Springer Transactions in Civil and Environmental Engineering

More information about this series at http://www.springer.com/series/13593

Alak De

Sedimentation Process and Design of Settling Systems

Springer

Alak De
Civil Engineering
Jadavpur University
Kolkata, West Bengal
India

ISSN 2363-7633 ISSN 2363-7641 (electronic)
Springer Transactions in Civil and Environmental Engineering
ISBN 978-81-322-3632-0 ISBN 978-81-322-3634-4 (eBook)
DOI 10.1007/978-81-322-3634-4

Library of Congress Control Number: 2016960753

Printed on acid-free paper

This Springer imprint is published by Springer Nature
The registered company is Springer (India) Pvt. Ltd.
The registered company address is: 7th Floor, Vijaya Building, 17 Barakhamba Road, New Delhi 110 001, India

Homage
To the noble soul of my wife Dr. Aditi De,
The sincerest and only friend of my life
Alak De

Preface

The subject of this monograph is sedimentation. The literature review on the subject presented herein will help researchers to select their field of study on the subject to carry forward; the subject matter being presented in lucid style will help undergraduate and postgraduate students of environmental engineering and other relevant fields to master the contents.

With the new concepts and latest information on sedimentation being presented, design engineers and consultants on the subject are likely to earn a skill in their activities with the subject.

Acknowledgements

The author is indebted to the great minds that illumined his vision and knowledge and also inspired him to carry with the subject.

Editing this type of manuscript itself is fairly difficult. Compared with the text-oriented manuscript, it is more difficult. Moreover, the author likes to take this opportunity to tender his humble submission that he did not present a tidy manuscript. Even then the editors discharged their meticulous endeavour to its finest presentation.

All credit for good presentation in the book goes to the editors, and for lapses, if there is any, the author remains liable.

The author expresses his profound thanks and regards to Swati Meherishi, Aparajita Singh, Joseph Daniel and S. Madhuriba for their hard work.

The author acknowledges his students who had to undertake hard work while doing their theses under his supervision at Jadavpur University.

Sanjay Biswas prepared the manuscript through sleepless nights. The author is extremely grateful for his support. The author acknowledges Parthajit Patra, who prepared some important sketches, and Mousumi Das, who collected data under trying conditions as shown in figures.

Thanks are also due to Journal of the Institution of Public Health Engineers, India (Jour.IPHE) and Journal of the Indian Institution of Chemical Engineers (Jour.IIChE) for necessary permissions.

Contents

Chapter 1
Introduction

Abstract This chapter introduces the term sedimentation, settling systems and the mode of presentation of varied aspects of subject of this book.

Keywords Settling system • Settling tank • Tube settling • Couette flow settling • Thickening

1.1 Sedimentation

Sedimentation is an indispensable operation in water and waste water treatment. It finds important application in chemical and metallurgical industries. It is a 'must do' operation in thickening of sludge during sludge handling. This has to be employed wherever settling has its role to play.

1.2 Settling System

Water and waste water may contain solids. These solids are settleable and non settleable. The settleables are removed through settling in a settler. The non-settleables are rendered settleable before they are removed through settling.

Non-settleables are rendered settleable by charge neutralisation of colloids with coagulants through rapid mixing by stirring or flash mixing followed by contacting between the particles with slow mixing when they form flocs under Vanderwaals' force of attraction.

The relative importance of coagulation-flocculation and sedimentation depends on the relative fraction of non-settleables and settleables present in water.

Where settleables are non-existent and only very small amount of colloids are present in water, the water may be subjected to polishing treatment through coagulation-flocculation followed by filtration.

With comparable fractions of settleables and non-settleables, solids are removed through coagulation-flocculation and sedimentation followed by filtration.

A. De, *Sedimentation Process and Design of Settling Systems*,
Springer Transactions in Civil and Environmental Engineering,
DOI 10.1007/978-81-322-3634-4_1

Solids with settleables only and negligible non-settleables are removed through settling and then filtration.

In the event of necessity of reducing the bulk of solid sludge, the slurry is subjected to settling, the object being the concentrating or thickening of the sludge solids in a thickener.

Hence, a settler in the form of a settling tank or tube settler together with or without 'rapid mixing' of coagulant in 'Flash mixer' followed by contacting process of slow mixing in a flocculator may form settling system.

The thickener is a settling system for sludge thickening.

1.3 Approach to the Study

The subject of 'Sedimentation Process and Design of Settling Systems' has diverse modes of applications.

'Sedimentation' comprehends the phenomenon through Introduction (Chap. 1), Developments in Settling Studies (Chap. 2) and Sedimentation Process (Chap. 4).

'Settling systems' differ in 'settling system with settling tank', 'settling system with tube settling', 'settling system with couette flow settling' and 'settling system of thickening'.

The studies of the first three settling systems employ a new concept of 'Velocity Profile Theorem' for solving settling velocity problems.

'Settling system with settling tank' reads through Velocity Profile Theorem (Chap. 3), Discrete Settling (Chap. 5), Flocculant Settling (Chap. 6), New Mode of Column Settling Data Analysis (Chap. 8), Analysis of Short Circuiting Phenomena (Chap. 9), In Quest of Parameter for Settling Comparison (Chap. 10), Design of Settling System (Chap. 11), Simulation of Real Settling System in Jar Testing (Chap. 12) and Compatible Design of Real Settling System (Chap. 13).

'Settling system with tube settling' develops through Velocity Profile Theorem (Chap. 3), Shallow Depth Settling (Chap. 14), Verification Tube Settling Theory (Chap. 15), Residual of the Assorted Solids Through Shallow Depth Settler (Chap. 16), Control Application on Design Parameters (Chap. 17) and Design of High-Rate Settlers (Chap. 18).

'Settling system with couette flow settling' is developed through Velocity Profile Theorem (Chap. 3) and Design of System Module for Couette Flow Setter (Chap. 19).

'Settling System of thickening' is contained in Zone Settling and Compression (Chap. 7) and Design of Thickeners (Chap. 20).

The study scheme of the book is depicted in the study tree and is presented in Fig. 1.1.

SEDIMENTATION PROCESS AND DESIGN OF SETTLING SYSTEMS				
SEDIMENTATION PROCESS	**SETTLING SYSTEM**			
	↓	↓	↓	↓
	WITH SETTLING TANK	**WITH TUBE SETTLING**	**WITH COUETTE FLOW SETTLING**	**OF THICKENING**
↓	↓	↓	↓	↓
Chapter 1 : Introduction	Chapter 3 : Velocity Profile Theorem	Chapter 3 : Velocity Profile Theorem	Chapter 3 : Velocity Profile Theorem	Chapter 7 : Zone Settling and Compression
↓	↓	↓	↓	↓
Chapter 2 : Developments in Settling Studies	Chapter 5 : Discrete Settling	Chapter 14 : Shallow Depth Settling	Chapter 19 : Design of System Module for Couette Flow Settler	Chapter 20 : Design of Thickeners
↓	↓	↓		
Chapter 4 : Sedimentation Process	Chapter 6 : Flocculant Settling	Chapter 15 : Verification of Tube Settling Theory		
	↓	↓		
	Chapter 8 : New Mode of Column Settling Data Analysis	Chapter 16 : Residual of The Assorted Solids Through Shallow Depth Settler		
	↓	↓		
	Chapter 9 : Analysis of Short Circuiting Phenomena	Chapter 17 : Control Application on Design Parameters		
	↓	↓		
	Chapter 10 : In Quest of Parameter for Settling Comparison	Chapter 18 : Design of High Rate Settlers		
	↓			
	Chapter 11 : Design of Settling System			
	↓			
	Chapter 12 : Simulation of Real System Settling in Jar Testing			
	↓			
	Chapter 13 : Compatible Design of a Real Settling System			

Fig. 1.1 Study tree of "Sedimentation Process and Design of Settling Systems"

Chapter 2
Developments in Settling Studies

Abstract Literature on settling studies has been reviewed since 1889, and the salient chronological developments on the subject are outlined.

Keywords Baffles • Contacts between particles • Camp's settling tank • Settling column analysis • Tube settling verification

2.1 Literature Review

Gravity separation of solids from its suspension has been in practice for a long time. The earliest study on the phenomenon, that could be traced, appears to have come through Sheddon. Subsequent understandings of its developments may be traced through the following:

1889: Sheddon (1889) recognised that continuous operation of settling tank gives as good a result as an intermittent operation. The fact that the use of baffle could help to reduce the volumetric capacity of the tank without the impairment of effluent quality was also noticed. Sheddon discussed the factors such as distribution of kinetic energy of the incoming liquid, temperature variation, action of wind (in open basin) that causes motion of the water and results in mixing, in detail. These factors are responsible for not allowing the settleable solids to fall through a quiescent liquid in the manner as expected.

Since Sheddon's publication in 1889, settling tanks were constructed using baffles, and they were put to continuous operation. Although considerable advancement was noticed in the performance of the settling tanks by the introduction of baffles, the cause of improved performance was not backed up by proper scientific analysis.

1888–1889: A falling particle sends out disturbances to the medium surrounding it. When its line of fall is in the vicinity of the wall of the container, the disturbances get reflected from the wall and modify the fall velocity of the particle. Munroe (1988–1989) developed empirical correction factor to the modified fall velocity due to this 'Wall Effect'.

A. De, *Sedimentation Process and Design of Settling Systems*,
Springer Transactions in Civil and Environmental Engineering,
DOI 10.1007/978-81-322-3634-4_2

1904: In 1904, Allen Hazen felt the necessity of theoretical analysis of the phenomenon for the better understanding of the subject. He based on Sheddon, as stated by him, and carried the analysis further.

In his classic attempt, he recognised the fact that what actually happens in the process is extremely complex and that 'first, conditions much simpler than those which actually exist must be assumed and from those simple assumptions the more complex conditions can be approached'.

Accordingly, he assumed (1) particle that hits the bottom stays removed and (2) all particles fall with the same settling velocity throughout their entire fall. Let

t_0 – time required to reach the bottom from the surface of water
t – sedimentation time in case of intermittent operation and the ratio:

$$\frac{\text{Volume of the tank } (V)}{\text{Flow rate } (Q)}$$

i.e., theoretical detention time in continuous operation
n – number of basins in series
y_0 – amount of suspended matters remaining in suspension at the commencement of time measurement t
y – amount of suspended matters settled in time t
$y_0 - y$ – amount of suspended matters remaining in suspension at time t
$\frac{y_0 - y}{y}$ – fraction of suspended matters remaining in the suspension
C – concentration of particles in the influent

Let us consider a basin full of water in absolutely quiescent condition. Particles are uniformly distributed throughout the entire volume. In shorter interval of time t, the fraction of solids removed is t/t_o and the fraction remained in suspension is

$$\frac{y_0 - y}{y_0} = 1 - \frac{t}{t_0} \tag{2.1}$$

Now we imagine the water in the basin kept mixed during the process in such a way that at any instant of time concentration of particles remains the same throughout. Over a sufficiently short interval of time τ during which the movement of water in the mixing can disregarded, the fraction of solids removed-

$$\frac{y_0 - y}{y_0} = 1 - \frac{\tau}{t_0}$$

At the end of next interval,

$$\frac{y_0 - y}{y_0} = \left(1 - \frac{\tau}{t_0}\right)\left(1 - \frac{\tau}{t_0}\right)$$

and at the end of $n\tau = t$

$$\frac{y_0 - y}{y_0} = \left(1 - \frac{\tau}{t_0}\right)^n \text{ i.e} \left(1 - \frac{1\,t}{nt_0}\right)^n \tag{2.2}$$

Next we consider the above basin in continuous operation. The mixing is continued so that at any instant of time, the particle concentration is the same throughout the volume. Accordingly the concentration of the particles in the suspension is the same as it is in the effluent, namely,

$$\frac{y_0 - y}{y_0}.C$$

In a sufficiently small interval of time τ to disregard mixing, solids entering into the tank are $Q\tau C$ and solids going out is

$Q\tau\left(\frac{y_0-y}{y}\right)C$ and the amount deposited is $-\frac{\tau}{t_0}v\frac{y_0-y}{y_0}C$ as shown in Fig. 2.1.

Q – flow rate
C – concentration of solids in the influent
τ – infinitely small interval of time
V – volume of the tank
t_0 – time required to reach the bottom from the surface of water
y_{0}- – amount of suspended matters remaining in suspension at the commencement of time measurement
y – amount of suspended matters settled in time t;

Then the mass balance equation can be written as

$$Q\tau C - Q\tau\left(\frac{y_0 - y}{y_0}\right)C = \frac{\tau}{t_0}\left(\frac{y_0 - y}{y_0}\right)CV \tag{2.3}$$

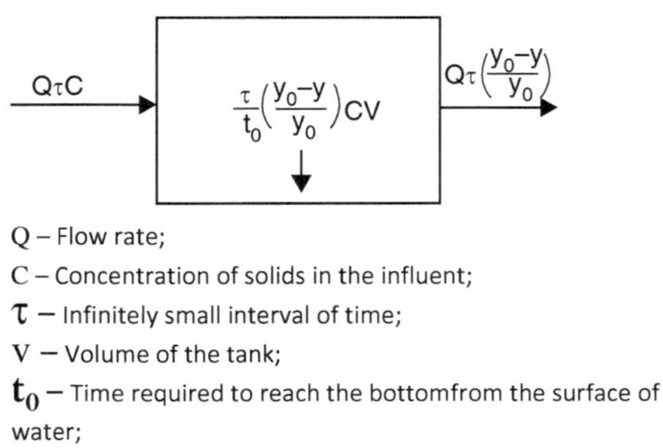

Fig. 2.1 Mass balance of solids

From which one obtains

$$\frac{y_0 - y}{y} = \frac{Q}{Q + \frac{V}{t_0} = \frac{1}{1+\frac{t}{t_0}}} \tag{2.4}$$

If we imagine the whole basin to be divided into n equal tanks connected in series, the flow through time through each is t/n. Fraction of settleable solids in the effluent through the first is

$$\frac{y_0 - y}{y_0} = \frac{1}{1 + \frac{1}{n} \cdot \frac{t}{t_0}}$$

In the effluent of the second tank

$$\frac{y_0 - y}{y_0} = \left(\frac{1}{1 + \frac{1}{n} \cdot \frac{t}{t_0}}\right)^2$$

and in the effluent leaving finally

$$\frac{y_0 - y}{y_0} = \left(\frac{1}{1 + \frac{1}{n} \cdot \frac{t}{t_0}}\right)^n$$

$$\frac{y_0 - y}{y_0} = \left(1 + \frac{1}{n} \cdot \frac{t}{t_0}\right)^{-n} \tag{2.5}$$

If we assume an infinite number of basins, this will mean absolutely complete baffling and continuous forward movement of the water at all points, mixing from top to bottom but with no mixing backwards or forwards.

Thus by assuming the basin to be divided into a number of hypothetical cells, Hazen attempted to make allowance for the departure of the basin from ideality. Redistribution is confined to one cell at a time.

Writing

$$\frac{y}{y_0} = \frac{t}{t_0} = \frac{V}{Q}\frac{v_0}{h_0} = \frac{v_0}{Q/A} \tag{2.6}$$

Equation 2.5 becomes – $\frac{y_0-y}{y_0} = \left(1 + \frac{1}{n} \cdot \frac{v_0}{Q/A}\right)^{-n}$

$$\frac{y}{y_0} = 1 - \left(1 + \frac{1}{n} \cdot \frac{v_0}{Q/A}\right)^{-n} \tag{2.7}$$

v_0 is the velocity of a particle moving through depth h_0 in time t_0 and A is the surface area of the tank. From the above equation, it is apparent that the greater is the number of such cells, the better is the damping of the factor that retard settling.

Equation 2.6 led Hazen to conclude that 'a shallow basin is as effective as deeper one as long as the bottom velocities do not prevent the deposition of solids'.

From Eq. 2.6,

$$\frac{y}{y_0} = \frac{t}{t_0} = \frac{v_0}{Q/A}$$

If we put the ratios $\frac{y}{y_0} = \frac{t}{t_0} = \frac{v_0}{Q/A} = 1$,

$\frac{y}{y_0}$ is 100 % removal for $t = t_0$ and $v_0 = \frac{Q}{A}$, i.e. ideal removal under quiescent settling.

In Hazen's expression if $n \to \infty$

$$\frac{y}{y_0} = 1 - \mathrm{Lt}_{n\to\infty}\left(1 + \frac{1}{n} \cdot \frac{v_0}{Q/A}\right)^{-n}$$
$$= 1 - \mathrm{e}^{-\frac{v_0}{Q/A}}$$

If we put $\frac{v_0}{Q/A} = 1$ for quiescent settling, Hazen's expression deduces.

Fractional removal $= 1 - \mathrm{e}^{-1}$, i.e. $1 - \frac{1}{2.7183}$ or 0.632 or 63.2 %.

Thus n-value cannot simulate ideal settling.

For any 75 % removal, say, ideal theory deduces $\frac{v_0}{Q/A} = 0.75$, i.e. $v_0 = 0 \cdot 75\frac{Q}{A}$.

Hazen deduces $0.75 = 1 - \mathrm{e}^{-\frac{v_0}{Q/A}}$, i.e. $\frac{v_0}{Q/A} = 1.386$ or $v_0 = 1.386\frac{Q}{A}$.

A factor 'n' has been introduced in the expression

$$\frac{y_0 - y}{y_0} = \left(1 + \frac{1}{n} \cdot \frac{v_0}{Q/A}\right)^{-n}$$

The expression shows that with increase in the value of 'n' which will mean an increase in the number of virtual baffles, the performance of the basis is better.

The value of 'n', i.e. the number of virtual baffles, which is indicative of the pattern of flow, more specifically the pattern of flow of the settleable particles inside the basin, depends on the weight fraction of the particles removed for the same value of $\frac{t}{t_0} = \frac{v_0}{Q/A}$. Again for the same weight fraction of particles removed the value of 'n' depends on $\frac{t}{t_0} = \frac{v_0}{Q/A}$. The flow rate, settling velocity of particles and the basin surface area being independent of each other the value of 'n' will also depend on them. The theory could not establish the direct relationship with the geometrical parameters of the tank, its inlet and outlet structures. 'n' is thus not a characteristic of an actual tank alone. To be of any practical utility, it is essential that tank characteristics should be related to the factors causing deviation of the flow from the ideal one to reflect the degree of deviation. In this regard, the utility of 'n' is questionable.

Let us consider that the influent to a settling tank contains particles with settling velocities $v_1, v_2, \ldots$etc. A question remains what should be the value of 'n'. If it is an experimentally determined value, let the value be n_1 for the particles with v_1 (settling velocity). Likewise n_2 is the value for particles with settling velocity $\mathbf{v}_2$ and so on. Unless all n -values $n_1, n_2, \ldots$ are equal, not other information can be deduced from the observations. In other words, 'n' should be the characteristic parameter of the actual tank alone.

If all the particles with a particular settling velocity v_0 are completely removed, we get

$$\left(1 + \frac{1}{n} \cdot \frac{v_0}{Q/A}\right)^{-n} = 0 \textbf{ and}$$

the value of 'n' in this case cannot be defined. In such cases deviation of the actual basin from an ideal one cannot be taken into account in accordance with the theory.

It has been assumed that all the particles with different settling velocities will have identical mixing patterns. This presumption is not beyond criticism. Consider a particle with settling velocity sufficiently large so as not to be affected by the turbulence created in the flow. Such particles will not be identically distributed.

1910: Newton (1910) deduced an expression for drag force on a falling particle:

$fA\rho v_s^2$ where f, a coefficient less than unity; A, projected area of the particle on a plane perpendicular to its line of fall; ρ, density of the medium; and v_s, instantaneous settling velocity of the particle.

Writing in terms of dynamic pressure, the above expression is written as $\frac{C_D A \rho v_S^2}{2}$ where $C_D$is Newton's drag coefficient and $C_D = 2f$.

1925: Oden (1925) could describe the settling velocity distribution among the particles in suspension during their batch settling.

1927: Capen (1927) studied a large number of settling tanks. Tracer studies in those tanks indicated that the ratio of the time to reach the centre of mass of the tracer response curve to its theoretical detention time could somewhat relate to the fractional removal of the solids through those tanks. In a large number of cases, the ratios were in the vicinity of 0.48.

1928–1929: A falling particle sends out disturbances to its surrounding medium. The distance, to which this disturbance reaches, is the extent of, what is called, the velocity field of the particle. For two closely falling particles, their velocity fields interfere, and their settling velocities are modified. This is 'hindered settling'.

Kermack et al. (1928–1929) observed in hindered settling of red blood corpuscles that the ratio

$$\frac{\text{Hindered settling velocity}}{\text{Settling velocity of the particle}}$$

increased with the increase in Reynolds' number.

1933: Francis (1933) studied 'Wall Effect' on falling particles like Munroe (1988–1989) and found out empirical correction factors.

1934: Rudolf and Lacy (1934) conducted experiments and indicated that the hindered settling velocity of 20,000 ppm activated sludge varied from 0.08 to 0.15 cm/s as against its free settling velocity of 0.14–0.27 cm/s.

1936: Shields (1936) in his paper entitled 'Application of Similitude of mechanics and Turbulence research to bed load movement (Translated)' had shown that particles on the surface of the sediment bed will not move if the quantity $0.1\boldsymbol{v}_S^2$ is less than $\frac{gd(\rho_s-\rho_L)}{\rho_L}$, where $\boldsymbol{v}_s$is settling velocity of the particle, g– acceleration due to gravity at the place of observation, $\boldsymbol{\rho}_s$ – density of solids and $\boldsymbol{\rho}_L$ – density of the liquid medium.

1936: Camp (1936) had shown that hydraulic characteristics of long narrow tanks are superior to those of wide low velocity tanks. The use of long narrow channel will minimize the effect of inlet and outlet disturbances etc. leading to the decrease in efficiency due to short circuiting. Common length to width ratios employed in design are from about 3.1 to 5.1.

1941: Peter Homack's (1941) experiment indicated that the free settling velocity of $\mathbf{caco}_3$ crystal aggregatates is about 0.06 cm/s. Settling is less rapid when Mg $(\mathbf{OH})_2$ is present than when $\mathbf{caco}_3$ is present alone.

1943: Camp and Stein (1943) deduced the number of contacts $\boldsymbol{N}_s$ (due to differential velocities) and $\boldsymbol{N}_v$ (due to velocity gradients) taking place per unit volume per unit time between particles of diameter $\boldsymbol{D}_1$ and $\boldsymbol{D}_2$, $\boldsymbol{n}_1$ and $\boldsymbol{n}_2$ being the numerical concentrations of particles of diameter $\boldsymbol{D}_1$ and $\boldsymbol{D}_2$, respectively, as

$$\boldsymbol{N}_s = \boldsymbol{n}_1\boldsymbol{n}_2\frac{(s-1)}{72}(\boldsymbol{D}_1+\boldsymbol{D}_2)^3(\boldsymbol{D}_1-\boldsymbol{D}_2);$$

s – Sp.Gr. of solids

due to differential settling and that due to velocity gradient as

$N_v = n_1 n_2 \frac{1}{6} G(D_1 + D_2)^3$, where G is mean temporal velocity gradient.

By computation with the help of the above equations, Camp remarked:

(i) Flocculation in deep tanks at low velocity ($R = 20$ ft and $v = 1$ ft/min) is due almost entirely to differential settling velocities.
(ii) The rates of flocculation by two processes are about the same in a tank 2 ft deep with a velocity of 10 ft/min.

By adjusting the turbulence mixing coefficient and the magnitude of G properly, Camp expected, both the effects of turbulent retardation and coagulation can be taken into account in settling test.

1944: Dobbins (1944) studied the effect of turbulence on settling. Turbulence delays the settling of particles. Camp transformed Dobbin's solution in terms of removal under no scour condition.

1946: Camp (1946) expressed that the main purpose of writing the paper was to collect in one compendium the known principles of sedimentation essential to the development of design theory and to present the theory of design developed to a stage which will permit its use in practice.

Accordingly he presented the paper under several subheadings as follows:

1. Settling velocities of individual particles:

Camp started with the drag force expression deduced by Sir Isaac Newton and deduced the settling velocity of a sphere. The solution of settling velocity equation involves trial computations. Camp suggested a method avoiding the same,

Camp pointed out that the particles to be removed from water and sewage by settling are usually irregular in shape and that the irregularities have greater influence upon the drag as the settling velocity increases. The settling velocity of a particle is also influenced by the presence of the walls of the container in the vicinity of the particle. Camp made a mention of the empirical correction factors developed by Francis (1933) and Munroe (1988–1989) for the 'Wall Effect'.

Camp commented that a theoretical analysis to find a correction factor in case of hindered settling was lacking, and the experimental data were not numerous. In his experiment with lucite spheres in still water and round sand grains suspended in a tube of rising water, it was observed that the correction factor

$$\frac{\text{Hindered settling velocity}}{\text{Settling velocity}}$$

increases with increase in Reynolds' number. This supports the observations by Kermack et al.

2. Nature of settling processes in water and sewage treatment:

Particles to be removed in water and sewage treatment plants consists of minerals, organic solids, grease with varying quantities of entrained water and occasionally gas.

Camp reported that with a rough approximation, the specific gravity of different substances such as fine sand grains, flocculated mud particles, suspended vegetable matters, alum floc $Al_2O_3 \cdot 20H_2O$, iron floc $Fe_2O_3 \cdot 20H_2O$, organic suspended solids like proteins and fats, may be taken to be 2.6, 1.5–1.0 (depending upon the quantity of entrained water), respectively. He recorded Homack's (1941) observation that the settling velocity of c_aco_3 crystal aggregates is about 0.06 cm/s, and it is reduced in the presence of $Mg(OH)_2$.

Practice indicated that a grit chamber removing sand grain 0.2 mm size and larger (settling velocity 2–2.4 cm/s) will protect pumps and/or other treatment units from undue abrasion and heavy deposits.

The free settling velocity of activated sludge, Rudolf and Lacy observed, was 0.14–0.27 cm/s and at concentration of 20,000 ppm, the hindered settling velocity was found to be varying from 0.08 to 0.27 cm/s.

3. Settling analysis of suspensions:

Camp advocated settling column analysis of suspensions for employing it to the efficient design of a settling tank and predicting or checking the performance of the same. He considered the analysis in detail (Camp 1936) for discrete suspension.

4. Clarification theory for ideal basin:

From the definition of an ideal settling basin, Camp characterized an ideal rectangular basin in continuous flow operation. He showed how to calculate the removal efficiency on the basis of settling column analysis of the influent suspension.

5. Tractive force and bed-load movement:

Camp deduced the channel velocity v_c required to start motion of particles of diameter D given by $v_C = \sqrt{\frac{8\beta}{f} g(s-1)D}$.

The value of constant β for fine nonuniform sand is 0.04 for impeding motion on smooth beds and has higher value 0.1–0.25 for impeding motion from sand ripples formed from smooth bed. 'g' is the acceleration due to gravity at the place of observation, 'S' is the Sp.Gr. of the material and f is the Darcy-Weisbach friction factor.

6. Effects of turbulence on sedimentation :

The nature of turbulent mixing process was discussed. Turbulence delays the settling of particles. Camp showed how to take the effect into account for discrete particles with two dimensional flows in rectangular tank.

7. Flocculent suspensions:

Flocculation occurs due to (i) differential settling and (ii) velocity gradients. Camp deduced N_S, N_v number of contacts taking place per unit volume per unit time due to differential velocities and velocity gradients respectively (Camp and Stein 1943). The equations helped Camp to make observations on the relative effects of flocculations due to differential settling and that due to velocity gradients.

8. Overflow rate, detention period, velocity and tank dimensions:

The theory of settling of discrete particles in an ideal basin indicates that the removal is a function of overflow velocity and independent of depth. The effect of turbulent mixing, though not independent of depth, can be shown to be influenced very little by it. This led Camp to suggest that the depth of the settling tank should be made as small as possible as is consistent with no scour condition. This conclusion, according to Camp, may also be drawn for flocculent suspensions. The magnitude of no scour velocity can be determined experimentally or with the help of the equation given under 'tractive force and bed-load movement'.

9. Short circuiting and stability:

Camp described tracer technique to get informations regarding the hydraulic characteristics of the tank such as the presence of dead space, short circuiting, etc.

10. Conclusions :

Finally, Camp illustrated the details of tentative design of one primary and one secondary sedimentation basin.

No doubt, the paper provided better understanding of the subject and paved the way for the further development.

1946: Eliassen (1946) was right when he remarked that 'Mr. Camp has accomplished one part of his announced objective – namely- to collect in one compendium the known principles of sedimentation essential to the development of design theory. However, he has only partly accomplished his other goal, which he announced was to present the theory of design, developed to a stage which will permit its use in practice'. Nevertheless it is true at the same time that 'one should not be condemned for starting the rationalisation of settling tank design on the ground that he has not presented a completed and fully tested theory and that the responsibility of progress belongs to the profession as a whole'. This is what Camp spoke in his defence.

Eliassen put forward a demonstration to show that if, according to Camp, the removal is governed by the overflow rate and not by detention, short circuiting does not decrease removal. This is intriguing.

1949: Schmitt and Voigt (1949) discussed an application of tray settling principle in the form of two storied settling tanks.

1951: Dresser (1951) reported a large increase of removal capacity of a settling tank with the introduction of trays.

1952: From a mathematical analysis of longitudinal mixing in settling tanks, Thomas and Achibald (1952) suggested that the value of 'n' in Hazen's theory is approximated by

$$\frac{t_{\text{mean}}}{t_{\text{mean}} - t_{\text{mode}}},$$

t_{mean} is the mean time required for the tracer to flow through the basin, and t_{mode} is the time required for the highest concentration to appear in the effluent. These times are identified on a typical tracer curve shown in Fig. 2.2. Here the flow pattern of water and that of the particles in suspension have been assumed identical.

1953: Camp (1953) again discussed an overflow rate and detention period. He demonstrated that tray in tank provides added floor area and increases the removal of solids and that the reduction of the tank depth does not increase the removal ratio. Thus he conclusively showed that the removal is independent of the tank depth in an ideal basin when free settling is concerned. He considered the various factors such as flocculation (both due to differential settling and velocity gradients), turbulence, short circuiting, etc. with reference to an ideal settling basin. Finally, he suggested the proposed design of primary and secondary sedimentation tanks.

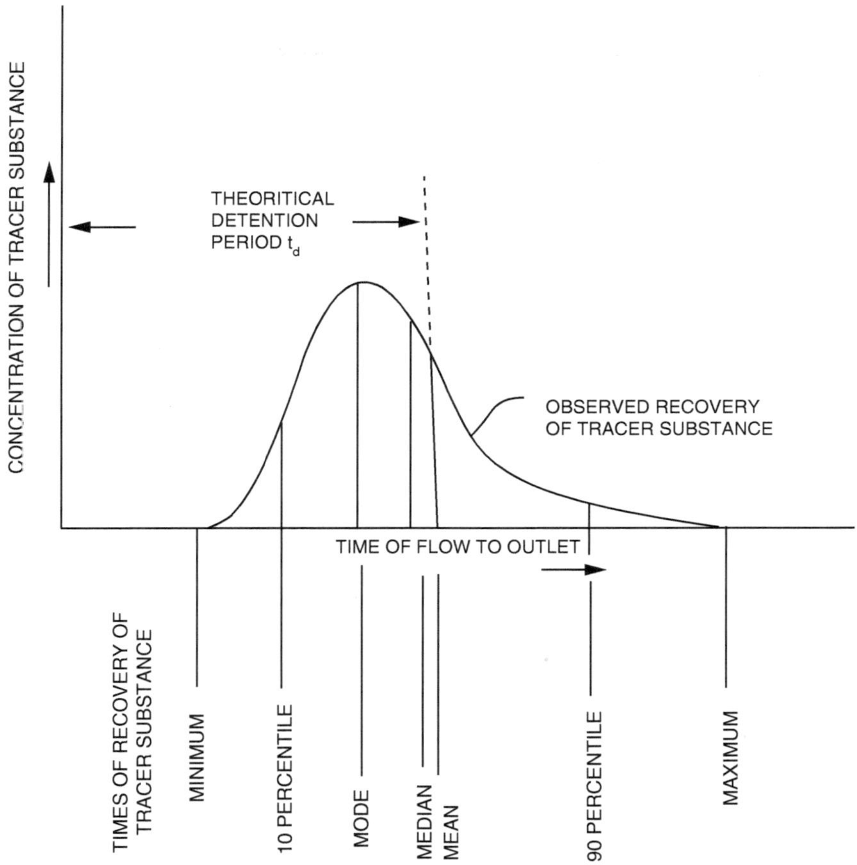

Fig. 2.2 Tracer response curve

This paper may be recognised as an attempt to present and explain the theory of design of settling tank in simple terms. This is, in fact, an extension and reconsideration of what Camp had tried to achieve in his previous paper (Camp 1936).

1955: Fischerstrom (1955) reported some successful applications of tray settling theory. He pointed out that for efficient removal it is necessary (i) to maintain proper hydraulic condition along with (ii) the proper overflow rate. Earlier experiences of others confirmed his feeling that attempts to use radial flow circular trays considered only the later factor. He realised that Reynolds' number 500 or less should be maintained in a basin for the most efficient performance.

The Reynolds' number could be reduced in a basin by increasing the wetted perimeter, i.e. by introducing longitudinal baffles. The baffles may be horizontal or vertical. The vertical baffles decrease the Reynolds' number but does nothing with the overflow rate. The horizontal baffle, on the other hand, not only reduces the Reynolds' number but also reduces the overflow rate and the vertical distance; the settling particles must fall through before striking the bottom surface.

According to him, the minimum spacing would be determined by the sludge removal problem and the difficulty of distributing the flow equally to a large number of trays.

He applied his theory to several cases. From the excellent performance of the operating installations, he could conclude that the tray settling was in no way only theoretical. Cost analysis revealed that the tray basins are less expensive in comparison to the conventional one.

1955: Talmadge and Fitch (1955) could relate the batch settling data to the determination of unit areas both as clarifier and thickener.

1956: Fitch (1956) expressed that an ideal basin must be one that can be analysed rigorously. According to him, the more closely the ideal conditions approach reality, the fewer will be the amendments necessary to predict the practical behaviour.

Fitch criticised Hazen and Camp because, Fitch stated, if the curvature of the flow path is considered, the flow may have upward velocity components at some points and downward components at others. It may also curve laterally. This makes the determination of the trajectory of the particle difficult.

He considered a vertical section of flownet of infinitesimal thickness extending from inlet to the outlet and bounded by flow lines as shown in Fig. 2.3. The width dw may vary from inlet to the outlet but will remain the same over the

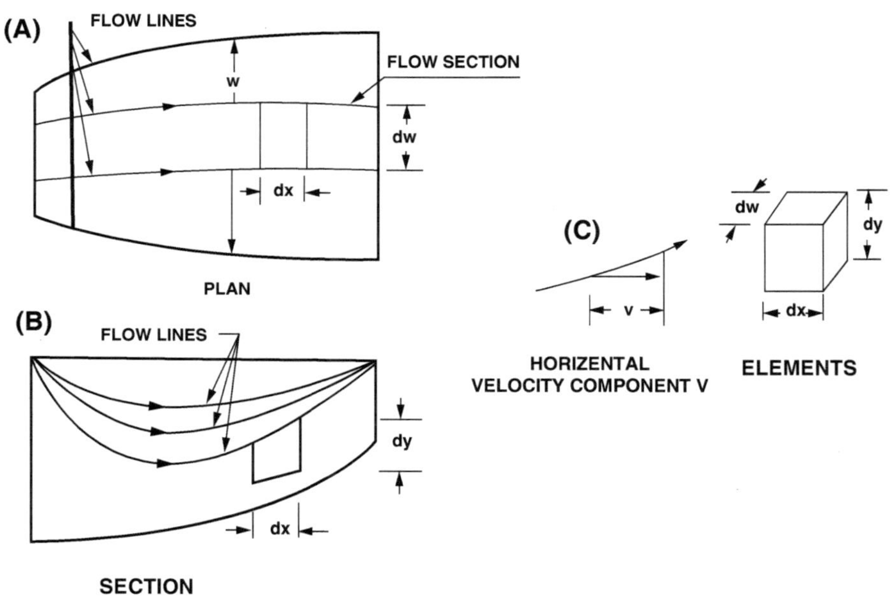

Fig. 2.3 Elements of flow section showing plan and section and the velocity components

entire depth of the basin. The flow through an infinitesimal cross section may be expressed as

$$dq = v\,\mathrm{d}w\,\mathrm{d}y \tag{2.8}$$

q = the flow through the section
V = the horizontal component of velocity assumed to vary from point to point
$\mathrm{d}w$ = the width of the flow stream at the cross section
$\mathrm{d}y$ = the vertical height of the filament of the flow passing through the cross section

$$\frac{dy}{dt} = v_s = \text{the settling velocity of the particle with respect to the fluid} \tag{2.9}$$

$\mathrm{d}t$ = small interval of time
$\mathrm{d}x$ = horizontal component of elemental distance along the flow line

$$v\,\mathrm{d}t = \mathrm{d}x \tag{2.10}$$

From Eqs. 2.9 and 2.10

$$\mathrm{d}y = v_s dt = \frac{v_s}{v} dx \tag{2.11}$$

$$\mathrm{d}q = v\,\mathrm{d}w\mathrm{d}y = v\,\mathrm{d}w\frac{v_s}{v}\mathrm{d}x \tag{2.12}$$

$$= v_s\,\mathrm{d}w\,\mathrm{d}x = v_s\,\mathrm{d}A \tag{2.13}$$

dA = projected area across which flow sweeps

Integrating over all filaments of flow traversed by a particle, assuming that the particle has traversed no flow at the time it has swept through no area.

$$\int_0^q dq = v_s \int_0^a dA \tag{2.14}$$

$$\text{i.e. } q = v_s a \tag{2.15}$$

Where a = the projected or surface area of the section swept by the flow.

Now q is the flow which the particle of given settling rate v_s can traverse.

$$v_0 = \text{Over flow velocity (by definition)} = \frac{q_s}{a} \tag{2.16}$$

i.e. $av_0 = q_s$ = total flow of the section.

If the solids are assumed to be uniformly distributed in the feed suspension initially, then the fraction of total section flow out of which particles of settling velocity v_s will settle is given by

$$\frac{q}{q_s} = \frac{v_s}{v_0} = F \tag{2.17}$$

Thus Fitch claimed to have deduced the same result arrived at by Hazen and Camp under less idealized condition.

It is really interesting and worthwhile to examine how far this claim is justified.

The assumption that the thickness of the section of flownet of infinitesimal thickness remains the same throughout the depth implies that there is no depth-wise variation of velocity. In writing Eq. 2.8, the horizontal component of velocity has been considered, and in writing Eq. 2.11, the vertical component of velocity has been neglected. This suggests that the velocity is horizontal and remains the same at every point on the vertical cross section at right angle to flow.

From the Eqs. 2.10 and 2.11 and 2.12,

dt = theoretical detention time in an elemental volume

$$= \frac{dx}{v} = \frac{\frac{dx}{dq}}{dw\,dy} = \frac{dwdydx}{dq} = \frac{\text{volume of the element}}{\text{flow through the section}}$$

The flow q_s of suspension down through and out of which a particle of settling velocity v_0 can settle is given by $q_s = v_0 a$. If the theoretical detention time corresponding to flow q_s is t, a particle which enters at the top will reach the bottom at a depth

$$v_0 t = \frac{q_s \cdot t}{a}$$

and will be removed. Thus the flow will be free from particles of settling velocity v_0.

This can be true only when a particle is removed from the suspension when it reaches the bottom of the settling zone and is not returned back to the suspension.

Again we consider the flow to be q_s containing particles of velocity v_s. In time t, a particle which enters at the top will reach a depth

$$v_s t = \frac{qt}{a}$$

If the concentration of the suspended particles of each size is the same at all points in the vertical cross section at the inlet end of the settling zone, the fraction of total flow from which particles of settling velocity v_s will be removed

$$F = \frac{\frac{qt}{a}}{\frac{q_s t}{a}} = \frac{q}{q_s} = \frac{v_s a}{v_0 a} = \frac{v_s}{v_0} \tag{2.18}$$

The claim that the results have been deduced under less idealised condition which Fitch extends is, therefore, untenable.

1956: Ingersoll et al. (1956) reviewed the fundamental concepts on sedimentation.

They pointed out the inadequacy of comparing the basin performances by finding out the total percentage removal of the suspended solids. A new method for comparison by comparing what they called 'overflow residual efficiencies' was proposed.

Their experiments with wax spheres and silica showed that the removal was independent of depth. At shallow depth and high displacement velocities, the removal decreased considerably. In the above experiment, the resuspension of fine light sediment was observed at horizontal velocities much lower than those required to start bed-load movement in accordance with critical channel velocity formula developed by Camp.

This might lead one to conclude that the resuspension which resulted in an inefficient removal was not due to scour but due to turbulent eddies. They also put forward a suspended load equation as a logical approach to the subject of limiting horizontal velocities to avoid scour by turbulent eddies.

The use of multiple inclined baffles to prevent scour in shallow tanks was suggested.

Dispersion studies revealed to them that the dispersion characteristics are largely governed by the inlets, and the outlets are of minor importance in such a case.

1956: Barham et al. (1956) reviewed the equipments employed in settling.

1957a: Fitch (1957a) discussed Eliassen's (1946) demonstration to show that the two assertions made by Camp such as:

1. 'The removal of suspension in a sedimentation basin is unaffected by the depth of the tank except through the influence of turbulence and bottom scour ', which is equivalent to stating that the removal is governed by overflow rate and not by detention, and
2. 'Short circuiting decreases removal' are not compatible to each other in explaining the settling phenomenon in a basin.

Let us assume a tank with uniform flow through the top third of the basin (as shown in Fig. 2.4 tank A). A particle settling to the bottom of the active one third will fall vertically through the stagnant two thirds of the basin.

Hence a particle entering at the top or identically into either of the tank A, tank B or tank D will reach the bottom at the same distance from the inlet end of the tank; tank A, tank C and tank D are identical.

Tank B has the depth one third of the other tanks and is otherwise identical with them. In case of tank A, the top third is active, and in case of tank D, its bottom third portion is active. Tank A, tank B and tank D will, therefore, accomplish identical removals.

If the assertion 1 is valid, then the tank A and tank D, both of which are short circuiting, will make identical removals with that of tank C and tank B. In other

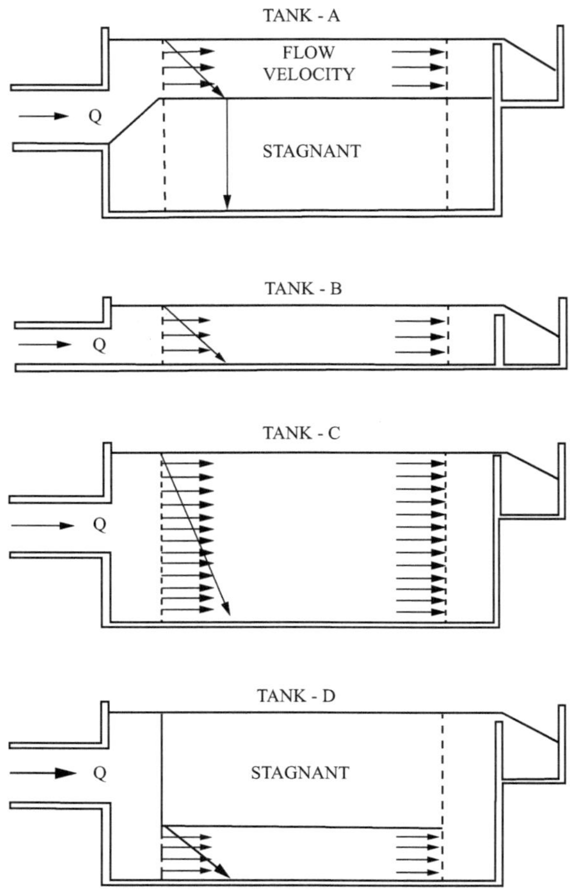

Fig. 2.4 Eliassen's demonstration

words, if the assertion 1 is to be valid, short circuiting should not affect the removal, i.e. assertion 2 must be invalid. This is anomalous.

Fitch wrote—'Eliassen's demonstration was countered with the statement: As commonly defined overflow rate is the discharge per unit surface area of the tank. For the purpose of this discussion the definition given by Stein is preferable i.e. overflow rate is equal to the ratio of depth to detention time. Since all particles of water do not have the same detention time they do not have the same overflow rate. In other words, short-circuiting affects the overflow rate in precisely the same manner as it does the detention time'.

'This answer', Fitch stated, 'does not in any way resolve the dilemma of Eliassen's demonstration'.

To the author it appears that Fitch was not correct either in concluding that 'short circuiting would not change removals if it is true that removal is not a function of tank depth'.

Eliassen's demonstration considered the short circuiting resulting from the depth-wise variation of flow through velocity. Indeed in such cases short circuiting does not appear to have any influence on the removal (Eliassen, discussion, 1946). But when one considers the widthwise variation of flow through velocity, the short circuiting thus resulted will decrease the removal (Chap. 9 of this book). In a tank short circuiting results from both the above variations.

Eliassen's demonstration is not a 'dilemma' as it was called so by Fitch. It exposes only a part of an entire picture. It is in no way in contradiction with the fact expressed by the statement made by Camp while answering Eliassen. The statement was 'The literature is full of experimental evidence that short-circuiting impairs the removal in settling tanks'.

Fitch appreciated that an ideal basin was conceived to translate the results of batch settling analysis and that the batch settling analysis should show the removal in it.

Fitch conducted the settling column analysis with c_aco_3 suspension in water. He plotted the removals in detention time and overflow rate coordinates in log-log paper. The curves were neither vertical nor horizontal but intermediate between the two. In fact, they were more horizontal than vertical.

Thus he concluded—'It would be imprudent to discard detention time as a design factor. There is evidence that for settling of class-2 suspensions, in which particles continue to coagulate or flocculate during the sedimentation period, detention can be of considerably greater significance'.

1957b: In a different paper, Fitch (1957b) described four characteristic types of settling phenomenon depending upon (1) the dilution of the suspension and (2) the relative tendency of the particles to cohere. They are (i) class-1 clarification, (ii) - class-2 clarification, (iii) zone settling and (iv) compression.

Removal of class-1 suspension is governed by the overflow rate.

In case of class-2 suspension, a demonstration was put forward by Camp (1936) to show that the flocculation by differential settling only is independent of depth. As shown in Fig. 2.5, the trajectories of two particles coalesce in an ideal basin. If the depth were decreased, the tank velocity would be increased proportionately, and the slopes of the trajectories would be proportionately decreased. Thus the particles A and B would be expected to coalesce at the same distance from the inlet end and to reach the bottom at an unchanged distance from the inlet. This was the argument.

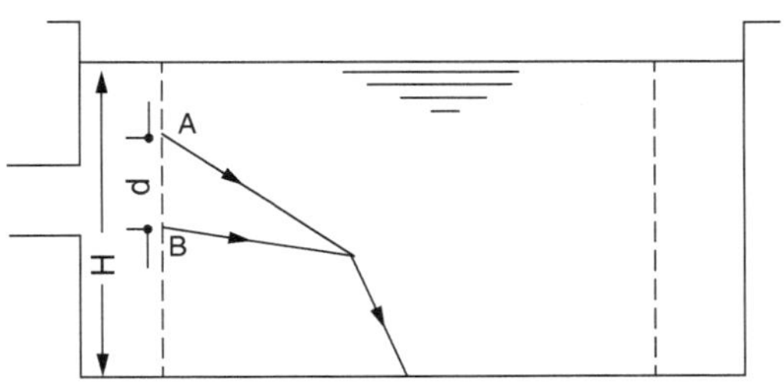

Fig. 2.5 Effect of particle agglomeration on settling

Now it is easy to see the following. If the settling velocity of A is $\boldsymbol{v}_1$ and that of B is $\boldsymbol{v}_2$ entering at '*d*' vertical separating distance one above the other in an ideal basin and Q be the flow rate, B and D are the width and depth of the basin, respectively, then the particle will collide at a distance from the inlet and given by

$$\frac{\boldsymbol{d}}{\boldsymbol{v}_1 - \boldsymbol{v}_2} \cdot \frac{\boldsymbol{Q}}{\boldsymbol{BD}}$$

This being dependent on the depth of the basin, the aforesaid argument seems not to be valid. This was pointed out and shown by Fitch in a somewhat different fashion.

So for class-2 suspension, the depth of the tank and hence the detention time are also a factor influencing its removal. As such the removal in this case is governed by (1) the overflow rate and (2) the detention time.

For zone settling, the capacity-controlling factor is the solids throughout per unit area per unit time, and for compression, the factors are solids detention and the sludge depth.

1957: Dallas (1957) described a tank with a triangular long section provided with sloped bottom. Effluent troughs were arranged at regular intervals on the surface. This arrangement resulted the stream lines extend obliquely upward. The shape of basin, according to Dallas, eliminated the so-called ineffective zone in an ideal basin. The basin was claimed to have certain advantages over flat bottom tanks. This resulted in an overall economic design of a settling tank.

1957: Katz and Geinopolos (1957) conducted tracer studies in two circular basins, one of which was centrally fed and the other one was peripheral fed. Studies indicated that the centrally fed basin had the tracer response better and was hydraulically better than the peripheral-fed circular basin.

1957: Lesperance (1957) looked upon the sedimentation unit both as clarifier and thickener. He concluded that unit area requirement for a thickener may often control the ultimate size of the unit.

1957: O'Connor and Eckenfelder (1957) performed settling column analysis with flocculant suspension as described by Camp. They showed how to utilise the results to determine the total removal of flocculent suspension in an ideal basin corresponding to a particular overflow rate. Subsequently they computed various percentages of removals and their associated detention times and overflow rates in order to determine the design criteria from the laboratory settling column analysis.

This paper bridges the gap in the literature by showing the analysis on class-2 suspension. But it requires critical evaluation.

1958: Bergman (1958) conducted tracer studies in order to determine and compare the hydraulic characteristics of basins. He observed the problems of instability, especially at low flow rates, are common in tracer analysis of sedimentation basins.

1967: Hansen and Culp (1967) made a detailed literature review to show that there had been many attempts to apply shallow-depth settling principles proposed by Hazen and Camp. Failure of the said attempts may be ascribed to two major reasons: (i) unstable hydraulic condition in a very wide shallow tray and (ii) the minimum tray spacing being limited by the vertical clearance required for mechanical sludge removal equipment.

The authors overcame all the difficulties by using small diameter (1–4 in.) tubes of 2–4 ft. length. The performance of the tubes, they concluded, would depend upon the tube length, diameter, flow rate, the nature of the incoming settleable material and the nature and quantity of the chemicals added.

1968: Tekippe and Cleasby (1968) conducted tracer studies in order to compare the hydraulic characteristics of basins like other workers and came out with the same observations made by Camp (1936), (1946), Bergman (1958) and Katz and Geinopolos (1957).

1968: Hansen et al. (1968) installed tube settling devices in many water treatment plants. They presented the operating experiences with those installations.

Over 20 water treatment plants employed horizontal tubes. Detention time was less than 10 min and capacities ranged from 20 gpm to 2000 gpm. A plant reduced the raw water turbidity of 1000 JU through the use of flocculation, tube sedimentation and mixed media filtration. The overall detention time was 16 min. In steeply inclined tubes, continuous cleaning of sludge took place. Test results in both the laboratory and the field indicated that at 60° inclination to the horizontal continuous sludge removal as well as settling took place very efficiently. The tube settlers coupled with mixed media filters reduced the size and cost of treatment.

1969: Hansen et al. (1969) reviewed their studies leading to the development of tube settlers. They discussed the results of pilot and plant scale installations and concluded that the capacity of an existing clarifier can be increased from two to four times by installing modules of steeply inclined tubes.

1969: Culp et al. (1969) concluded by conducting studies on various tube clarifiers already installed in different plants that such installations were highly efficient and economic (Camp 1946).

1970: Hernandes and Wright (1970) evaluated the performances of tubes having cross sections of rectangular, circular, square and hexagonal. They studied the effect of flow rate on tube settler efficiency and proposed that for any water or waste water to be clarified by sedimentation in a tube settler, there exists a relationship between the percent turbidity removal and logarithm of the ratio $\frac{v_0^2 R}{L}$,

where v_0 – flow-through velocity
R – hydraulic radius
L – length of tube

For any particular water or waste water suspensions to be used in tube settlers for clarification, there exists a particular value of

Log $\frac{v_0^2 R}{L}$, beyond which rapid deterioration of tube performance occurred. However, this parameter does not include the effect of variables like settling velocity of particles, inclination of the tube, etc.

1970: Culp and Culp (1970) reported in their book about plants where tube modules were installed in conventional sedimentation basins and they worked efficiently. They worked with all kinds of tube cross sections, namely, rectangular, square and circular. They also reported that tubes with 'chevron-shaped' (vee-shaped) cross section were already in operation in some plants. The firms producing such tubes claimed that such cross section helps optimum sludge compaction along with uniform flow condition.

The authors also reported about plants where closely spaced plates were used in place of tubes. They observed that these plate settlers, i.e. 'lamella clarifiers', were equally good.

1970: Yao (1970) (Chap. 11 of this book) made a basic theoretical study of the characteristics and governing physical properties of high-rate settling system. He realised that the concept of overflow rate and its significance with the high rate settlers were not defined. He also felt the necessity of extensive generalisation of Camp's model for being applied to the tube settling system. To provide information as general guidance for practical design, he developed a parameter '*S*' given by

$$s = \frac{\boldsymbol{v_s}}{\boldsymbol{v_0}} (\mathbf{sin}\,\boldsymbol{\theta} + \boldsymbol{L}\,\mathbf{cos}\,\boldsymbol{\theta})$$

Where, $\boldsymbol{v_s}$ is – settling velocity of the particles,
v_0– mean flow through velocity
$\boldsymbol{\theta}$ – inclination of the tube
L – length/diameter of the tube

For a particular type of settler, there is a critical S value S_c. Theoretically all particles having S values greater than or equal to S_c will be removed completely. He discussed on design considerations and illustrated the applications of the equations developed.

1971: Slechta and Conley (1971) reported the experiences in plant scale application of the tube settlers with reference to the primary clarification and secondary clarification of activated sludge and trickling filter solids. He concluded that tube settler in clarification of activated sludge should be considered as a device for protecting clarifier efficiency under peak flow condition. In the clarification of raw and trickling filter solids the tube settlers increased the performance.

1972: Beach (1972) put forward an empirical relationship between maximum flow rate and tube dimensions. He claimed that this relationship took into account the effect of transition from turbulent to laminar flows on relative length and settling efficiency of particles in laminar flow.

1973: Yao (1973) reported that the removal efficiency of raw water turbidity decreases with increases in overflow rate. He observed 85 % removal of turbidity at an overflow rate of 1500 gpd/sq.ft (41.7 $m^3/d/m^2$). This efficiency of tube settlers of circular cross section exceeded that of conventional settling tanks of best performance.

1975: Zanoni and Bolomoquist (1975) considered flocculant solids to present a methodology for settling column test with a graphical procedure for data analysis and interpretation.

1976: De's (1976) studies on settling had the following outcomes:

1. He established a general framework (set of assumptions) under which settling in settling tanks of any shape can be ideally analysed.
2. He evaluated the inadequacies of conventional settling column analysis for discrete suspension and flocculant suspension and proposed a single method for the analysis of suspension irrespective of its nature (De 1998).
3. He also performed settling analysis under short circuiting to show that short circuiting from the depth-wise variation of velocity does not affect settling but that resulting from the widthwise variation of velocities impairs the same (De 1990, 2009c).
4. De discussed the disadvantages of all the measures for the comparison of settling performances of different tanks and proposed a new parameter for the same (De 1983b).
5. He employed the design criteria and the laboratory settling data to the process design.
6. De established complete theory of tube settling by deducing (i) the general equation of a particle trajectory through a tube settler, (ii) the critical fall velocity equal to or beyond which all particles having the settling velocities equal to and more than critical fall velocity will be removed completely and (iii) the fraction of particles settled, having settling velocity less than the critical fall velocity and demonstrated the application of the theory.
7. De also considered the tube settler as an ideal basin to design a real tube settling system.

1976: Krishnan (1976) followed Zanoni and Bolomoquist and dispensed with the drawing of iso-removal curves in accordance with the conventional method of analysis. At different constant times the removals of the collected samples from all the ports of settling column were determined and from these values the average solids in the column were determined directly.

1978: Fischer (1978) reported the use of shallow-depth sedimentation theory in designing 'lamella clarifier'.

1978: Grimes and Nyer (1978) presented some design considerations for 'lamella clarifiers'.

1978: Wills (1978) suggested maximum diameter of tubes in relation to the ratio or flow to the cross-sectional area of the tube module. He claimed these criteria would ensure laminar flow condition. He suggested that the maximum overflow rate should be within 85–400 gpd/sq.ft (3.5–16.3 $m^3/d/m^2$). But for better results, his recommendation is that the overflow rate should not exceed 250 gpd/sq.ft (10.2 $m^3/d/m^2$).

1979: Verhoff (1979) worked on parallel plate settlers based on the assumptions made by Yao (1970). He formulated an equation for critical settling velocity as a function of settler plate length and angle of inclination and optimised this critical velocity. He reported that this critical velocity should be minimum so that particles having settling velocity greater than this would be removed completely. He also suggested that to achieve the object of high-rate sedimentation, knowledge of settling velocity distribution of particles should be known. He worked with this by placing plate settlers in both rectangular and circular tanks.

For circular tanks, he concluded that the angle should be 25°–45° and upflow gives better results.

In case of a rectangular tank, he suggested that no optimisation for angle is required. So, for an assumed angle, only the length of settlers has to be optimised.

1980: Mendis and Benedek (1980) studied plant scale secondary clarifiers both with and without tubes. They concluded that when separation process is clarification, the tube settlers permit overflow rate up to 4 m/h (96 $m^3/d/m^2$) at solids loading rate up to 12 $kg/d/m^2$. This is 50–100 % greater than for basins without tube. When the separation process is thickening, they concluded that the tube did not provide additional capacity but improved the quality of effluent.

1980: Mazumdar (1980) analysed the effect of bending the tube settler in vertical plane on the settling of particles through it.

It was observed that bending reduces the length of the tube required without bending for a critical fall velocity of particle. This implies that bending a tube in vertical plane increases its removal efficiency.

1982: Berthouex and Stevens described the concentration profile of solids by a mathematical model as

$C(z,t) = a + bz + ct + dt^2 + ezt$, where $C(z,t)$ is the concentration of solids at depth z and time t.

1984: It appeared from the study that by providing curvature to a parallel plate settling system in vertical plane, called 'bent plate settling system', the system may be made more efficient with regard to the settling of particles and the continuous draining of sludge through the same.

Sinha (1984) and Sinha and De (1984) worked out the theory of bent plate settling system and plotted the particle trajectory through the system with the help of the theory in the way of illustration.

1985: Ong (1985) used Berthouex and Stevens model to use least square technique for the analysis of discrete settling data.

1986: A particle entering through the topmost point of a tube settler travels through a length to settle to the bottom. This length is 'critical length' for the particle. Mullick (1986) had studied the variation of the measured critical length with the variation of characteristic parameters of tube settling and compared with the values computed from theory using activated carbon and marble dust particles.

Mullick observed that the variation of critical length with the rate of flow, angle of inclination and Reynolds' number appeared to have been in accordance with the theory throughout the entire range of studies made by him up to Reynolds' number 645. So-called additional, transition, initial length mentioned in the theory (1986) was found non-existent.

1986: Roy (1986) made studies similar to that made by Mullick (1986) using sand, and fly ash up to Reynolds' number 5602. His conclusions were similar to that made by Mullick.

1988: Roy (1988) repeated the experimental studies made by Mullick (1986) and Roy (1986). He used plaster of paris, kaolin, chalk dust and colour pigment as particles and studied up to Reynolds' number 305. His conclusions did not differ from that of Mullick (1986) and Roy (1986).

1989: Hasan Ali (1989) also used mathematical description for column settling data analysis.

1989: Mehera (1989) studied and concluded similar to the studies made by Mullick (1986), Roy (1986) and (1988). He extended his studies up to Reynolds' number 3797.

1989: Ghosh (1989) investigated into the impairment of settling in a tube settler. He observed that ideal performance in tube settler was obtained even at Reynolds' number 1707. Ideal performance implied that the largest settling velocity of particle in the effluent was less than the critical fall velocity. Even then the settling performance was impaired. The impairment was due to scouring.

1990: Nandi (1990) studied upflow clarification through vertical tubes and found the mechanism of removal of settleable solids through them distinctly different from that through inclined tube system. The largest settling velocity (v_s) in the effluent through a vertical tube, the flow through velocity being v_0, is given by

$$\boldsymbol{v}_s = 1 \cdot 81 \boldsymbol{v}_0^{0.71}$$

1990: In the literature, removal of solids through a settling tank is described in terms of 'overflow velocity'. In number of cases, the 'weir loading' is also mentioned. Dependence of solids removal in a settling tank on both overflow velocity and weir flow velocity has been studied.

Acharya (1990), Acharya and De 1994) undertook a study with a laboratory-scale sedimentation tank to ascertain the dependence on the 'overflow velocity (v_0)' and weir loading, more specifically the 'velocity through the weir flow area (v_w)' as parameters for the description of settling performance.

From the analysis of experimental data, the removal of solids through a sedimentation tank was found to be a function of both 'weir flow velocity' and 'overflow velocity'. It appears, therefore, that removal through a settling tank cannot be described in terms of either overflow velocity (v_0) or weir flow velocity (v_w) only.

Both the parameters should be taken care of in the design of settling tank.

1992: Bhaskar et al. (1992) considered flocculant settling in column to evaluate removal efficiency.

1993: An upflow clarification system employing vertical baffles can provide tremendous flexibility in its operation by changing the distance of separation between the baffles.

Deb (1993) investigated such system to find the criteria for designing the same by finding out the relationship between the particle with the largest settling velocity (v_s) escaping with the effluent and the upward flow through velocity (v_0). He obtained the relationship

$$\boldsymbol{v}_s = 1 \cdot 79 \boldsymbol{v}_0^{0.71}$$

1998: For the efficient operation, maintenance and economic design of settling tank, the characteristics of settleable solids in raw water suspension should be related to their removal by a settling tank in its plant scale performance. The characteristics of settleable solids are studied by analysing the column settling data collected in laboratory.

To predict the removal in a plant scale settling tank, Camp advocated 'settling column analysis' and described an analysis for discrete suspension. The suspended solids encountered in domestic and industrial waste waters are usually flocculant in nature. O'Connor and Eckenfelder, Jr employed a different mode of analysis for flocculant suspension. They based their method on a conclusion that is valid for discrete particles only. Since then all the standard text on the subject describes two modes of analysis—one for discrete suspension and the other for flocculant one. Literature on the subject about the analysis for suspension that is a combination of both of them is silent.

Conventional modes of such analyses are based on assumptions and differ widely for discrete and flocculant suspensions. Inadequacies of such analyses are pointed out. De (1976) put forward a direct rational mode of analysis regardless of the nature of suspension, i.e. discrete or flocculant and independent of any assumption. Laboratory test data have been analysed to illustrate the mode of analysis.

2002: De (2002) investigated on the instantaneous velocities of a settling particle employing method of successive approximation. He employed a new method that produces direct solutions to settling velocity determination.

2005: De (2005) worked out a methodology to set the bases for setting the speed and duration of rotations of the paddles during 'flash mixing ', 'slow mixing' and also the 'settling time' in the jar test procedure.

Till date the question of compatibility of operating a real settling system in accordance with jar test results has been left out without the recognition of its significance. For the compatible operation of a real settling system according to the jar testing procedure, *Gt* values in the jar for flash mixing and slow mixing should be equal to those values in the real settling system.

In order to exemplify the design of jar testing procedure for compatible operation of a real settling system, the 'settling system' of Serampore Water Treatment Plant, processing 5000 m^3 per hour of water, was taken into consideration. Jar testing procedure was designed for the compatible operation of the real settling system.

Thus the methodology may serve towards the standardisation of the procedure that is practised with indiscriminate arbitrariness throughout the globe so far.

2006: Overcamp (2006) carried out analysis on flocculant settling data.

2009: Velocity Profile Theorem (De 2009a) is a new concept. It is simple and can help solving the settling problem analysis through any settling system. Velocity Profile Theorem has been employed to deduce the 'theory of ideal settling' and establish the complete 'theory of tube settling'. Application of the theorem to solve numerical problems has been demonstrated by solving a numerical problem.

2009: The cost-saving potentiality of shallow-depth settling system has been known for a long time. For controlling the parameters of tube settling presented herein is a procedure to control the design parameters of tube settling system to fix their coordinated values for optimised design (De 2005).

Quantitative changes in the critical fall velocity for small changes in one or more of the design parameters have been worked out. This provides solution for adjustment for small changes in the values of the parameters.lt has been shown that increase in the angle of inclination and also the mean velocity of flow through the tube settler and its radius settling performance deteriorates. With the increase in the length, performance of the tube settler improves. Limitations of the values of the design parameters for optimised design of the tube settler have been worked out. An example has been solved to demonstrate how to control the values of the parameters for the optimised combination of the same.

2009: Yao published his theoretical study on tube settling in 1970. He deduced the expression for the 'critical fall velocity' through tube settler under laminar flow condition. It was suggested that the designed length of the tube settler need to include an initial length for the development of laminarity in the flow. These being the basic to the design of the tube settling need experimental verification.

The experimental verification was undertaken (De 2009) at Environmental Engineering Laboratory of CE Dept, Jadavpur University, Kolkata. Different suspensions were sent through inclined tubes of varying lengths and varying diameters. The critical lengths traversed by particles of varying settling velocities were measured. The following observations were recorded.

1. Even within a turbulent flow with high value of Reynolds' number settling of particles in the tube takes place in accordance with the theory deduced under laminar flow condition.
2. The settling of particles in tube settler takes place according to the theory without the necessity and provision of an additional or transition length for the development of laminarity in the flow. This length has been found to be redundant and may be done away with. It appears that no need is there to include the so-called transition length in the designed length of the tube settler.
3. Settling of particles in the tube settler is impaired, while the particles settle according to the theory even at very high value of Reynolds' number. This impairment is due to scouring.

In the design of tube settler, the scouring should be the main consideration and not the Reynolds' number of flow.

2009: 'Velocity Profile Theorem' has been applied to study the effect of short circuiting (De 2009b) on settling. The study reveals that short circuiting arising out of the variation of flow velocity along with the width of the settler impairs the settling performance. For the design of an efficient settler, the inlet width should be made narrow. The flow velocity variation along the depth of the settler does not affect the settling performance in any way. This leaves a scope for redistribution of velocity component vectors, in the direction of flow, along the depth to the convenience of calculating the actual removal of solids, the redistribution having maintained the same rate of flow through the settler.

2010: Based on the settling data of raw water suspension, a methodology was developed to compute the residual concentration of solids through the tube settler. Laboratory settling data have been employed to illustrate the numerical application of the methodology (De et al. 2009) to work out the effluent concentration of solids through a given tube settler carrying the raw water suspension at a given rate.

2010: Couette flow settler (De 2010b) can be a successful application of shallow-depth sedimentation. The theory of couette flow settling has been worked out and presented. The system adjustment for the minor quantitative changes in the system parameters can be controlled as indicated by the expressions derived herein. The basis and the procedure of design of couette flow settling module have been presented and illustrated by working out a problem.

2011: Pise and Halkude (2011) modified Krishnan's (1976) method of analysis by averaging the removal values of the settling solids along the depth of the column to compute the total removal in the column.

2.2 Developments

From the foregoing review of literature, the following salient developmental steps may be noted:

1. Sheddon's (1889) three observations were very significant understandings in the development of the application of settling phenomenon. The observations are as follows:
   (i) Continuous flow operation in settling tank can perform as good as intermittent operation.
   (ii) Baffles could reduce the volumetric capacity of the settling tank.
   (iii) Incoming momentum, mixing resulting from temperature variation and wind action were factors affecting settling performance.
2. Hazen (1904) was the first to initiate the theoretical study on the phenomenon of settling in a tank in continuous operation. Hazen evolved

   the concept of 'ideal settling tank' working under hypothetical conditions to deduce settling performance in terms of ratio of settling velocity of particles and 'overflow velocity ' or surface loading. He conceived of virtual baffles to take into account the effect of different patterns of flow in his theory. In totality different flow patterns give rise to different patterns of unequal times of flow of the elements through the tank. This is referred as a phenomenon of 'short circuiting'.
3. Newton (1910) deduced expression of drag on a falling particle. This led to the expression for settling velocity of a particle.
4. Oden (1925) could describe the settling velocity distribution among the particles in suspension during their batch settling.

5. Shields (1936) had shown that the particles on the sediment bed will not move if the quantity 0.1 v_s^2 is less than $\frac{gd(\rho_s - \rho_l)}{\rho_l}$
   Where,

   v_s – settling velocity of a particle
   g – acceleration due to gravity
   d – diameter of the particle
   ρ_l – density of the liquid
   ρ_s – density of the solid

6. Camp (1936) observed that long narrow tanks $\left(\frac{\text{Lenth}}{\text{width}} \text{ratios } 3:1 \text{ to } 5:1\right)$ are better settlers.
7. Camp and Stein (1943) deduced expressions for number of contacts between two types of particles per unit volume per unit time due to velocity gradients within suspension.
8. Camp (1946) described and advocated settling column analysis for efficient design of settling tank.
   He considered the analysis in detail for discrete suspension only. He deduced clarification theory for ideal basin on the basis of settling column analysis of discrete influent suspension. Camp also deduced channel velocity required to start motion of particles towards the design of settling tank.
9. Eliassen (1946) criticised Camp with the help of a demonstration that according to Camp's ideal basin theory, the phenomenon of short circuiting does not affect removal.
10. Fischerstrom (1955) advocated maintaining proper hydraulic condition at Reynolds' number 500 or less and also proper overflow rate for efficient removal of solids. This can be achieved with the help of longitudinal baffles.
11. Talmadge and Fitch (1955) could relate batch settling data to the determination of unit areas both as clarifier and thickener.
12. Ingersoll et al. (1956) proposed 'overflow residual efficiency' as parameter for the comparison of settling tank performances.
13. Fitch (1957b) described four characteristic types of settling as discrete settling or class-1 clarification, flocculant settling or class-2 clarification, zone settling and compression.
    This classification paved towards the development of the rationalised theory of settling.
14. O' Connor and Eckenfelder (1957) put forward the method of settling column analysis with flocculant suspension for the computation of removal of flocculant solids through settling tank.
15. Hansen and Culp (1967) pointed out that shallow-depth sedimentation using trays could not be implemented because unstable hydraulic condition resulted and the minimum spacing of trays were limited by the sludge removal mechanism.
    The problem was overcome by using small diameter (1–4 in.) of tubes of 2–4 f. length.

16. Yao (1970) attempted deduction of shallow-depth sedimentation theory.
17. De (1976) studied settling column analysis, analysis of phenomenon of short circuiting, measures of settling performance comparison and tube settling theory. He:

    (i) Pointed out the inadequacies of 'settling column analyses' made by both Camp and also that by O'Connor and Eckenfelder. A method of analysis, irrespective of the nature of suspension and not based on any assumption, was proposed.
    (ii) Showed that short circuiting resulting from depth-wise variation of velocities does not affect removal of solids but that resulting from the widthwise variation of velocities impairs the same
    (iii) Proposed a new measure for the settling performance comparison
    (iv) Established complete theory of tube settling

18. De et al. (2009) conducted experimental verification of tube settling theory in 1986, 1988, 1989.
19. Ghosh (1989) investigated into the impairment of settling in a tube settler.
20. Nandi (1990) studied upflow clarification through vertical tubes and obtained largest settling velocity v_s of particle in the effluent through overflow velocity v_0 as

$$v_s = 1 \cdot 81 v_0^{0.71}.$$

21. Acharya (1990) through laboratory-scale study could show that in the design of settling tank, both 'overflow velocity' and 'weir flow velocity ' or 'weir loading' should be considered the removal of solids being dependent on them.
22. Deb (1993) investigated an upflow clarification system employing vertical baffles and observed the largest settling velocity of particle escaping with the effluent and the upflow through velocity v_0 are related as

$$v_s = 1 \cdot 79 v_0^{0.71}.$$

23. De (2002) investigated on the instantaneous velocities of a settling particle employing method of successive approximation. A new method that produces direct solution to settling velocity calculation avoiding trial was proposed.
24. De (2005) – For the compatible operation of a real settling system, a methodology was worked out to set the bases for setting the speed and duration of rotations of the paddles during 'flash mixing', 'slow mixing' and also 'settling time' in 'jar test' procedure.
25. De (2009a, b, c) devised and established 'Velocity Profile Theorem' that can be employed to solve any settling velocity problem.
26. De (2009a, b, c) worked out quantitative changes in the critical fall velocity for small changes in one or more of design parameters for shallow-depth sedimentation system.

27. De (2009c) analysed the phenomenon of short circuiting through 'Velocity Profile Theorem'.
28. De (2010a) presented computational methodology for calculating residual solids through tube settler using settling column test data with influent suspension.
29. De (2010b) worked out 'theory of couette flow settling' and presented a design procedure for the same.

References

Acharya TK (1990) Dependence of solids removal through a settling tank on overflow velocity and weir flow velocity. M.C.E, thesis, Jadavpur University, Kolkata, India

Acharya TK, De A (1994) Overflow velocity and weirflow velocity – a study for their significance – Recent researches in Ecology. Environ Pollut 9:257–269

Ali San H (1989) Analytical approach for evaluation of settling data. J Environ Eng Div ASCS 115(2):455–461

Barham WR et.al (1956) Clarification, sedimentation and thickening equipment patent review. Bulletin No.54, published by Engineering Experiment Station, Louisiana State University, Baton Rouge La

Beach WA (1972) Fundamentals of tube settler design. In: 27th Industrial waste conference, Purdue University, p 805

Bergman BS (1958) An improved sedimentation design. J Proc Ins Sewage Purif South African Branch, Part I, p 50–67

Berthouex PM, Stevens DK (1982) Computer analysis of settling data. J Environ Eng ASCE 108 (5):1065–1069

Bhaskar PU, Chuadhuri S, Jawed M (1992) Type II sedimentation – removal efficiency from column settling test. J Environ Eng ASCE 118(3):757–760

Camp TR (1936) A study of rational design of settling tanks. Sewage Work J 8:742–758

Camp TR (1946) Sedimentation and the design of settling tasks. Trans Am Soc Civ Eng 111:895

Camp TR (1953) Sedimentation basin design. Sewage Ind Waste 25:1

Camp TR, Stein PG (1943) Velocity gradients and internal work in fluid motion. J Boston Soc Civ Eng 30:219

Capen CH (1927) Study of sewage settling tank design. Eng News Rec 99:833

Culp GL, Culp RL (1970) New concepts in water purification. Van Nostrand Co Inc., New York

Culp GL, Hsiung KY, Conley WR (1969) Tube clarification process – operating experiences. ASCE San Eng Div SA5, p 829

Dallas JL (1957) Uniflow tank – an improved settling tank. Biological treatment of sewage and industrial wastes, vol 2. Reinhold Pub. Corp.,New York

De A (1976) Conceptual studies on discrete and flocculent settling. PhD (Eng) thesis, Jadavpur University, Calcutta, India

De A (1982) Critical appraisal of Hazen's theory on sedimentation. J IPHE India, no.2, 43

De A (1983a) Application of ideal basin concept to the design of a real tube settling system. J IPHE, India, no.3, 29

De A (1983b) Parameter for settling tank performance comparison. J IPHE, India, no.4, 21

De A (1990) Effect of short circuiting on the basin efficiency. J IPHE 2, 37

De A (1998) Revised mode of analysis of column settling data. Indian Chem Eng Section B 40, No 4, Oct–Dec

De A (2002) New methods of solutions of settling velocity problems. J IPHE India, Vol 2002 No 2

De A (2005) Design of Jar testing procedure and compatible operation of a real settling system. J IPHE India 2005 No 5

De A (2009a) Velocity profile theorem – Concept for solving settling problem analysis. J IPHE India vol 2009–10, No 1
De A (2009b) Theoretic study on the control of design parameters for tube settling. J IPHE India 2009–10 No 2
De A (2009c) Analysis of the effect of short – circuiting on settling – an application of velocity profile theorem. J IPHE India 2009–10 No 4
De A (2010a) Computational methodology for residual solids through tube settler. J IPHE India 2010–2011 No 1
De A (2010b) Design of simple couette flow module for removal of solids. J IPHE India 2010–11 No 2
De A et al (2009) Experimental verification of the theory of tube settling. JIPHE India 2009–10, No. 3
Deb B (1993) Upflow clarification through vertical baffles – its design and mechanism of removal. M.C.E thesis, Jadavpur University, Kolkata, India
Dobbins WF (1944) Effect of turbulence on sedimentation. Trans ASCE 109:629
Dresser HG (1951) Trays nearly triple settling tank capacity. Eng News Rec p 32
Eliassen R (1946) Discussion, sedimentation and the design of settling tank. Trans ASCE 111:895
Fischer MC (1978) Wet scrubber water treatment in the iron and steel industry using Lamella gravity settlers. In: Proceedings of 33rd industrial wastes conference, Purdue University, p 808
Fischerstrom CNH (1955) Sedimentation in rectangular basins. Proc ASCE J San Eng Div
Fitch EB (1956) Flow path effect on sedimentation. Sewage Ind Waste 28:1
Fitch EB (1957a) The significance of detention in sedimentation. Sewage and Ind Waste 29:1123
Fitch EB (1957b) Sedimentation process fundamentals. Biological treatment of sewage and industrial wastes, vol 2. Reinhold Pub Corpn., New York
Francis AW (1933) Wall effect in falling bell method for viscosity. J Appl Phys 4:403–406
Ghosh A (1989) Impairment of settling in a tube settler. MCE thesis, Jadavpur University, Calcutta, India
Grimes CB, Nyer EK (1978) Lamella clarification: Design application. In: Proceedings of 33rd industrial waste conference, Pardue University, p 950
Hansen SP, Culp GL (1967) Applying shallow depth sedimentation theory. J AWWA 59:1134
Hansen SP, Culp GL, Richardson G (1968) High rate sedimentation in water treatment work. J AWWA 60:681
Hansen SP, Culp GL, Richardson G, Stukenberg JR (1969) Practical application of idealised sedimentation theory in waste water treatment. J WPCF 41:1421
Hasan A (1989) Analytical approach for evaluation of settling data. J Environ Eng Div ASCS 115(2):455–461
Hazen A (1904) On sedimetation. Trans ASCE 53:45
Hernandes JW, Wright JR (1970) Design parameters for tube settlers. In: Industrial waste conference, Purdue University, p 805
Homack P (1941) A study of the precipitates formed in line softening. MS thesis, submitted to Mass. Inst. Tech., Cambridge
Ingersoll AC, Mckee JE, Brooks NH (1956) Fundamental concepts of rectangular settling tanks. Trans ASCE 121:1179
Katz WJ, Geinopolos A (1957) A comparative study of the hydraulic characteristics of two types of circular solids separation basins. Biological treatment of sewage and industrial wastes, vol 2, Reinhold Pub.Corpn., New York
Kermack WO, Mckenderick AG, Ponder E (1928–1929) The stability of suspensions. The velocities of sedimentation and cataphoresis of suspensions in various fluid. In: Proceedings of the Royal Society of Edinburgh, vol XLIX
Krishnan P (1976) Column settling test for flocculant suspension. J Environ Div EI:227–229
Lesprance TW (1957) Application of fundamental to waste treatment sedimentation design. Biological treatment of sewage and industrial wastes, vol 2, Reinhold Pub.Corpn., New York

Mazumdar K (1980) Theoretical study of bent tube settling system. MCE thesis, Jadavpur University, Calcutta, India
Mehera AK (1989) an investigation into the extent of adherence of tube settling performance to its theory. MCE thesis, Jadavpur University, Calcutta, India
Mendis JB, Benedek A (1980) Tube settlers in secondary clarification of domestic waste water. J WPCF 52:1893
Mullick S (1986) Critical length determination for a tube settling system using activated carbon and marble dust.-M.C.E thesis, Jadavpur University, Calcutta, India
Munroe HS (1988–1989) The English versus continental system of jigging – is close sizing advantageous?. Trans AIMME 117:637–659
Nandi S (1990) Upflow tube clarification – its design and mechanism of removal. MCE thesis, Jadavpur University, Calcutta, India
Newton I (1910) Mathematical principles of natural philosophy. Vol 2, Sections 2 and 7
O'Connor DJ, Ecknfelder WW Jr (1957) Evaluation of laboratory settling data for process design. Biological treatment of sewage and industrial wastes, vol 2, Reinhold, Pub.Corpn., New York
Oden S (1925) The size distribution of particles in soils and the experimental methods of obtaining them. Soil Sci 19:1
Ong SL (1985) Least square analysis of Settling data under discrete settling conditions. Water SA 11(4)
Pise CP, Halkude SA (2011) A modified method for settling column data analysis. Int J Eng Sci Technol 3(4):3177–3183
Roy T (1986) Sedimentation of sand and fly ash in light of tube settling theory. MCE thesis, Jadavpur University, Calcutta, India
Roy K (1988) An experimental study of tube settling parameters and their relationship. M.C.E thesis, Jadavpur University, Calcutta, India
Rudolf W, Lacy IO (1934) Settling and compacting of activated sludge. Sewage Work J VI(4):647
Schmitt EA, Voigt OD (1949) Two storey flocculation sedimentation basin for the Washington Aqueduct. J Am Water Works Assoc 41:837
Sheddon (1889) Cleaning water by settlement. J Assoc Eng Soc p 477
Shields A (1936) An Wendung der Aehnlich Keitsmechanik under Turbulenz forschung auf die Geschiebebewe – gung. Mitteilungen der Preussischen Versuchsanstalt fur Wasserban und Sehiffbau,Heft 26, Berlin
Sinha A (1984) Theory of bent plate settling and design of the system module. MCE thesis, Jadavpur University, Calcutta, India
Sinha A, De A (1984) Theory of bent plate settling system. J IPHE India No 4, 67
Slechta AP, Conley WR (1971) Recent experiences in plant scale application of the settling tube concept. J WPCF 43:1725
Talmadge WP, Fitch EB (1955) Determining thickener unit areas. Ind Eng Chem 47:38
Tekippe RJ, Cleasby JL (1968) Model studies of a peripheral feed settling tank. J Sanit Eng Div 94 (1):85–102
Thomas HA, Archibald RS (1952) Longitudinal mixing measured by radioactive tracers. Trans ASCE 117:839
Verhoff (1979) Optimal design of high rate sedimentation devices. J ASCE Environ Eng Div 105:199
Willis RM (1978) Tubular settlers – a technical review. J AWWA 70:531
Yao KM (1970) Theoretical study of high rate sedimentation. J WPCF 42:218
Yao KM (1973) Design of high rate settlers. ASCE Environ Eng Div 99:621
Zanoni A, Bolomoquist M (1975) Cloumn Setling test for floculant suspension. J Environ Eng ASCE 101(B):309–318

Chapter 3
Velocity Profile Theorem

Abstract Velocity Profile Theorem is a new concept. It is simple and can help solving the settling problem analysis through any settling system.

The simple concept has been introduced. The theorem has been employed to deduce ideal settling theory and establish the complete theory of tube settling.

Keywords Velocity Profile Theorem • Velocity Profile diagram • Ideal settling theory • Tube settling theory • Solids removal

3.1 Velocity Profile Theorem and Its Application to Deduce Settling Theories

3.1.1 Velocity Profile Theorem

X, Y and α are three mutually perpendicular axes. Consider a flow section at distance α_i from X-Y plane. The flow lines are parallel to the X-axis and inclined at an angle θ with the horizontal (Fig. 3.1).

A particle having settling velocity v_s entering through the point $(0, y, \alpha_i)$ will start moving forward in the direction of X–axis with velocity $\phi(y)$ and the resolved component of v_s in the direction of X,$-v_s \sin\theta$ and that in the direction of Y,$-v_s\cos\theta$.

In time dt it falls through dy moving through a distance dx given by

$$dx = (\phi(y) - v_s \sin\theta)(-)\frac{dy}{v_s \cos\theta}$$

It falls from y_1 to y_2 while moving from x_1 to x_2 and accordingly

$$\int_{x_1}^{x_2} dx = \int_{y_1}^{y_2} (\phi(y) - v_s \sin\theta)(-)\frac{dy}{v_s \cos\theta}$$

A. De, *Sedimentation Process and Design of Settling Systems*,
Springer Transactions in Civil and Environmental Engineering,
DOI 10.1007/978-81-322-3634-4_3

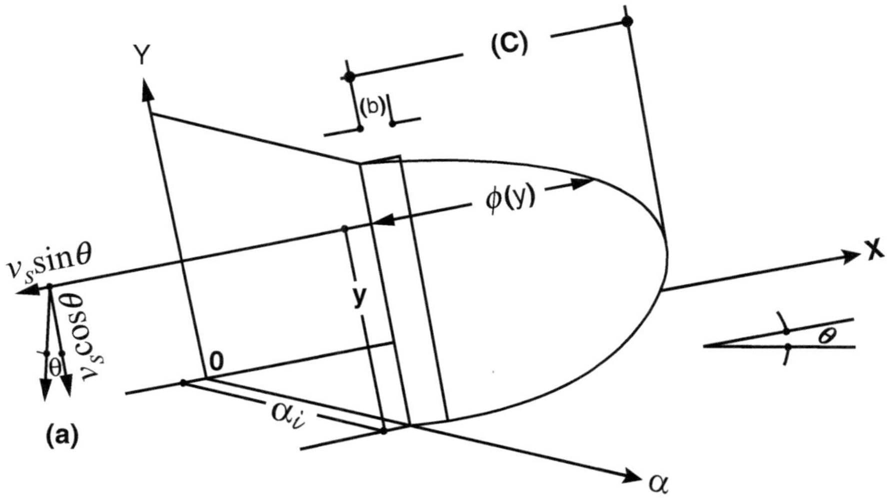

Fig. 3.1 Velocity Profile diagram

$$\text{i.e}\,(x_2 - x_1) = \frac{\int_{y_2}^{y_1} \phi(y)dy - (y_1 - y_2)v_s \sin\theta}{v_s \cos\theta} \tag{3.1}$$

In the above equation,

$$\begin{aligned}&\text{Distance moved through in the direction of X} - \text{axis}\\ &= \frac{(\text{Area of flow velocity diagram} - \text{Area of particle velocity diagram})\text{ over change of Y}}{\text{Particle velocity component in the} -\text{ve direction of Y}}\\ &= \frac{\text{Area of velocity profile diagram over the change of Y}}{\text{Particle velocity component in the} -\text{ve direction of Y}}\end{aligned}$$

Area of flow velocity diagram between y_1 and y_2

$$= (x_2 - x_1)v_s \cos\theta + (y_1 - y_2)v_s \sin\theta \tag{3.2}$$

Staring from a point (x_1, y_1, α_i), the coordinate x_2 where the particle with settling velocity v_s falls to a chosen depth y_2 can be computed from Eq. 3.1 using area of flow Velocity Profile diagram between y_1 and y_2.

From the set of such values of (x_2, y_2), the trajectory of the particle may be drawn through them.

These simple computations in Eqs. 3.1 and 3.2, herein after, are to be known as 'Velocity Profile Theorem'. This theorem can deduce complicated settling problem analysis through any settling system.

3.1.2 *Computation of Area of Velocity Profile Diagram*

Computation of the area of Velocity Profile diagram between y_1 and y_2 requires such computation for the flow velocity diagram and the particle velocity diagram.

Such computation for the flow velocity diagram may be done:

1. By graphical integration of the flow velocity diagram and the particle velocity diagram
2. By simple integration if the flow velocity diagram is defined by equation
3. From the expression

   Area of flow velocity diagram between y_1 and y_2

$$= (x_2 - x_1)v_s \cos\theta + (y_1 - y_2)v_s \sin\theta$$

$$= (\text{Increase in } X)(\text{Particle vel. component along } -\text{ve } Y) + (\text{Decrease in } Y) \times (\text{Particle vel. component along } -\text{ve } X)$$

Computation of area of particle velocity diagram is simple and is

$$= (y_1 - y_2)v_s \sin\theta$$

$$= (\text{Decrease in } Y)(\text{Particle vel. component along } -\text{ve } X).$$

3.2 Application to Deduce Settling Theories

3.2.1 *Ideal Settling Theory*

Ideal settling tank works under hypothetical assumptions never realised in practice. Even then the concept is important since the settling efficiency is described in terms of overflow velocity till date.

L (length) $\times$ B (width) $\times$ D (depth) is the settling zone of an ideal settling tank. It is fed with flow rate Q, carrying solids concentration C_s consisting of identical particles as regards their settling velocities.

Total solids entering into the zone per second $= QC_s$.

A critical particle having critical settling velocity v_c entering at the top falls through D travelling the length L of the settling zone.

By Velocity Profile Theorem (Fig. 3.2),

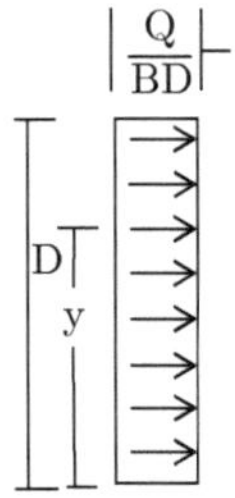

Fig. 3.2 Flow velocity diagram, also Velocity Profile diagram

$$L = \frac{(D)\left(\frac{Q}{BD}\right) - (D) \cdot 0}{v_c}$$

$$\text{i.e}\, v_c = \frac{Q}{BL}$$

= overflow velocity (v_0), surface loading or critical velocity

Obviously all particles having $v_s \geq v_0$ will be removed completely. For particles having settling velocity $v_s < v_0$, a particle entering at height y from the bottom will fall through y travelling the distance L.

By Velocity Profile Theorem (Fig. 3.2),

$$L = \frac{(y)\left(\frac{Q}{BD}\right)}{v_s}\ \text{i.e}\ y = \frac{LBDv_S}{Q}$$

i.e. all such particles entering through this bottom depth y will be completely removed.

The removal through the bottom depth y is $(y)(B)\left(\frac{Q}{BD}\right)C_s$ per second.

$$= \frac{(LBDv_S)}{Q}\left(\frac{Q}{BD}\right)(B)C_s\ \text{i.e}\,(Lv_s)(B)C_s;$$

Total solids entering into the zone per second – QC_s.

Such solids are removed in the ratio:

$$= \frac{(Lv_s)(B)C_S}{QC_s}\ \text{i.e}\, v_s/\frac{Q}{BL},\ \text{i.e}\frac{v_s}{v_0}.$$

3.2.2 *Theory of 'Tube Settling'*

During the early and mid-1960s of the last century, employing detention time of few minutes through inclined tubes, trays, etc., the so-called high-rate settling system became popular.

Yao (1970) deduced the trajectory equation of a settling particle entering through the vertical diameter of an inclined tube, which he claimed to be a general equation. This equation not being a general one could not be utilised to deduce the complete theory of 'tube settling system'.

De (1976) deduced the general equation and established the complete theory of 'tube settling system'; Velocity Profile Theorem is employed here to deduce the complete theory of 'tube settling system' in a simpler way.

Flow Velocity Through 'Inclined Tube' in Terms of Three Mutually Perpendicular Axes

The end areas of circular tube cross sections have their centres at (0, R, 0) and (L, R, 0) in Fig. 3.3. Flow velocity $f_{y,\alpha}$ through any point $(0, y, \alpha)$ can be written from any standard textbook on the subject for the flow rate Q through the tube as

$$f_{y,\alpha} = \frac{2Q}{\pi R^4}\left(2yR - y^2 - \alpha^2\right) \tag{3.3}$$

Any particle having settling velocity v_s entering through $(0, y, \alpha)$ has its settling velocity components

$-v_s \sin\theta$ and $-v_s \cos\theta$ in the direction of X-axis and Y-axis, respectively.

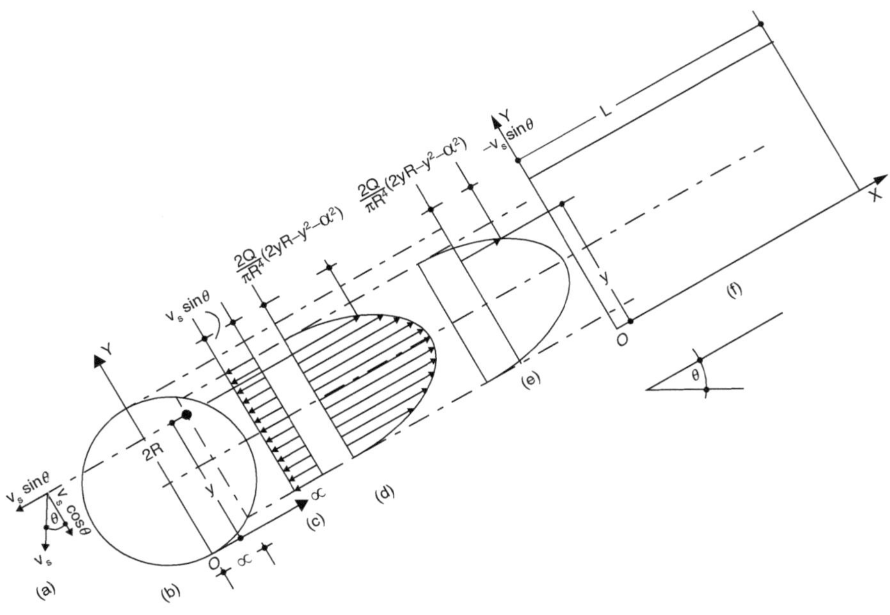

Fig. 3.3 Tube Settling (*a*) Particle velocity components; (*b*) Tube cross section and the entry point of the particle (0,*y*,∝); (*c*) Particle velocity diagram; (*d*) Flow velocity diagram; (*e*) Velocity Profile diagram; (*f*) Inclined tube

The particle will start moving with velocity:

$$u_{y,\,\alpha} = \frac{2Q}{\pi R^4}\left(2yR - y^2 - \alpha^2\right) - v_s \sin\theta \tag{3.4}$$

This will give the velocity distribution profile on any chord {Fig. 3.3e} on the tube cross section at distance $\propto$ from X-Y plane.

Critical Fall Velocity

The critical fall velocity v_{ci} through the i-th chordal section is the fall velocity of a particle entering through the point $\left(O, y_{1i} = R + \sqrt{R^2 - \alpha_i^2}, \alpha_i\right)$ and just reaching $(L, y_{2i} = R - \sqrt{R^2 - \alpha_i^2}, \alpha_i)$ travelling through a distance L,

By Velocity Profile Theorem,

$$L = \frac{\frac{2}{3}(2\sqrt{R^2 - \alpha_i^2})\left(\frac{2Q}{\pi R^4}\right)(R^2 - \alpha_i^2) - (2\sqrt{R^2 - \alpha_i^2})v_{ci}\sin\theta}{v_{ci}\cos\theta}$$

$$\text{i.e}\, v_{ci} = \frac{8Q\left(R^2 - \alpha_i^2\right)^{3/2}}{3\pi R^4\left(L\cos\theta + 2\sqrt{R^2 - \alpha_i^2}\sin\theta\right)} \tag{3.5}$$

All particles through the i-th chordal section having $v_s \geq v_{ci}$ will be completely removed.

The nearer this chord is to the centre of the section, the more is the magnitude of v_{ci} through the chordal section. The magnitude of v_{c0} on the diametral section is the maximum and it is

$$v_{c0} = \frac{8Q}{3\pi R(L\cos\theta + 2R\sin\theta)} \tag{3.6}$$

Hence all particles having settling velocity $v_s \geq v_{c0}$ will be removed through the section.

Solids Removal

Flow rate Q at solids concentration C_S will carry solids QC_S per second into the tube. To compute the removal through the settler, the cross section may be imagined to be divided into n strips each of width $w = 2R/n$. The number of strips n should be evenly distributed on either side of the vertical diameter for the covenience of computational work. The strips are marked by their central chord.

Critical Fall Velocity Through the *i*-th Chordal Section

$$v_{ci} = \frac{8Q\left(R^2 - \alpha_i^2\right)^{3/2}}{3\pi R^4\left(L\cos\theta + 2\sqrt{R^2 - \alpha_i^2}\sin\theta\right)}$$

$$= \frac{Q(\text{Chordlength})^3}{3\pi R^4\left(L\cos\theta + (\text{Chordlengh})\sin\theta\right)} \quad (3.7)$$

3.2.2.1 Computation of Solids Removal for Particles Having $v_s < v_{ci}$

All those particles having $v_s < v_{ci}$ entering through the i-th chordal section will move through the length from $(0, y_{1i}, \alpha_i)$ to $\left(L, y_{2i} = R - \sqrt{R^2 - \alpha_i^2}, \alpha_i\right)$ to be removed.

$$\text{i.e} \int_{y_{1i}}^{y_{2i}} \left[\frac{2Q}{\pi R^4}\left(2yR - y^2 - \alpha_i^2\right) - v_s\sin\theta\right](-)\frac{dy}{v_s\cos\theta} = L$$

$$i.e\, ay_{1i}^3 - by_{1i}^2 + cy_{1i} - d = 0 \quad (3.8)$$

where $a = \frac{2Q}{3\pi R^4 v_s\cos\theta}$; $b = 3aR$; $c = \left(3a\alpha_i^2 + \tan\theta\right)$;

$$d = \left(ay_{2i}^3 - by_2^2 + cy_{2i} - L\right)\,\text{and}\, y_{2i} = R - \sqrt{R^2 - \alpha_i^2} \quad (3.9)$$

y_{1i} is obtained by solving the cubic Eq. 3.8.

Now, the area of flow diagram between y_{1i} and y_{2i}

$$= \left[Lv_s\cos\theta + (y_{1i} - y_{2i})v_s\sin\theta\right]$$

Solids removal through this section $= C_s w\left[Lv_s\cos\theta + (y_{1i} - y_{2i})v_s\sin\theta\right]$ (3.10)

Solids removal through all such strips

$$= C_s w\Sigma\left[Lv_s\cos\theta + (y_{1i} - y_{2i})v_s\sin\theta\right] \quad (3.11)$$

3.2.2.2 Computation of Solids Removal for Particles Having $v_s \geq v_{ci}$

Here the area of flow diagram between y_{1i} and y_{2i}

$$= [Lv_{ci}\cos\theta + 2\sqrt{R^2 - \alpha_i^2}\ v_{ci}\sin\theta]$$

i.e. the solids removal through the strip

$$= C_s w\left(Lv_{ci}\cos\theta + 2\sqrt{R^2 - \alpha_i^2}\ v_{ci}\sin\theta\right) \quad (3.12)$$

Solids removal through all such strips

$$= C_s w\Sigma(Lv_{ci}\cos\theta + 2\sqrt{R^2 - \alpha_i^2}\ v_{ci}\sin\theta) \quad (3.13)$$

Total solids removal through the tube cross section is the sum of Eqs. 3.11 and 3.13

$$= C_s w\Sigma\left(Lv_s\cos\theta + (y_{1i} - y_{2i})v_s\sin\theta\right) + C_s w\Sigma(Lv_{ci}\cos\theta + 2\sqrt{R^2 - \alpha_i^2}\ v_{ci}\sin\theta)$$
$$= C_s w\Sigma(Lv_s\cos\theta + (y_{1i} - y_{2i})v_s\sin\theta),$$

Where $v_s = v_{ci}$, $y_{1i} = R + \sqrt{R^2 - \alpha_i^2}$, $y_{2i} = R - \sqrt{R^2 - \alpha_i^2}$ for section when $v_s \geq v_{ci}$ and

$v_s = v_s$, y_{1i} is calculated from Eq. 3.8, $y_{2i} = R - \sqrt{R^2 - \alpha_i^2}$ for section where $v_s < v_{ci}$.

3.2.3 Application to Numerical Problem

Problem A 50-cm-long tube of diameter 5 cm, inclined at an angle of 30° with the horizontal, is employed for the removal of solids from a flow rate of 0.06 litres/sec with concentration of solids of 100 mg/l consisting of particles that are all identical as regards their settling velocities of 0.3 cm/sec. Calculate the concentration of solids in the effluent.

Solution
Divide the cross section into 10 strips each of width 0.5 cm, 5 strips being on either side of the vertical diameter, marked by the identification numbers of their central chords. The problem is worked out in steps that are tabulated in Table 3.1

Table 3.1 Calculation of solids in the effluent

1. No.	Steps	Chord identification numbers $i =$ (1)	(2)	(3)	(4)	(5)
1.	Distance of the chord from the centre α_i cm	0.25	0.75	1.25	1.75	2.25
2.	Chord length= $2\sqrt{R^2-\alpha_i^2}$cm	4.97494	4.76970	4.33013	3.57071	2.17945
3.	The bottom point of the chord $-y_{2i}=R-\sqrt{R^2-\alpha_i^2}$cm	0.01253	0.11515	0.33494	0.71464	1.41028
4.	Critical velocity v_{ci} through the chordal section in cm/sec. Calculated from Eq. 3.7	0.43824	0.38708	0.29102	0.16456	0.03801
5.	y_{1i} cm (calculated from Eq. 3.8. where$v_s(=0\cdot 3\,\text{cm/s})$ less than v_{ci} and $y_{1i}=R+\sqrt{R^2-\alpha_i^2}$ where $v_s \geq v_{ci}$) (Appendix)	3.07486	3.37267	4.66506	4.28536	3.58972
6.	Velocity v_s with which particle falls from $y_{1i}$to y_{2i} travelling through the tube length	0.3	0.3	0.29102	0.16456	0.03801
7.	$Lv_s\cos\theta\text{cm}^2/\text{sc}$	12.99038	12.99038	12.60154	7.12566	1.64588
	$\Sigma Lv_s\cos\theta\text{cm}^2/\text{sc}$	47.35384				
8.	$(y_{1i}-y_{2i})v_s\sin\theta\text{cm}^2/\text{sc}$	0.45935	0.48863	0.63008	0.29380	0.4142
	$\Sigma(y_{1i}-y_{2i})v_s\sin\theta\text{cm}^2/\text{sc}$	2.28606				
9.	Solids removed $=2C_s w\Sigma[Lv_s\cos\theta+(y_{1i}-y_{2i})v_s\sin\theta]=2\left(\frac{100\,\text{mg}}{1}\right)\left(\frac{11}{1000\text{cm}^3}\right)(0.5\text{cm})(47.35384+2.28606)\frac{\text{cm}^2}{\text{s}}$ $=4.96399\text{mg/s}$					
10.	Concentration of solids in the effluent= $\frac{(\text{Solids entering per sec.}-\text{solids removed per sec.})}{\text{Flow rate}}$ $=\frac{\left(\frac{100\,\text{mg}}{1}\right)\left(\frac{0.061}{\text{s}}\right)-(4.96399\,\text{mg/s})}{\left(\frac{0.061}{\text{s}}\right)}$ i.e 17.3mg/l					

Appendix

Calculation of $(y_{11} - y_{21})$;

$$a = \frac{2Q}{3\pi R^4 v_s \cos\theta} = \frac{2 \times 60}{3\pi(2.5)^4 \times 0.3\cos 30^\circ} = 1.25458;$$

$$b = 3aR = 3 \times 1.25458 \times 0.25 = 9.40935;$$

$$\mathrm{c} = 3a\alpha_1^2 + \tan\theta = 3 \times 1.25458 \times 0.25^2 + \tan 30^\circ = 0.81258;$$

$$d = ay_{21}^3 - by_{21}^2 + cy_{21} - 50 = -49.99129;$$

$$1.25458\, y_{11}^3 - 9.40935\, y_{11}^2 + 0.81258\, y_{11} + 49.99129 = 0$$

Solving the above Eqn. $y_{11} = 3.07486; y_{21} = 0.01253;$
Calculation for $(y_{12} - y_{22})$:

$$a = 1.25458; b = 9.40935;$$

$$c = 3 \times 1.25458 \times 0.75^2 + \tan 30^\circ$$
$$= 2.69445;$$
$$d = 1.25458 \times 0.11515^3 - 9.40935 \times 0.11515^2 + 2.69445 \times 0.11515 - 50$$
$$= -49.81258;$$

$$1.25458y_{12}^3 - 9.40935y_{12}^2 + 2.69445y_{12} + 49.81258 = 0$$

i.e $y_{12} = 3.37267$; and $y_{22} = 0.11515$;
and

$$y_{12} - y_{22} = 3.25752;$$

for the rest of the strips $(y_{1\mathrm{i}} - y_{2\mathrm{i}}) = \text{Chord length};$

Notations

$\boldsymbol{x}, \boldsymbol{y}$,	**Coordinates**
$\boldsymbol{\phi}(\boldsymbol{y})$	Flow velocity at coordinate y
$\boldsymbol{v}_s$	Settling velocity of particle
$\boldsymbol{\theta}$	Inclination of the tube with horizontal
$\boldsymbol{y}_1, \boldsymbol{y}_2$	Particle falls from y_1 to y_2
$\boldsymbol{Q}$	Flow rate

$\boldsymbol{L}, \boldsymbol{B}, \boldsymbol{D}$	Length, width, depth of settling zone
v_0	Overflow velocity
$\boldsymbol{C}_s$	Concentration of solids
$\boldsymbol{R}$	Radius of the tube cross section
v_{ci}	Critical fall velocity through the i-th chord
y_{1i}, y_{2i}	Particle falls from y_{1i} to y_{2i} on the i-th chord

References

De Alak Kumar (1976): Conceptual studies on discrete and flocculent settling. PhD thesis (Engg). Jadavpur University, Kolkata

De Alak: Velocity profile theorem: concept for solving settling problem analysis. J IPHE India 2009–2010(1)

Yao (1970) Theoretical study of high rate sedimentation. J WPCF 42:218

Chapter 4
Sedimentation Process

Abstract Settling of solids is discussed, and its characteristic classification is presented.

Keywords Settleables and nonsettleables • Discrete settling • Flocculant settling • Zone settling • Compression settling

Sedimentation/settling is a process of gravity separation of solids from their suspension. When such separation aims at the clarified effluent, the process is called 'clarification'. The process is termed 'thickening' when such separating process promotes thickening of the sludge produced.

4.1 Settleables and Non-settleables

For a settling tank of depth 4 m and detention time of 4 h, particles with settling "velocities 0.0278 cm/s or more will reach the bottom and they are 'settleables'".

Smaller particles have settling velocities lesser than 0.0278 cm/s and cannot reach the bottom. Such particles are 'poorly settleable solids'.

Smaller particles have larger surface area per unit volume of solids. By virtue of this large surface area, the solids experience unbalanced impacts from the surrounding molecules while they are executing kinetic heat motion.

So long the particle mass is such as not to be affected by the transfer of momentum to it by the kinetic heat motion of the surrounding molecules, the particles will be settling with their settling velocities however small it may be.

When the particles assume colloidal dimensions, i.e., of the order of 10^{-4} cm and tend to move downwards, the small mass of the particle moves in a helter-skelter fashion, a motion executed being known as 'Brownian movement', under the larger number of unbalanced impacts from the surrounding molecules on larger surface area. The particles cannot, therefore, follow their line of fall with their computed settling velocities however small. As such they will never fall through even a very very small depth to touch the bottom. These are 'non-settleable solids'.

Particles less than 10^{-4} cm in size enter into dissolved state.

A. De, *Sedimentation Process and Design of Settling Systems*,
Springer Transactions in Civil and Environmental Engineering,
DOI 10.1007/978-81-322-3634-4_4

Poorly settleable and non-settleable solids are rendered settleable through coagulation-flocculation, in a settling system, when particles conglomerate into a bigger size with larger settling velocity and are removed.

4.2 Characteristic Classification of Sedimentation Process

Fitch (1958) described four characteristic types of sedimentation:

(i) Class-I clarification or discrete settling
(ii) Class-II clarification or flocculant settling
(iii) Zone settling
(iv) Compression settling

Discrete settling: Here all the individual settling particles fall with their respective settling velocities unimpaired throughout their entire fall.

Flocculant settling: In flocculant settling, flocculant particles go on forming bigger flocs with higher settling' velocities.

Zone settling: In zone settling by virtue of their flocculant nature, the particles bind one another to form latticed structure to form into zone.

Any layer in the zone receives at its top the latticed particles from the layer just overlying it. The layer itself also releases latticed particles at its bottom to the layer just underlying it. Thus, the zone settles as a whole.

Compression settling: In compression settling underlying particle carries the weight of the particles lying above. The sharing of the proportion of weight of the particles increases when the pore pressure in the interstices is reduced due to oozing out of pore water and the compression settling consolidates the sludge mass.

The phase changes of the four types of settling processes may be demonstrated with the help of 'paragenesis diagram' (Fig. 4.1).

In the diagram, the ordinate downward indicates increasing concentration of solids from low solids to high solids. The abscissa rightward indicates the higher degree of flocculating tendency of the particles from particulates that do not form floc to highly flocculant particles that form flocs with large number of particles. Higher degree of flocculating tendency means a particle can form floc with larger number of particles.

Along the ordinate from low solids to high solids with particulates, settling is 'discrete'.

At any intermediate concentration of solids such as at A, if the degree of flocculant tendency of the particles increases along AB, the settling becomes more and more flocculant meaning thereby that flocs are formed with more particles.

At B, structured lattices are established to form layers of zone settling. With further increase in the degree of flocculating tendency of particles, the flocs become

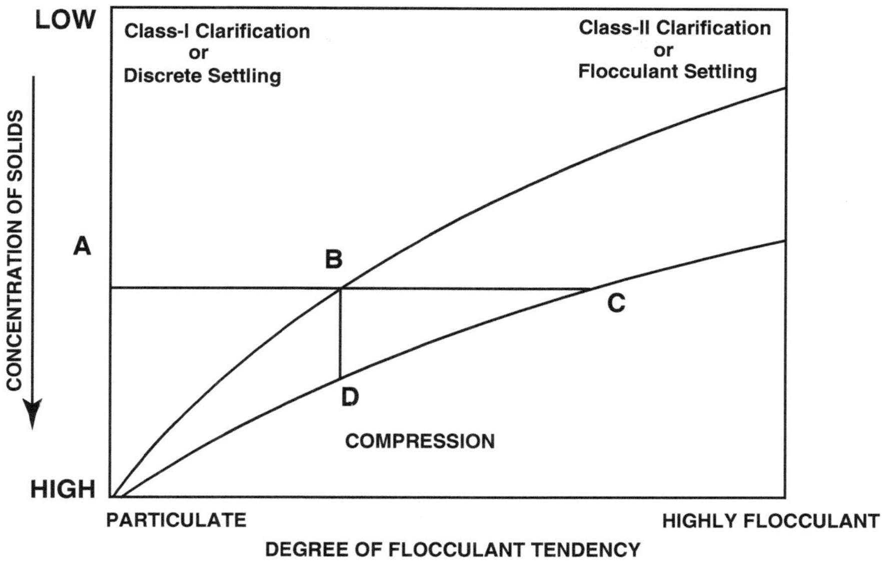

Fig. 4.1 Paragenesis diagram

more and more compact, and ultimately at C compression settling sets in. The same thing happens when degree of flocculating tendency at B is held, and concentration of solids increases from B to D.

Reference

Fitch EB, In McCabe J, Eckenfelder WW Jr (1958) Biological treatment of sewage and industrial wastes, vol 2: anaerobic digestion and solid – liquid separation. Reinhold Publishing Corpn., New York, p 160

Chapter 5
Discrete Settling

Abstract Settling velocity expression for a settling particle is deduced. Several methods for finding the setting velocity of a particle from the diameter and vice versa are presented. Computation of the ideal removal with the help of settling column test data has been demonstrated.

Keywords Settling velocity • Trial solution • Direct solution • Ideal settling theory • Settling column test

5.1 Class-I Clarification or Discrete Settling

In discrete settling, all settling particles fall with their individual settling velocities same throughout their entire fall, i.e. each particle falls through equal depth in equal time. If the trajectory of a particle with settling velocity v_s is plotted in depth-time coordinates, it will give a straight line OB as shown in Fig. 5.1.

If the particle is at depth D_t. at time t, the settling velocity v_s of the particle $= \frac{D_t}{t}$.

When a particle is just immersed in a fluid, it starts gaining momentum under the gravitational field, and the rate of gain of this momentum is the gravity force. Simultaneously it has to lose momentum to resist buoyancy. The rate of loss of such momentum is buoyant force.

Under the balance of the two forces, the particle accelerates. With its movement comes into play the fluid friction on the surface of the particle that increases with the increase in settling velocity of the particle. The rate of loss of momentum to the surrounding fluid mass due to friction is the 'drag force'.

The rate of gain of momentum under the gravitational field and the rate of loss of momentum to resist buoyancy remain the same, while the rate of loss of momentum due to fluid friction increases with the increase in settling velocity. A situation appears when the rate of gain of momentum by the particle under the gravitation field is the same as the rate of loss of momentum by it to the surrounding fluid mass. Under this dynamic equilibrium, the particle falls with constant momentum, i.e. the constant settling velocity v_s. This velocity is the characteristic settling velocity or simply the settling velocity (v_s) of the particle. In settling a particle is identified by its settling velocity (v_s) and no other parameter.

A. De, *Sedimentation Process and Design of Settling Systems*,
Springer Transactions in Civil and Environmental Engineering,
DOI 10.1007/978-81-322-3634-4_5

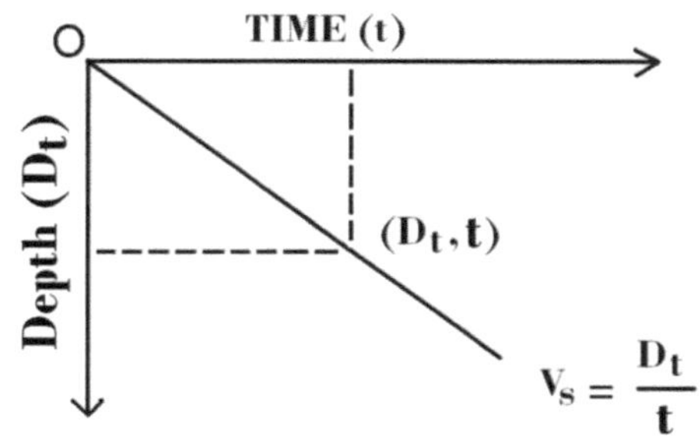

Fig. 5.1 Settling trajectory of a discrete settling particle in depth-time coordinates & Settling velocity of a particle

5.1.1 Derivation of Settling Velocity (v_s) Equation

Under dynamic equilibrium,

$$\text{Gravity force} - \text{Buoyant force} = \text{Drag force}$$

For a particle of diameter d, mass density of the particle ρ_s, mass density of fluid ρ_l, and acceleration due to gravity at the place of observation g,

$$\frac{\pi d^3(\rho_s - \rho_1)}{6} = \text{Drag Force} \tag{5.1}$$

Drag Force

With the instantaneous settling velocity v_s, the particle sweeps out $\frac{\pi d^2}{4} v_s$ volume of fluid per second, containing mass of fluid $\frac{\pi d^2 v_s \rho_l}{4}$. If each of the elements of this mass would move with velocity v_s,the rate of loss of momemtum to the fluid mass would be $\frac{\pi d^2}{4} \rho_l v_s^2$. Since to all the fluid elements the settling velocity v_s could not be communicated, the rate of loss of momentum to the fluid mass is

$f \frac{\pi d^2}{4} \rho_l v_s^2$ where $f < 1$

i.e. Drag force $= f \frac{\pi \mathrm{d}^2}{4} \rho_l v_s^2$

$= C_D \left(\frac{\pi d^2}{4}\right) \frac{\rho_l v_s^2}{2}$,writing in terms of dynamic pressure head $\frac{\rho_l v_s^2}{2}$;

$= C_D \frac{A \rho_l v_s^2}{2}$, where A is the projected area of the particle on a horizontal plane and C_D is Newton's drag coefficient

Equation 5.1 can be written:

$$\frac{\pi}{6} d^3(\rho_s - \rho_l) g = C_D \frac{\pi d^2}{4} \frac{\rho_l v_s^2}{2} \tag{5.2}$$

$$\text{i.e.}\, v_S = \sqrt{\frac{4}{3}\frac{gd}{C_D}\left(\frac{\rho_s - \rho_l}{\rho_l}\right)} \tag{5.3}$$

$$v_S = \sqrt{\frac{4}{3}\frac{gd}{C_D}(s-1)} \tag{5.4}$$

Equations (5.3) and (5.4) are Newton's law for falling bodies and are valid for all values of Reynolds' number R providing values of C_D in accordance with Eq. (5.5) as

$$C_D = \frac{24}{R} + \frac{3}{\sqrt{R}} + 0 \cdot 34 \tag{5.5}$$

Graphical plot of Eq. (5.5) is shown in Fig. 5.2.

If C_D written as

$$C_D = \frac{24}{R}$$

$$\text{i.e}\log C_D = \log 24 - \log R$$

i.e. $\log C_D$ versus $\log R$ plot is straight line. In Fig. 5.2 such plot extends up to $R = 1$.

This indicates that (Appendix 1) if C_D is written as $\frac{24}{R}$, at $R = 1$, $R = 0.5$, and $R = 0.1$, the computed settling velocity is more than the actual value by 6.1 %, 4.4 %, and 1.96 %, respectively.

Allowing up to 2 % increased values for the settling velocities of particles for $R \leq 0 \cdot 1$,

$$C_D = \frac{24}{R}$$

$$= \frac{24\nu}{v_s d}$$

ν – Kinematic viscosity

μ – Coeff. of viscosity

and with Eq. 5.3

$$v_s = \frac{gd^2}{18^{\mu}}(\rho_s - \rho_l) \tag{5.6}$$

$$\text{and} = \frac{gd^2}{18}\frac{(s-1)}{\nu} \tag{5.7}$$

Equations (5.6) and (5.7) are Stoke's law.

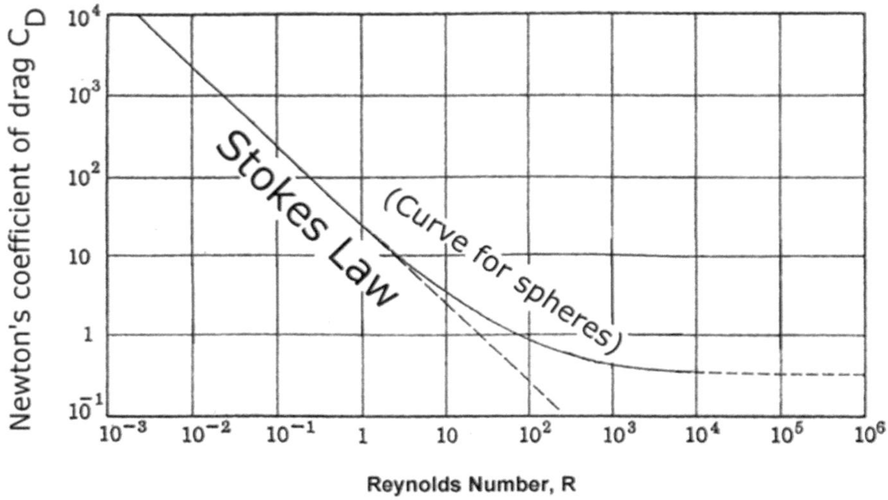

Fig. 5.2 Drag coefficient versus Reynolds' number curve for spheres

5.1.2 *Settling Velocity Calculations*

In the following, different methods and types of settling velocity calculations are presented.

1. Method 1: Trial solution

v_s from given value d:

$$\text{Newton's law} - v_s = \sqrt{\frac{4gd}{3C_D}(s-1)} \tag{5.8}$$

$$\text{Stoke's law} - v_s = \frac{gd^2}{18\nu}(s-1) \tag{5.9}$$

$$R = \frac{v_s d}{\nu} \tag{5.10}$$

$$\text{Newton's drag coefficient } C_D = \frac{24}{R} + \frac{3}{\sqrt{R}} + 0.34 \tag{5.11}$$

Step 1: Find v_s from Eq. (5.9).
Step 2: Use v_s found in step 1 to find R from Eq. (5.10).
Step 3: Use R found in step 2 to find c_D from Eq. (5.11).
Step 4: Use c_D found in step 3 to find v_s from Eq. (5.8).

With repetitive use of steps 2, 3 and 4, the values of v_s are successively approximated till it converges reasonably.

Problem 5.1 Calculate the settling velocity of a particle of 0.5 mm dia. and of material specific gravity 2.65 falling through water at 20 °C. Kinematic viscosity of water at 20 °C = 1.004×10^{-6} m²/s.

Solution

1. Find from Eq. 5.9. $v_s = \dfrac{(9.81\text{m/s}^2)(5 \times 10^{-4}\text{m})^2(2.65 - 1)}{18 \times 1.004 \times 10^{-6}\text{m}^2/\text{s}}$
$= 22.391 \times 10^{-2}\text{m/s}$

2. Find from Eq. 5.10. $R = \dfrac{(22.391 \times 10^{-2}\text{m/s})(5 \times 10^{-4}\text{m})}{1.004 \times 10^{-6}\text{m}^2/\text{s}}$
$= 111.5$

3. Find from Eq. 5.11. $C_D = \dfrac{24}{111.5} + \dfrac{3}{\sqrt{111.5}} + 0.34$
$= 0.8393$

4. Find from Eq. 5.8. $v_s = \sqrt{\dfrac{4 \times (9.81\text{m/s}^2)(5 \times 10^{-4}\text{m})(2.65 - 1)}{3 \times 0.8393}}$
$= 11.338 \times 10^{-2}\text{m/s};$

5. $R = \dfrac{(11.338 \times 10^{-2})(5 \times 10^{-4})}{1.004 \times 10^{-6}}$
$= 56.46$

6. $C_D = \dfrac{24}{56.46} + \dfrac{3}{\sqrt{56.46}} + 0.34$
$= 1.162$

7. $v_s = \sqrt{\dfrac{4 \times (9.81\text{m/s}^2)(5 \times 10^{-4})(2.65 - 1)}{3 \times 1.162}}$
$= 9.637 \times 10^{-2}$

8. $R = \dfrac{(9.637 \times 10^{-2})(5 \times 10^{-4})}{1.004 \times 10^{-6}}$
$= 47.993$

9. $C_D = \dfrac{24}{47.993} + \dfrac{3}{\sqrt{47.993}} + 0.34$
$= 1.273$

10. $v_s = \sqrt{\dfrac{4 \times (9.81)(5 \times 10^{-4})(2.65 - 1)}{3 \times 1.273}}$
$= 9.207 \times 10^{-2}$

11. $R = \dfrac{(9.207 \times 10^{-2})(5 \times 10^{-4})}{1.004 \times 10^{-6}\text{m}^2/\text{s}}$
$= 45.851$

12. $C_D = \dfrac{24}{45.851} + \dfrac{3}{\sqrt{45.851}} + 0.34$
$= 1.306$

13. $v_s = \sqrt{\dfrac{4 \times (9.81)(5 \times 10^{-4})(2.65 - 1)}{3 \times 1.306}}$
$= 9.09 \times 10^{-2}$

14. $R = \frac{(9.09 \times 10^{-2})(5 \times 10^{-4})}{1.004 \times 10^{-6}}$
$= 45.269$

15. $C_D = \frac{24}{45.269} + \frac{3}{\sqrt{45.269}} + 0.34$
$= 1.316$

16. $v_s = \sqrt{\frac{4 \text{ x } (9.81)(5 \times 10^{-4})(2.65 - 1)}{3 \times 1.316}}$
$= 9.055 \times 10^{-2}$

17. $R = \frac{(9.055 \times 10^{-2})(5 \times 10^{-4})}{(1.004 \times 10^{-6})}$
$= 45.09$

18. $C_D = \frac{24}{45.09} + \frac{3}{\sqrt{45.09}} + 0.34$
$= 1.319$

19. $v_s = \sqrt{\frac{4 \times (9.81)(5 \times 10^{-4})(2.65 - 1)}{3 \times 1.319}}$
$= 9.045 \times 10^{-2}$

20. $R = \frac{(9.045 \times 10^{-2})(5 \times 10^{-4})}{(1.004 \times 10^{-6})}$
$= 45.04$

21. $C_D = \frac{24}{45.04} + \frac{3}{\sqrt{45.04}} + 0.34$
$= 1.319$

22. $v_s = \sqrt{\frac{4 \times (9.81)(5 \times 10^{-4})(2.65 - 1)}{3 \times 1.319}}$
$= 9.045 \times 10^{-2} \text{m/s}$

Diameter d from the given value of v_s:

$$\text{From Newton's law} - d = \frac{3v_s^2 C_D}{4g(s-1)} \tag{5.12}$$

$$\text{From Stoke's law} - d = \sqrt{\frac{18\nu v_s}{g(s-1)}} \tag{5.13}$$

$$\text{Reynolds' number R} = \frac{\text{v}_\text{s}\text{d}}{\nu} \tag{5.14}$$

$$\text{Newton's 'Drag Coeff.'}\ C_D = \frac{24}{R} + \frac{3}{\sqrt{R}} + 0.34 \tag{5.15}$$

1. Find D from Eq. 5.13.
2. Find R from Eq. 5.14 with d obtained in step 1.

3. Find C_D from Eq. 5.15 with the value of R obtained in step 2.
4. Find v_s from Eq. 5.12 with C_D obtained in step 3.

'd' is successively approximated with repetitive use of steps 2, 3 and 4 till the value of 'd' converges reasonably.

Problem 5.2 Calculate the diameter of a particle of material Sp.Gr. = 2.65 falling through water at 20 °C with settling velocity $9.03 \times 10^{-2}\text{m/s}$. Kinematic viscosity of water at 20 °C = $1.004 \times 10^{-6}\text{m}^2/\text{s}$.

Solution

1. Find from Eq. 5.13 $d = \sqrt{\frac{18 \times (1.004 \times 10^{-6}\text{m}^2/\text{s})(9.03 \times 10^{-2}\text{m/s})}{(9.81\text{m/s}^2)(2.65 - 1)}}$, i.e. $1.004 \times 10^{-4}\text{m}$.
2. Find from Eq. 5.14. $R = \frac{(9.03 \times 10^{-2}\text{m/s})(3.175 \times 10^{-4}\text{m})}{(1.004 \times 10^{-6}\text{m}^2/\text{s})}$, i.e. 28.6.
3. Find From Eq. 5.15. $C_D = \frac{24}{28.6} + \frac{3}{\sqrt{28.6}} + 0.34$, i.e. 1.74.
4. Find from Eq. 5.12. $d = \frac{3 \times (9.03 \times 10^{-2}\text{m/s})^2 \times 1.74}{4 \times (9.81\text{m/s}^2)(2.65 - 1)}$, i.e. $6.57 \times 10^{-4}\text{m}$.
5. $R = \frac{(9.03 \times 10^{-2})(6.57 \times 10^{-4})}{(1.004 \times 10^{-6})}$, i.e. 59.1.
6. $C_D = \frac{24}{59.1} + \frac{3}{\sqrt{59.1}} + 0.34$, i.e. 1.14.
7. $d = \frac{3 \times (9.03 \times 10^{-2})^2 \times 1.14}{4 \times (9.81)(2.65 - 1)}$, i.e. 4.31×10^{-4}.
8. $R = \frac{(9.03 \times 10^{-2})(4.31 \times 10^{-4})}{(1.004 \times 10^{-6})}$, i.e. 38.76.
9. $C_D = \frac{24}{38.76} + \frac{3}{\sqrt{38.76}} + 0.34$, i.e. 1.44.
10. $d = \frac{3 \times (9.03 \times 10^{-2})^2 \times 1.44}{4 \times (9.81)(2.65 - 1)}$, i.e. 5.44×10^{-4}.
11. $R = \frac{(9.03 \times 10^{-2})(5.44 \times 10^{-4})}{(1.004 \times 10^{-6})}$, i.e. 48.9.
12. $C_D = \frac{24}{48.9} + \frac{3}{\sqrt{48.9}} + 0.34$, i.e. 1.26.
13. $d = \frac{(3)(9.03 \times 10^{-2})^2 \times 1.26}{4 \times (9.81)(2.65 - 1)}$, i.e. 4.76×10^{-4}.
14. $R = \frac{(9.03 \times 10^{-2})(4.76 \times 10^{-4})}{(1.004 \times 10^{-6})}$, i.e. 42.8.
15. $C_D = \frac{24}{42.8} + \frac{3}{\sqrt{42.8}} + 0.34$, i.e. 1.36.

16. $d = \frac{3 \times (9.03 \times 10^{-2})^2 \times 1.36}{4 \times (9.81)(2.65 - 1)}$, i.e. 5.14×10^{-4}.
17. $R = \frac{(9.03 \times 10^{-2})(5.14 \times 10^{-4})}{(1.004 \times 10^{-6})}$, i.e. 46.2.
18. $C_D = \frac{24}{46.2} + \frac{3}{\sqrt{46.2}} + 0.34$, i.e. 1.3.
19. $d = \frac{3 \times (9.03 \times 10^{-2})^2 \times 1.3}{4(9.81)(2.65 - 1)}$, i.e. 4.9×10^{-4}.;
20. $R = \frac{(9.03 \times 10^{-2})(4.9 \times 10^{-4})}{(1.004 \times 10^{-6})}$, i.e. 44.
21. $C_D = \frac{24}{44} + \frac{3}{\sqrt{44}} + 0.34$, i.e. 1.34.
22. $d = \frac{3 \times (9.03 \times 10^{-2} \mathrm{m/s})^2 \times 1.34}{4(9.81 \mathrm{m/s^2})(2.65 - 1)}$, i.e. 5.06×10^{-4}m
    i.e. 5×10^{-4}m.

Diameter of the particle is 5×10^{-4}m.

In working out the problems, all the steps are presented to reveal the monotony and time-consuming affair of trial solution.

2. Method 2: Semigraphical method

Finding v_s from the given value of d

Eliminating v_s between Newton's law and Reynolds' number one gets

$$C_D = \frac{4gd^3(s-1)}{3\nu^2} \cdot \frac{1}{R^2}$$

$$\text{i.e. } \log C_D + 2\log R = \log \frac{4gd^3(s-1)}{3\nu^2} \qquad (5.16)$$

This is a straight line equation for C_D versus R on log-log plot. The straight line passes through point A at coordinates

$$C_D = 1, R = \sqrt{\frac{4gd^3(s-1)}{3\nu^2}};$$

and also through the point B at coordinates

$$C_D = \frac{4gd^3(s-1)}{3\nu^2}, R = 1$$

The value of R of the falling particle lies on Eq. 5.16 and also on C_D versus R plot, on log-log paper, of the Eq.

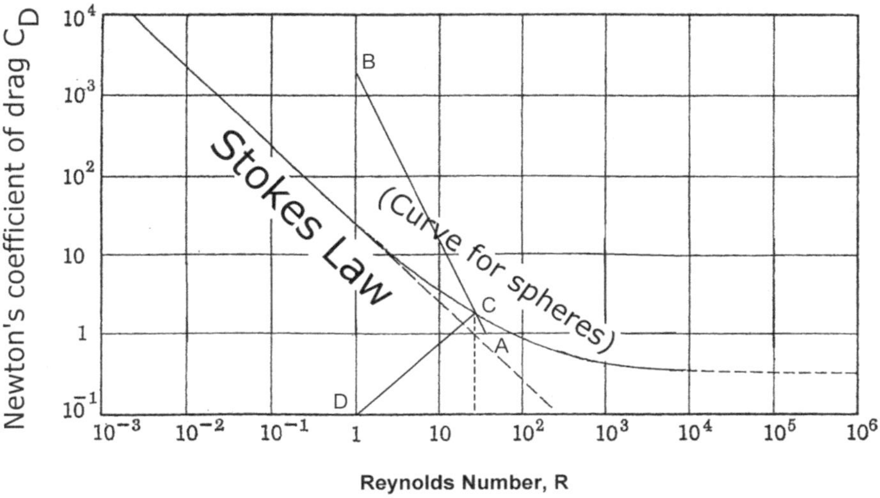

Fig. 5.3 Semigraphical solution of Problems 5.1 and 5.2

$C_D = \frac{24}{R} + \frac{3}{\sqrt{R}} + 0.34$, i.e. the value of the R of the falling particle corresponds to the intersection of the above curves.

This method is applied to solve Problem 5.1.

The diameter of the particle = 5×10^{-4}m.

The coordinates of point A are

$$C_D = 1, \quad R = \sqrt{\frac{4 \times (9.81\text{m/s}^2))(5 \times 10^{-4}\text{m})^3(2.65 - 1)}{3 \times (1.004 \times 10^{-6}\text{m}^2/\text{s})^2}}$$

$$= 50 \text{ (approx.)}$$

And the coordinates of B are

$$\begin{aligned} C_D = & \frac{4 \times (9.81\text{m/s}^2)(5 \times 10^{-4}\text{m})^3(2.65 - 1)}{3 \times (1.004 \times 10^{-6}\text{m}^2/\text{s})^2} \\ = & \ 2700 \text{ (approx.)}, \quad R = 1 \end{aligned}$$

A and B are plotted on Fig. 5.2 and connected by a straight line as shown in Fig. 5.3. The point of intersect ion of the two curves corresponds to Reynolds' number 45.

Hence settling velocity v_s of the particle

$$= \frac{45 \times (1.004 \times 10^{-4}\text{m}^2/\text{s})}{(5 \times 10^{-4}\text{m})}, \text{ i.e. } 9.03 \times 10^{-2}\text{m/s}$$

Finding 'd' from the given value of 'v_s'

Eliminating '*d*' between Newton's law and Reynolds' number

$$\log C_D - \log R = \log \frac{4\nu g(s-1)}{3v_s^3} \quad (5.17)$$

This is a straight line equation on log-log plot. It passes through C:

$$\left(C_D = 1, R = \frac{3v_s^3}{4\nu g(s-1)}\right) \text{and} D \left(C_D = \frac{4\nu g(s-1)}{3v_s^3}, R = 1\right)$$

R of the falling particle lies on CD and C_D versus *R* plot in Fig. 5.2 on log-log paper of the equation

$C_D = \frac{24}{R} + \frac{3}{\sqrt{R}} + 0.34$, i.e. the *R* of the falling particle is the R of the point of intersection of the two curves. This method is applied to solve Problem 5.2. The settling velocity of the particle v_s is $= 9.03 \times 10^{-2}\text{m/s}$. Then the coordinates of the point C are

$$C_D = 1, R = \frac{3 \times \left(9.03 \times 10^{-2}\text{m/s}\right)^3}{4 \times (9.81\text{m/s}^2)\left(1.004 \times 10^{-6}\text{m}^2/\text{s}\right)(2.65-1)}, \text{ i.e. } 34$$

Similarly the coordinates of D are

$$\begin{aligned} C_D &= \frac{4 \times (9.81\text{m/s}^2)\left(1.004 \times 10^{-6}\text{m}^2/\text{s}\right)(2.65-1)}{3 \times \left(9.03 \times 10^{-2}\text{m/s}\right)^3} \\ &= 2.9 \times 10^{-2}; \text{ and } R = 1. \end{aligned}$$

The points C and D are plotted on Fig. 5.3 shown. The points are connected by straight line to find the point of intersection. The Reynolds' number R corresponds to the point = 45.

Hence the diameter '*d*' of the particle

$$= \frac{\nu R}{v_s}, \text{ i.e. } \frac{\left(1.004 \times 10^{-6}\text{m}^2/\text{s}\right) \times 45}{\left(9.03 \times 10^{-2}\text{m/s}\right)}, \text{ i.e. } 5 \times 10^{-4}\text{m}$$

3. Method 3: New methods of solutions of settling velocity problems (De 2002).

In settling settleable particles are characterised by their terminal velocities termed settling velocities of particles. Two types of problems may be there with settling velocity. One may have to find out the settling velocity of a particle from its given diameter and vice versa.

In Stoke's range ($R \leq 0.1$), settling velocity problems may be solved by using Stoke's law. There are trial, semigraphical and graphical (presented elsewhere) solutions to the problems in Newton's range. The difficulties and shortcomings of such solutions are well revealed. In the following the development of velocity of settling particle has been investigated, and the direct solutions to the settling velocity problems are presented.

Development of Settling Velocity (v_s)

A particle of volume V and material density ρ_s, just immersed in a liquid (of density ρ_l, coefficient of viscosity μ), will start accelerating under the balance of gravity force $V\rho_s g$ and buoyant force $V\rho_l g$, i.e. $V(\rho_s - \rho_l)$. Its velocity will be increasing. Due to fluid friction, the drag on its surface will also be increasing with the increasing velocity. This may be defined as 'unsteady state' of motion in which the particle will be gaining momentum under the gravitational field at constant rate and losing its momentum, to the surrounding fluid mass, the rate of which will be increasing with the increasing velocity of the particle.

The 'steady state' of dynamic equilibrium will be reached when the rate of gain of momentum just balances the rate of loss of momentum, and the particle will be settling with constant terminal velocity that is the settling velocity of the particle.

Equation in 'Unsteady State'

At any instantaneous velocity v_s with the projected area, on a horizontal plane, A, the coefficient of drag on the particle C_D and the differential equation of motion of the particle may be written as

$$V\rho_s \frac{dv_s}{dt} = V(\rho_s - \rho_L)g - \frac{C_D A \rho_l v_s^2}{2}$$

$$\text{i.e. } \frac{dv_s}{dt} = \frac{(\rho_s - \rho_l)}{\rho_s} g - \left(C_D = \frac{24}{R} + \frac{3}{\sqrt{R}} + 0.34 \right) \frac{1}{2} \cdot \frac{\rho_l}{\rho_s} \cdot \frac{A}{V} \cdot v_s^2 \tag{5.18}$$

Equation (5.18) may be rewritten with $(\rho_s/\rho_l) = s$, $R = v_s d\rho_l/\mu$, $\nu = \mu/\rho_l$, for spherical particle of diameter 'd', i.e. (A/V)=(3/2d) as

$$\frac{dv_s}{dt} = \left(1 - \frac{1}{s}\right) g - \left(\frac{18\nu v_s}{d^2 s} + \sqrt{\frac{81\nu v_s^3}{16 d^3 s^2}} + 0.255 \frac{v_s^2}{sd} \right) \tag{5.19}$$

writing Eq. 5.19 in finite form as

$$\Delta v_s = \left[\left(1 - \frac{1}{s}\right) g - \left(\frac{18\nu v_s}{d^2 s} + \sqrt{\frac{81\nu v_s^3}{16 d^3 s^2}} + 0.255 \frac{v_s^2}{sd} \right) \right] \Delta t \tag{5.20}$$

The instantaneous velocity of a particle in 'unsteady state' may be determined by successive applications of Eq. (5.20). This may also be employed to calculate the settling velocity in 'steady state'.

Equation in 'Steady State'

In state of dynamic equilibrium $\frac{dv_s}{dt} = 0$ and from Eq. (5.18), it may be written as

$$\left(\frac{24}{R}+\frac{3}{\sqrt{R}}+0.34\right)v_s^2=\frac{2V}{A}\cdot\frac{\rho_l}{\rho_s}\frac{(\rho_s-\rho_l)}{\rho_s}g \tag{5.21}$$

$$=\frac{4gd}{3\rho_l}(\rho_s-\rho_l) \tag{5.22}$$

Equation (5.22) is written for spherical particle,
writing v_s in terms of Reynolds' number

$$0.34R^2+3R^{1.5}+24R=\frac{4gd^3(\rho_s-\rho_l)\rho_l^2}{3\mu^2\rho_l}$$

$$=\frac{4gd^3(s-1)}{3\nu^2}$$

writing $R=10^x$ in the above equation,

$$\text{i.e.}\, 0.34\times10^{2x}+3\times10^{1.5x}+24\times10^x=\frac{4gd^3}{3\nu^2}(s-1) \tag{5.23}$$

again, $R=10^x$ and expressing d in terms of Reynolds' number in Eq. (5.22):

$$0.34\,R^{-1}+3\,R^{-1.5}+24\,R^{-2}=\frac{4\mu g}{3v_s^3\rho_l^2}(\rho_s-\rho_l)$$

$$=\frac{4}{3}\frac{\nu g}{v_s^3}(s-1)$$

$$\text{i.e.}\, 0.34\times10^{-x}+3\times10^{-1.5x}+24\times10^{-2x}=\frac{4}{3}\frac{\nu g}{v_s^3}(s-1). \tag{5.24}$$

If the RHS of the Eq. (5.23) can be calculated and the calculated value is compared with the tabulated values of- $\phi(x)=0.34\times10^{2x}+3\times10^{1.5x}+24\times10^x$ for varying values of x then from the corresponding value of x obtained from comparison, the settling velocity v_s can be calculated as

$$v_s=\frac{(\mu10^x)}{\mathrm{d}\rho_l}$$

Similarly the diameter of the particle may be calculated from the given value of v_s from Eq. (5.24) by comparing the calculated value of its RHS with the tabulated values of

$$\phi_2(x)=0.34\times10^{-x}+3\times10^{-1.5x}+24\times10^{-2x}$$

$$\text{as } d=\frac{(\mu10^x)}{\mathrm{v_s}\rho_l}$$

Tabulated values of $\phi_1(x)$ and $\phi_2(x)$ are obtained as computer printout as given in Appendices 2 and 3, respectively, for $x=-3.0\,\text{to}\,x=+3.0$, the intervals being so chosen as to permit interpolation with sufficient accuracy.

Illustrative Examples

Problem 5.1 Investigate on the development of settling velocity of a particle of 0.5 mm dia. and of material Sp.Gr. = 2.65, falling through water at 20 °C. Calculate the time required and also the distance traversed by the particle till dynamic equilibrium is attained.

Solution

Diameter of the particle	$= 5 \times 10^{-4}$m
Density of water at 20 °C	$= 998.2\,\mathrm{kg/m^3}$
Specific gravity	= 2.65
Material density	$= 2650\,\mathrm{kg/m^3}$
Coefficient of viscosity at 20 °C	$= 1.002 \times 10^{-3}\,\frac{\mathrm{N \cdot s}}{\mathrm{m^2}}$
Calculated from the above data	
Kinematic viscosity at 20 °C	$= 1.004 \times 10^{-6}\,\frac{\mathrm{m^2}}{\mathrm{s}}$
The ratios, i.e. (ρ_s/ρ_l)	= 2.65

Equation 5.20 may be written for the problem:

$$\Delta v_s = \left[6.108 - \left(27.278\, v_s + 76.094\, v_s^{1 \cdot 5} + 192.453\, v_s\right)\right]\Delta t$$

The instantaneous velocities of the particle may be calculated and tabulated in the following table (Table 5.1).

The above tabulation traces the development of the settling velocity of the particle. It shows that the settling velocity of the particle is 0.0902 m/s and it is developed in 0.045 s. During this time, this may be calculated to have traversed through a distance of 3.27×10^{-3}m only.

Problem 5.2 Calculate the settling velocity of the particle mentioned in Problem 5.1.

Solution

The RHS of Eq. (5.23) for the above particle

Table 5.1 The instantaneous velocities of the particle

No.	Instantaneous velocity $(v_s)\,\mathrm{m/s}$	Time (t) s	Time increment (t) s	Increment of velocity $(v_s)\,\mathrm{m/s}$
1	0	0	0.01	0.06108
2	0.06108	0.01	0.01	0.02575
3	0.08683	0.02	0.005	0.00171
4	0.08854	0.025	0.005	0.00090
5	0.08943	0.03	0.005	0.00050
6	0.0899	0.035	0.005	0.00020
7	0.0901	0.04	0.005	0.00015
8	0.0902	0.045	–	–

$$= \frac{4gd^3}{3\nu^2}(s-1)$$
$$= \frac{4 \times 9.81\left(5 \times 10^{-4}\right)^3}{3 \times \left(1.004 \times 10^{-6}\right)^2}(2.65-1)$$
$$= 2676.30$$

From the table presented in Appendix 2, ϕ_1 (1.649998) = 2646.025635 and ϕ_1 (1.699998) = 3121.307861, ϕ_1 = 2676.30 can be interpolated at $x = 1.653182$.

Hence the settling velocity of the particle

$$= \frac{\nu \times 10^x}{d}$$
$$= \frac{1.004 \times 10^{-6} \times 10^{1.653182}}{5 \times 10^{-4}}, \text{ i.e. } 0.0904\ \text{m/s}$$

Problem 5.3 Calculate the diameter of a particle of material Sp.Gr. = 2.65 falling through water at 20 °C with velocity 0.0904 m/s.

Solution

The RHS of the Eq. (5.24) for the above particle

$$= \frac{4}{3}\frac{\nu g}{v_s^3}(s-1)$$
$$= \frac{4 \times 1.004 \times 10^{-6} \times 9.81}{3 \times 0.0904^3}(2.65-1)$$
$$= 0.029331$$

From the table presented in Appendix 3, $\phi_2(1.649998) = 0.029689$ and $\phi_2(1699998) = 0.024794$. $\phi_2(x) = 0.029331$ may be interpolated at $x = 1.65365$.

Hence the diameter of the particle is

$$\frac{\nu \times 10^x}{v_s}$$
$$= \frac{1.004 \times 10^{-6} \times 10^{1.65365}}{0.0904}, \text{ i.e. } 5 \times 10^{-4}\text{m}$$

4. Limiting diameter 'd' and settling velocity v_s of the particle for the application of Stoke's law

Limiting diameter

Stoke's law – $v_s = \frac{gd^2(s-1)}{18\nu}$,

$$\text{i.e. } R = \frac{v_s d}{\nu} = \frac{g d^3 (s-1)}{18\nu^2}$$

$$\text{i.e.} \, d = \left[\frac{18\nu^2 R}{g(s-1)}\right]^{1/3}$$

Limiting velocity v_s

Stoke's law – $v_s = \dfrac{g d^2 (s-1)}{18\nu}$,

$$\text{i.e.} \, R = \frac{v_s d}{\nu}$$

$$\text{i.e.} \, v_s = \left[\frac{g R^2 \nu (s-1)}{18}\right]^{1/3}$$

If the computed settling velocity of the particle with Stoke's equation is acceptable within the limits of 6.1 %, 4.4 % and 1.96 % +ve error (Appendix 1), the limiting Reynolds' numbers of the falling particle are R = 1, R = 0.5 and R = 0.1, respectively.

Limiting diameter '<u>d</u>' at $R = 1$, $R = 0.5$ and $R = 0.1$,

$$d = \left[\frac{18 \times \left(1.004 \times 10^{-6}\text{m}^2/\text{s}\right)^2 \times 1}{(9.81\text{m/s})(2.65-1)}\right]^{\frac{1}{3}}, \text{ i.e.} \, 1.04 \times 10^{-4} \text{m at} R = 1$$

$$d = \left[\frac{18 \times \left(1.004 \times 10^{-6}\text{m}^2/\text{s}\right)^2 = 0.5}{(9.81\text{m/s}^2)(2.65-1)}\right]^{\frac{1}{3}}, \text{ i.e.} \, 0.82 \times 10^{-4} \text{m at} R = 0.5$$

$$d = \left[\frac{18 \times \left(1.004 \times 10^{-6}\text{m}^2/\text{s}\right)^2 \times 0.1}{(9.81\text{m/s})(2.65-1)}\right]^{\frac{1}{3}}, \text{ i.e.} \, 0.48 \times 10^{-4} \text{m at} R = 0.1$$

Limiting settling velocity 'v_s' at $R = 1$, $R = 0.5$ and $R = 0.1$,

$$v_s = \left[\frac{(9.81\text{m/s}^2)(1^2)(1.004 \times 10^{-6}\text{m}^2/\text{s})(2.65-1)}{18}\right]^{\frac{1}{3}}$$
$$= 0.967 \times 10^{-2}\text{m at } R = 1$$

$$v_s = \left[\frac{(9.81\text{m/s}^2)(0.5^2)(1.004 \times 10^{-6}\text{m}^2/\text{s})(2.65-1)}{18}\right]^{\frac{1}{3}}$$
$$= 0.609 \times 10^{-2}\text{m at } R = 0.5$$

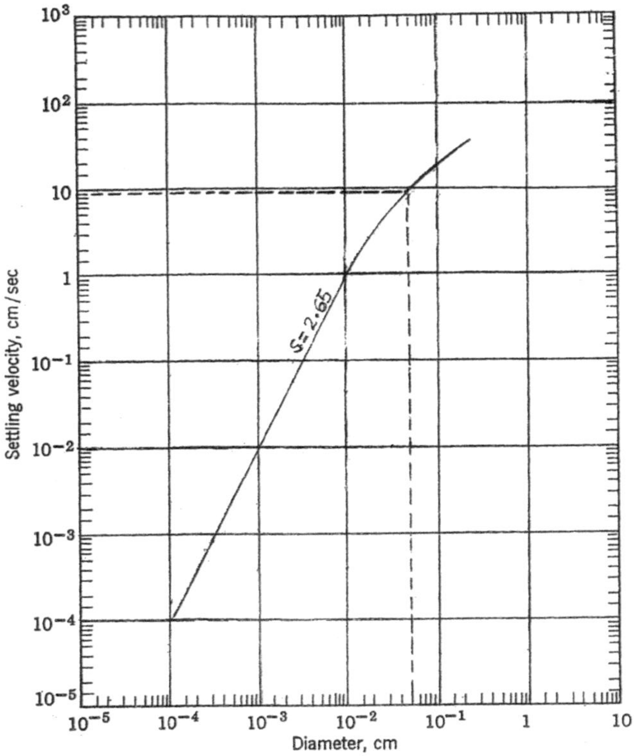

Fig. 5.4 Settling velocity versus diameter of the particle at water temperature 20 °C and material Sp.Gr. = −2.65

$$v_s = \left[\frac{(9.81\text{m/s}^2)(0.1^2)(1.004 \times 10^{-6}\text{m}^2/\text{s})(2.65 - 1)}{18}\right]^{\frac{1}{3}}$$
$$= 0.208 \times 10^{-2}\text{m at } R = 0.1$$

Needless to say that from the limiting diameter of the particle, the limiting velocity can be deduced from Stoke's equation and vice versa.

5. Method 4: Graphical method

For a particle of diameter '*d*' and temperature T° C of water it is falling through, a graph may be prepared as follows.

Stoke's law, $v_s = \frac{gd^2(s-1)}{18\nu}$, suggests that within the limit of its application v_s versus d plot on log-log paper is straight line.

Let us draw the curve in Fig. 5.4 for material Sp.Gr. = 2.65. Let us choose two values of diameter ($> 10^{-4}$ cm colloidal dimension), ν for the water at 20 °C and compute the settling velocities v_1 and v_2 from Stoke's law.

Plot (v_1, d_1) and (v_2, d_2) on log-log paper. Straight is run through those two points. It is extended to $(v,\ d)$ such that $vd = \nu(\text{for} R = 1)$ if 6,1 % error is acceptable with computation with Stoke's law.

The straight enters into curve in Newton's range. The curved portion is traced out though few plotted points with their coordinates being computed with Newton's law through trial solution.

Set of such curves may be prepared, as required, for its application.

In Stoke's range, v_1 determined from a particular curve can give the value v_2 for changed parameters in accordance with the relation:

$$\frac{v_1}{v_2} = \left(\frac{d_1}{d_2}\right)^2 \left(\frac{s_1 - 1}{s_2 - 2}\right)$$

Application

Problem 5.1: Find the settling velocity v_s of a particular of diameter 5×10^{-4} m and material Sp.Gr. = 2.65, falling through water at temperature 20 °C.

From Fig. 5.4 corresponding to diameter 5×10^{-4} m, the settling velocity of the particle $v_s = 0.09\,\text{m/s}$.

Note: It is obvious that from the give n value of $v_s = 0.09\text{m/s}$, the diameter $d = 5 \times 10^{-4}\text{m}$ can be found out from Fig. 5.4.

5.2 Ideal Settling Theory

Ideal settling theory aimed at translating the results of batch settling to the settling in tanks in continuous operation.

5.2.1 Ideal Settling Tanks

T.R. Camp (1946) hypothetically conceived four functional zones—(1) inlet zone, (2) settling zone, (3) sludge zone and (4) outlet zone in settling tanks in continuous operation.

1. Rectangular tank

Let us consider a rectangular settling tank with four hypothetical zones (Fig. 5.5). Flow enters into inlet zone at the rate Q containing settleable solids of concentration C_S consisting of identical particles as regards their settling velocities v_S. In infinitesimally small interval of time τ, volume of water that enters into the tank is $Q\tau$ carrying with it solids $Q\tau\ C_S$.

$Q\tau$ and $Q\tau\ C_S$ on entering into the inlet zone will go on distributing themselves perpendicular to the direction of flow. This distribution is complete where inlet zone ends, and uniform distribution is obtained over the entire cross-sectional area.

Hence $Q\tau$ and $Q\tau\ C_S$ enter into the settling zone (Length $L \times$ Breadth $B \times$ Depth D) from the inlet zone forming a slab of thickness $t = \frac{Q\tau}{BD}$. The slab moves forward

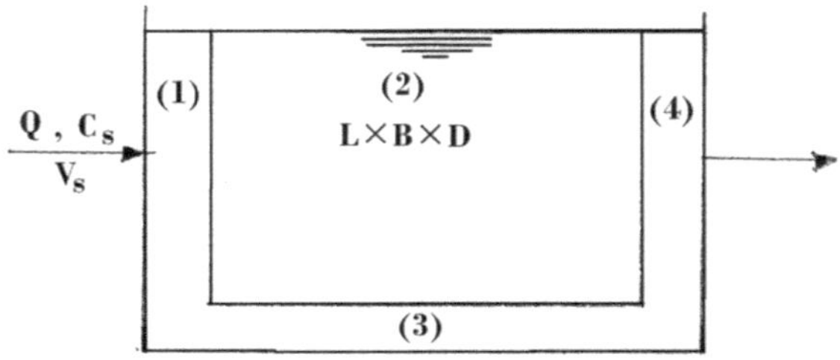

Fig. 5.5 Ideal rectangular settling tank

with velocity $v_w = \frac{Q}{BD}$ called 'flow-through velocity' and reaches the end of the zone at the lapse of time called 'theoretical detention time':

$$T = \frac{L}{\frac{Q}{BD}}, \text{ i.e.} \frac{LBD}{Q}$$

While the slab moves forward, the particles will be settling as if they are settling through a quiescent column of liquid. Since all settling particles have same settling velocity v_S, they will maintain invariable relative position as they settle. As such they will be present at concentration C_s wherever they are present.

A particle that enters into the settling zone at the top of the slab will move through a vertical distance $v_S T$ when the slab reaches the end of the settling zone, and the vertical length $v_S T$ will be free from particles. Outlet zone extends over a distance from a point to the end of the tank over which no particle settles and are carried into the effluent. The particles $B v_S T t C_s$ contained in vertical distance $v_S T$ are, obviously, settled.

Hence, the fraction of solids settled

$$= \frac{B v_S T t C_s}{Q \tau C_s}$$

$$= \frac{B v_S T}{Q \tau C_s} \frac{Q \tau}{BD} C_s$$

$$= v_s/(D/T) \tag{5.25}$$

$$= v_s/\left(D/\frac{LBD}{Q}\right), \text{ i.e.}\, v_s/\frac{Q}{BL} \tag{5.26}$$

D/*T* is settling velocity of a particle that enters at the top and just reaches the bottom at the lapse of theoretical detention time. This is 'critical settling velocity' of the particles as it makes a sharp division between the removal ratios of the settling particles.

All particles having $v_s < v_{cr} \left(= \frac{D}{T}\right)$ that are travelling through the length of the settling zone will not touch the sludge zone (formed by the deposition of sludge) and will be removed in the ratio $v_s/v_{cr} \ldots$. Equation (5.25) provided the particles that touched the sludge zone do not get back into the suspension and stay removed. Particles having $v_s \geq v_{cr}$ will be removed completely.

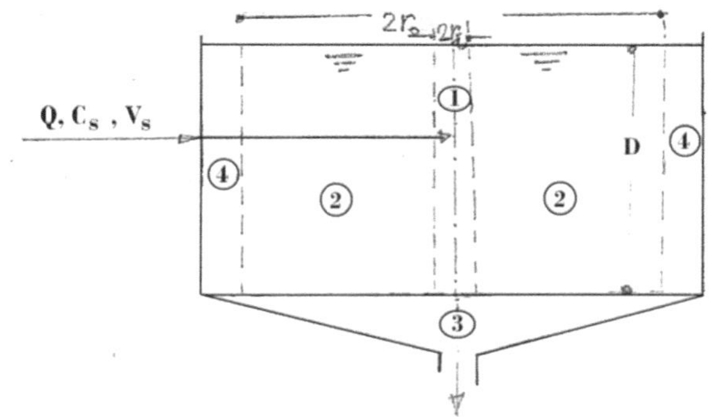

Fig. 5.6 Ideal circular settling tank

$Q/(BL)$ is the velocity at which the flow Q would come out through the surface of the settling zone, i.e. 'overflow velocity' v_0. Incidentally critical velocity is same as overflow velocity (Eq. (5.26)) in this case.

2. Circular settling tank

Circular settling tank may be centrally fed or peripherally fed. In peripheral-fed tank, the flow converges to the outlet resulting in very high outlet velocity that impairs the removal efficiency, and, as such, they are not to be used.

In the following is analysed an ideal centrally fed circular tank in the light of ideal settling theory.

Let us consider a circular settling tank (Fig. 5.6). It has four hypothetical zones. Flow enters into the tank at the rate Q carrying concentration of solids C_s consisting of identical particles as regards their settling velocity v_s.

In infinitesimally small interval of time τ, a volume of water enters into the tank $Q\tau$ carrying solids $Q\tau\ C_s$.

The flow moves forward in radial direction. In the inlet zone (1), the flow along with solids distributes themselves at right angle to the direction of flow, i.e. on the concentric surfaces. The uniform distribution is complete where inlet zone ends.

The flow enters into the settling zone (2) (extending from r_1 (radius of inlet zone), r_0 (radius to the outlet zone) and D (depth of the settling zone)) forming a concentric cylindrical shell of thickness:

$\alpha_i = \frac{Q\tau}{2\pi r_i D}$ just on entering into the settling zone.

Each of the particles at any instant of time moves with horizontal component of velocity equal to the flow-through velocity of water through a concentric cylindrical surface containing that particle. At any time t at distance r from the centre, the particle will have a horizontal component of velocity in radial direction:

$$= \frac{Q}{2\pi rD}$$

A further distance dr will be moved through in time dt given by

d$t = \frac{2\pi rD}{Q}$dr, and each particle will reach the end of the settling zone at the lapse of time:

$$\begin{aligned} T &= \text{Theoretical detention time} \\ &= \int_{r_i}^{r_0} \frac{2\pi rD}{Q} \mathrm{d}r \\ &= \frac{\pi\left(r_0^2 - r_i^2\right)D}{Q} \\ &= \frac{\text{Volume of the settling zone}}{\text{Discharge rate}}. \end{aligned}$$

At any instant of time t, a particle moving from the outside face of the cylindrical shell will be at a distance r_1 from the centre given by

$$r_1 = \sqrt{\frac{Qt}{\pi D} + (r_i + \alpha_i)^2}$$

Similarly a particle just opposite to the previous particle and on the inside face of the cylindrical shell will be at a distance r_2 from the centre at the same time t given by

$$r_2 = \sqrt{\frac{Qt}{\pi D} + r_i^2}$$

Time measurement commences as soon as Q enters and the cylindrical shell is formed.

So at the instant of time t, the distance between those two particles, i.e. the thickness of the cylindrical shell,

$$\begin{aligned} \alpha &= r_1 - r_2 \\ &= \sqrt{\frac{Qt}{\pi D} + (r_i + \alpha_i)^2} - \sqrt{\frac{Qt}{\pi D} + r_i^2} \\ &= \sqrt{\frac{Qt}{\pi D} : +r_i^2}\left[\left(1 + \frac{2r_i\alpha_i + \alpha_i^2}{\frac{Qt}{\pi D} + r_i^2}\right)^{\frac{1}{2}} - 1\right] \\ &= \sqrt{\frac{Qt}{\pi D} : +r_i^2}\left[1 + \frac{1}{2}.\frac{2r_i\alpha_i + \alpha_i^2}{\frac{Qt}{\pi D} + r_i^2} - 1\right] \end{aligned}$$

$= \sqrt{\frac{Qt}{\pi D} : +r_i^2}\,.\frac{r_i\alpha_i + \frac{1}{2}\alpha_i^2}{\frac{Qt}{\pi D} + r_i^2}$ expanding binomially and rejecting the terms of higher order of smallness

$= \frac{r_i\alpha_i}{r_2}$, neglecting α_i in comparison with r_i,

$$\text{i.e. } r_2\alpha_2 = r_i\alpha_i$$

i.e.

$$2\pi D r_2\alpha_2 = 2\pi D r_i\alpha_i$$
$$= Q\tau$$
$$\text{i.e. } \alpha_2 = \frac{Q\tau}{2\pi D r_2} \tag{5.27}$$

The same is the result for other pairs. The cylindrical shell of the solids, therefore, expands in radius, and from Eq. (5.27), it is seen that the thickness of the shell at any time t at distance r from the centre will be obtained by distributing uniformly the volume over the new cylindrical surface.

Since all the particles settle with the same velocity, they maintain the same relative position with respect to each other, and since the volume containing the solids neither contracts nor dilates the concentration of solids, they will remain at C_s wherever they are present.

So at time t when the radius of the shell is r, in further interval of time dt, amount of solids settled

$$= \frac{Q\tau}{2\pi D r} \cdot 2\pi r \cdot v_s \mathrm{d}t \cdot C_s$$
$$= \frac{Q\tau}{2\pi D r} \cdot 2\pi r \cdot v_s \cdot C_s \frac{\mathrm{d}r}{Q/(2\pi r D)}$$

and by the time the cylindrical shell expands in radius to a value $= r_0$, total solids settled

$$= \int_{r_i}^{r_0} \tau v_s C_s 2\pi r dr$$
$$= \tau v_s C_s \pi \left(r_0^2 - r_i^2\right)$$

So total fraction of solids settled $= \frac{\tau v_s C_s \pi \left(r_0^2 - r_i^2\right)}{Q\tau C_s}$

$$= \frac{\frac{v_s}{Q}}{\pi\left(r_0^2 - r_i^2\right)}, \text{ i.e.} \frac{v_s}{v_0} \tag{5.28}$$

where v_0 is overflow velocity,

$$= \frac{\frac{v_s}{QD}}{\pi\left(r_0^2 - r_i^2\right)D} \text{ i.e.} \frac{v_s}{D/T}, \text{ i.e.} \frac{v_s}{v_{cr}} \tag{5.29}$$

where v_{cr} is critical velocity.

Velocity of a particle that enters at the top of the settling zone and reaches the bottom at the lapse of the theoretical detention time T is the 'critical settling velocity' of the particles. Particles having critical settling velocity will be just completely removed. This is incidentally also the 'overflow velocity' in this case.

Equation (5.29) suggests that the particles having settling velocities $v_s < v_{cr}$ will be removed in the ratio v_s/v_{cr}, and particles having settling velocities $v_s \geq v_{cr}$ will be completely removed.

5.2.2 Framework of Assumptions in Ideal Settling Theory

T.R. Camp set up a framework with functional assumptions to deduce the foregoing settling theory.

Four functional zones needed to be conceived. They are (1) inlet zone, (2) settling zone, (3) sludge zone and (4) outlet zone.

1. *Inlet zone*: Just on entering into the tank is required a space at the end of which uniform distribution of solids over the cross section of flow should be obtained where inlet zone ends.
2. *Settling zone*: In the settling zone, the flow with the solids moves forward with same velocity, and the particles will be settling as if they are settling through a quiescent column of liquid.
3. *Sludge zone*: It is formed by the deposition of sludge. Particles that once touch this zone will not get back into the suspension again.
4. *Outlet zone*: Particles from the suspension are dragged into the outlet from a distance from the end of the tank. This distance sets the extent of the outlet zone and limits the end of the settling zone from the end of the inlet zone.

The particles that did not touch the sludge zone during the travel through the settling zone and carried into the outlet zone will not be removed and are carried with the effluent.

Evaluation of the Framework The word 'evaluation' connotes weighing the framework of assumptions in the light of real situation. It is true that it remained unknown to none that none of the assumptions made in deducing ideal settling in continuous operation conforms to real situation. Still it needs careful and critical understanding. For, the original intention behind proposing the ideal settling theory was to find out factors to cater for taking into account the deviations in real situation.

Assumption 1: Conceptually it is easy to divide the tank into inlet zone, settling zone, sludge zone and outlet zone. But it is impossible to consider geometrical partitioning as pictured in deducing the theory, if not for anything else, due to shear exerted by the contact surfaces of the liquid with tank.

Even for same rate of flow, the length of the inlet zone may not be same. This is so because of the variation in the distribution of incoming momentum pushes the limit of the zone at one point and pulls in the limit at the other. Depending upon the flow rate, the length of the zone may even extend to the end of the tank making settling zone conceived under the assumption to be non-existent.

Assumption 2: Every element cannot move forward with same forward velocity, at least, due to frictional resistance from the contact surface leaving aside stagnant pockets of dead space, phenomenon of short circuiting, etc. The detention time distribution of the flow elements with incoming momentum distribution among them is very sensitive to changes.

No element can move forward with same velocity if the shape geometry of the tank changes. A circular settling tank is a case in point.

Particles start settling as soon as they enter into the tank. They will go on settling till and until they are dragged into the effluent by the upward velocity component near the end, i.e. outlet zone of the tank. Settling process, thus, continues over distance that extends from the entry points into the tank to the outlet zone.

Assumption 3: Sludge zone is formed by the deposition of sludge. Solids from the sludge zone do get back into suspension by scouring, the quantity of these solids being dependent on the scouring velocity on the surface of the deposited sludge that can be controlled by reducing the velocity. If left uncontrolled, they may be thrown into suspension either to settle again or to be carried with the effluent.

Assumption 4: Particles that take more time to be dragged through the vertical distances for escape from their depths at which they enter into the outlet zone than that required by them to travel through the length of the outlet zone will not be carried with the effluent.

All particles of outlet zone cannot escape with the effluent. The particles other than that just mentioned will settle.

5.2.3 *Critical Velocity, Overflow Velocity, Surface Loading*

Critical velocity: If a particle entering into the settling zone at its surface spends theoretical detention time T within the zone to reach just the bottom of the zone at its outlet, it has critical settling velocity:

$$v_{Cr} = \frac{D(\text{Depth})}{T(\text{Theoretical detention time})}$$

This is critical because according to ideal settling theory, all particles having settling velocity $v_s \geq v_{Cr}$ will be completely removed, and particles having velocity $v_s < v_{Cr}$ will be removed fractionally in the ratio v_s/v_{Cr}.

This is true for all settling tanks irrespective of their shape, size and geometry.

Overflow velocity: Overflow velocity of all settling tanks

$$= \frac{Q(\text{Flow rate})}{A(\text{Surface area})}$$

$$= \frac{QD}{AD}$$

$$= \frac{D}{\frac{V\,(\text{Volume of the settling zone})}{Q}}$$

$$= \frac{D}{T}, \text{ i.e. } v_{Cr} \text{ or critical velocity}$$

In writing AD = volume of the settling zone, it is implied that both the surface area of the settling zone and the surface area of its bottom are identical and parallel to each other.

This shows that only when both the areas of the surface and the bottom of the settling zone are identical and parallel and the particles are in discrete settling critical fall velocity equals the overflow velocity or surface Ioading and the removal is independent of the depth of the tank.

5.2.4 Removal of Solids

Having known the direction as regards which particle will be removed and at what proportion through an ideal settling tank settling velocity distribution of solids in the influent need be known for the computation of total removal of solids through the same.

This can be accomplished by performing settling column analysis.

Settling Column Analysis[1]

Settling column test is carried out in a cylinder provided with ports at its various depths (Fig. 5.7).

1. *Settling column*: The ports are stoppered. The stoppers are fitted with simple devices for the collection of samples from various depths.

These may be simple collection pipes or fitted hypodermic needles.

The height of the cylinder may go up to 3 m depending upon the nature, concentration and distribution of solids in the samples.

The cross section of the cylinder should depend upon the volume of the collected samples such that the drawing of samples should not lower the water surface in the cylinder to such an extent as to introduce unacceptable error in the settling velocity calculations.

[1]Settling column analysis for discrete suspension, advocated by Camp, and settling column analysis for flocculant suspension, advocated by Eckenfelder and O'Connor, differ in their modes of analysis. Both the methods have inadequacies in their own terms. Elimination of these inadequacies ends up in a single method. To avoid repetition, the inadequacies in both of the methods and a single mode of analysis for both are presented in Chap. 8.

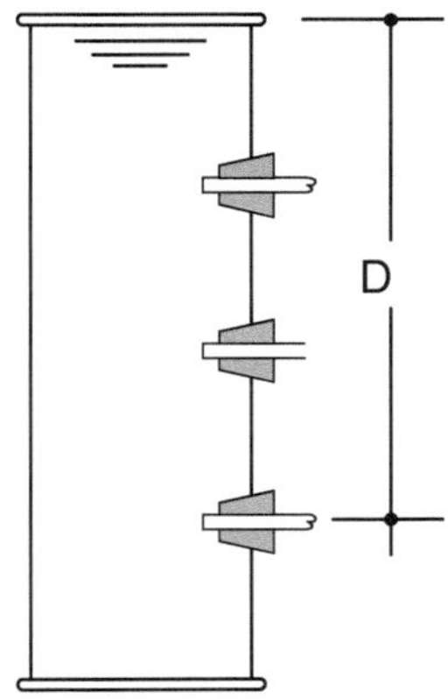

Fig. 5.7 Settling column

2. *Settling column test*: Water samples containing solids are poured into the cylinder, while the stoppers remain closed. The sample is to be stirred up and down vertically to make the sample uniform throughout the depth. The stirring should be very gentle, either with a stirrer or pneumatically, so as not to change the settling velocity distribution among the particles.

The moment when the stirring is discontinued marks the zero time of the time measurement. From this zero time, samples are to be collected through the ports at different times from various depths from the surface of the liquid. The concentration of solids in the samples is determined.

If $C_{D,t}$ is the concentration of solids in the sample collected from depth D at time t, $C_{D,t}$ will contain all particles having settling velocities $v_s \leq D/t$. At zero time $C_{D,0} = C_0$ (initial concentration of solids) same throughout the depth. Then $f = C_{D,t}/C_0$ is the fraction of particles in the sample having $v_s \leq D/t$.

3. *Analysis*

Mode 1 The coordinates (f, D/t) are plotted in f versus D/t plot (Fig. 5.8) to obtain what is known as cumulative frequency distribution diagram for settling velocities of the particles.

For a settling tank with top and bottom surfaces geometrically similar and parallel, fed with a flow rate Q, containing discrete settling solids,

$$\text{Overflow velocity} = \frac{Q}{A}$$

$$= \frac{D(\text{Depth of the tank})}{T(\text{Theoretical detention time})}$$

$$= v_{cr}$$

From the cumulative frequency distribution diagram of the settling velocities among the solids, the total removal of solids may be calculated.

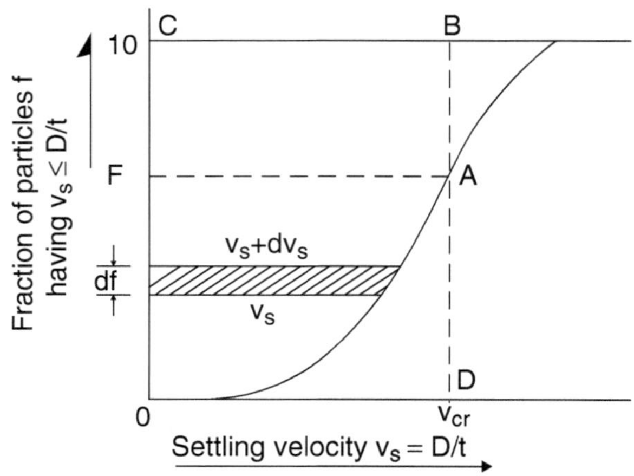

Fig. 5.8 Cumulative frequency distribution diagram among the settling particles

'F' is fraction corresponding to settling velocity v_{cr}. (1−F) is fraction of solids having settling velocities $v_s \geq v_{cr}$, and they will be removed completely.

If df is the fraction of particles having $v_s < v_{cr}$ and settling velocities lying between v_s and $v_s + \mathrm{d}v_s$, they will be removed in the ratio v_s/v_{cr}, and the total removal of such particles

$$= \int_0^{v_{cr}} \frac{v_s df}{v_{cr}}$$

The total removal through the tank $= (1 - F) + \int_0^{v_{cr}} \frac{v_s df}{v_{cr}}$

$$= \frac{\text{shaded area 0ABC}}{\text{0D}}$$

$$= \text{Average ordinate of the shaded area}$$

Problem 5.1 Settling column test was performed with discrete suspension with initial concentration of solids of 1000 mg/l. The following observations were made.

Samples collected at		Concentration of solids
Depth	Time	
25 cm	0 min 50 s	800 mg/l
25 cm	4 min 10 s	300 mg/l
25 cm	8 m 20 s	100 mg/l
50 cm	2 min 5 s	650 mg/l
50 cm	3 min 20 s	500 mg/l
50 cm	41 min 40 s	50 mg/l

An ideal settling tank was fed with the above water and critical fall velocity was 0.2 cm/s. Find the percentage removal expected in accordance with ideal settling theory.

Solution
From the observations recorded in the table, the following table is prepared.

Concentration in mg/l	Fraction	Settling velocity v_s cm/s less than equal to
800	0.8	0.5
650	0.65	0.4
500	0.5	0.25
300	0.3	0.1
100	0.1	0.05
50	0.05	0.02

The above values are plotted to obtain cumulative frequency distribution diagram for the settling particles as shown in Fig. 5.9.

In Fig. 5.9 OD measures the critical settling velocity 0.2 cm/s. At D the ordinate DB is raised to meet the fraction 1.0 line at B.

From the figure,

one small square measures

$$= 0.02 \times 0.01\,\text{cm/s, i.e.}\,0.0002\,\text{cm/s}.$$

Total fractional removal of solids in accordance with ideal settling theory

$$= \frac{\text{shaded area 0ABC}}{\text{0D}}$$

$$= \frac{739\,\text{squares} \times 0.0002\,\text{cm/s}}{0.2}$$

$$= 0.739,\ \text{i.e.}\ 73.9\%$$

Mode 2 The observations made in settling column test are the concentration, of solids, C at depths (D) at times t. These concentration values may be plotted in depth (D)-time (t) coordinates.

Isoconcentration curves $C_1, C_2 \cdots\cdots\cdot C_n$ may be drawn through them.

Nature of curve for discrete suspension:

From the cumulative frequency distribution diagram for settling velocities, the concentration of solids $v_s \leq v_{cr}$

$$C_1 = C_0 \int_0^{v_1} v_s df,\ \text{similarly}\ C_2 = C_0 \int_0^{v_2} v_s df \ldots\ldots\ldots\ldots\ldots$$

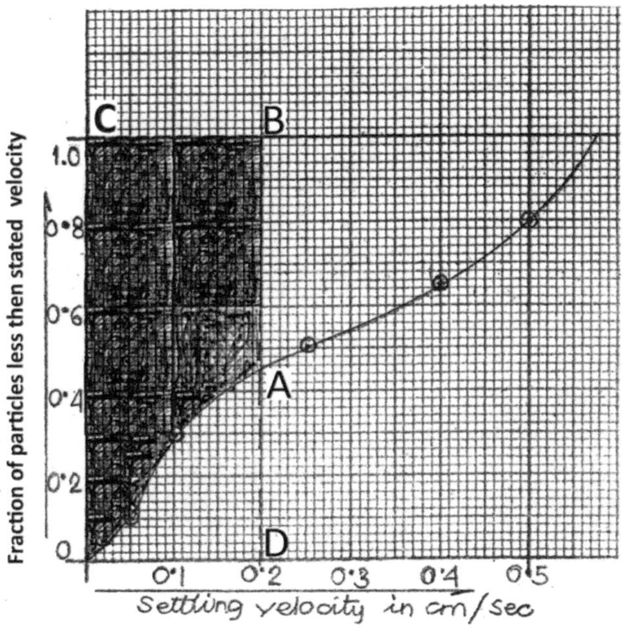

Fig. 5.9 Cumulative frequency distribution diagram for the settling velocities of the particles

$C_n = C_0 \int_0^{v_n} v_s df$, where C_0 is the initial concentration of solids and $C_1, C_2 \ldots\ldots \ldots C_{n-1}, C_n$ settle with $v_1, v_2 \ldots\ldots\ldots v_{n-1}, v_n$, respectively.

In Fig. 5.10, isoconcentration curves $C_1, C_2 \ldots\ldots\ldots C_n$ have been drawn. C_n is to be so chosen that beyond this concentration, the removal may be considered insignificant.

If Q be the flow rate into the tank of surface area A and depth D,

Overflow velocity v_0

$$= \frac{Q}{A} \cdot \frac{D}{D}$$

$$= \frac{D}{\frac{V(\text{Volume})}{Q}}$$

$$= \frac{D}{T(\text{Theoretical detention time})}$$

$$v_{Cr} = (\text{Critical settling velocity})$$

A vertical is drawn at $t = T$ up to the depth D, the depth of the tank.

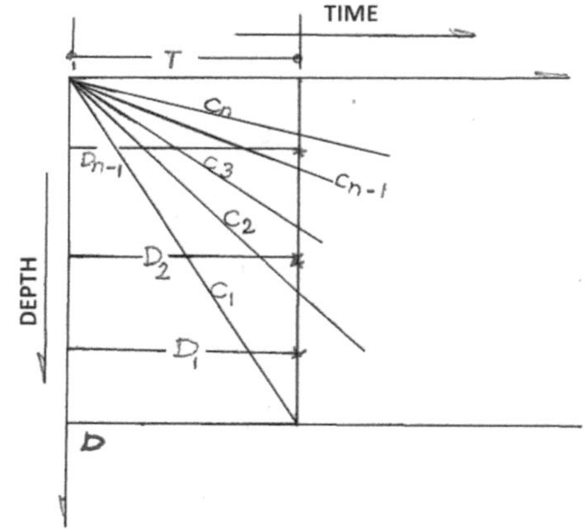

Fig. 5.10 Drawing of isoconcentration curves in depth-time coordinates

From Fig. 5.10, concentration $C_0 - C_1$ particles had settling velocities $v_s \geq d/t$ and will be completely removed. $C_1 - C_2$ concentration of particles had average velocities $v_s = (D_1/T)$, and they will be removed in the ratio $(D_1/T)/(D/T)$, i.e. $\frac{D_1}{D}$ and so on.

So the total solids removed

$$= (C_0 - C_1) + (C_1 - C_2)\frac{D_1}{D} + (C_2 - C_3)\frac{D_2}{D} + \ldots\ldots\ldots\ldots + (C_{n-1} - C_n)\frac{D_{n-1}}{D}$$

Hence total fractional removal

$$= \frac{1}{C_0}\left[(C_0 - C_1) + (C_1 - C_2)\frac{D_2}{D} + (C_2 - C_3)\frac{D_2}{D} + \ldots\ldots\ldots\ldots + (C_{n-1} - C_n)\frac{D_{n-1}}{D}\right]$$

Problem 5.2 A settling tank is fed with water containing discrete settling solids of concentration 1000 mg/l. The critical fall velocity is 0.4 cm/s and its depth is 2 m. Find the expected removal in accordance with the ideal settling theory.

In Fig. 5.11, the observed concentrations of solids at different depths at different times from the ports of settling column are plotted in depth-time coordinates.

Isoconcentration lines were drawn as shown. Detention time of the tank is 200/0.4 s, i.e. 500 s. At t=500 s, a vertical is drawn. The total removal of solids expected

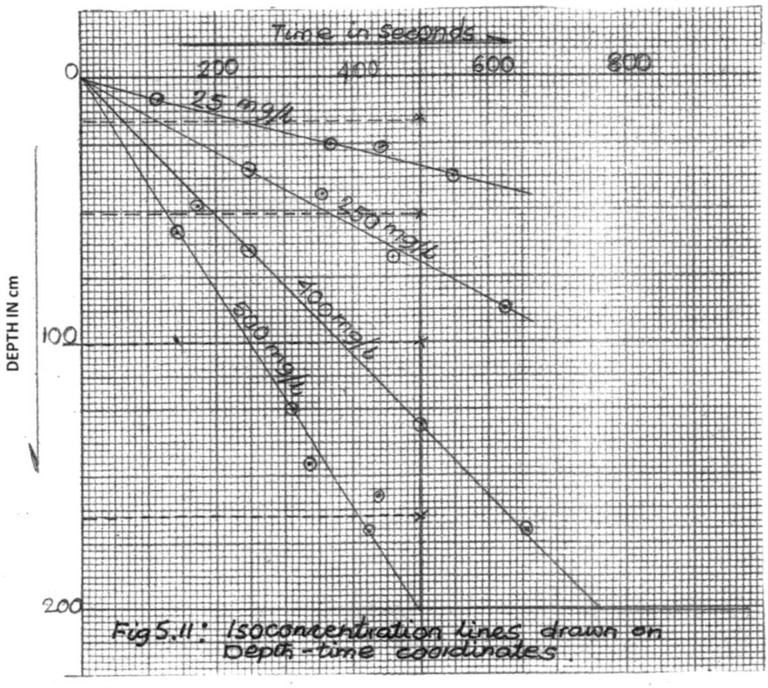

Fig. 5.11 Isoconcentration lines drawn on depth-time coordinates

$$= \frac{1}{1000}\left[(1000-500)+(500-400)\frac{66}{80}+(400-250)\frac{40}{80}\right.$$

$$\left.+(250-25)\frac{20.5}{80}+(25-0)\frac{6.5}{80}\right]$$

$$= 0.717, \text{ i.e. } 72\%$$

Appendices

Appendix – 1

1, The expression for C_D may be studied in two parts. At lower values of R, the first part $24/R$ predominates, while increasing the value of R, the second part $\frac{3}{\sqrt{R}} + 0.34$ predominates.

Settling velocity of a particle may be written — $v_s = \frac{K}{\sqrt{C_D}}$, where K is constant.

$$\log v_s = \log K - \frac{1}{2}\log C_D$$

Differentiating with respect to C_D,

$$\frac{1}{v_s}\cdot\frac{dv_s}{dC_D} = (-)\frac{1}{2}\frac{1}{C_D}$$

Writing in finite form and multiplying both sides by 100,

$$100\frac{\Delta v_s}{v_s} = (-)\frac{1}{2}\frac{\Delta_{C_D}}{C_D}\cdot 100$$

Calculated from above:

at $R = 1.0$ $C_D = 27.34$ $\Delta C_D = -3.34$ $100\frac{\Delta v_s}{v_s} = +6.1\%$

at $R = 0.5$ $C_D = 52.58264$ $\Delta C_D = -4.58264$ $100\frac{\Delta v_s}{v_s} = +4.4\%$

at $R = 0.1$ $C_D = 249.82683$ $\Delta C_D = -9.82683$ $100\frac{\Delta v_s}{v_s} = +1.96\%$

Appendix – 2

$$\phi_1(x) = 0.34 \times 10^{2x} + 3 \times 10^{1.5x} + 24 \times 10^x$$

x	ϕ_1 (x)	x	ϕ_1 (x)
−3.000000	0.024095	0.049998	30.921852
−2.950000	0.027042	0.099998	34.990536
−2.900000	0.030349	0.149998	39.615532
−2.850000	0.034061	0.199998	44.877071
−2.800000	0.038228	0.249998	50.867779
−2.750000	0.042905	0.299998	57.694756
−2.700000	0.048155	0.349998	65.482018
−2.650000	0.054049	0.399998	74.373421
−2.600000	0.060665	0.449998	84.536072
−2.550000	0.068093	0.499998	96.164482
−2.500000	0.076431	0.549998	109.485390
−2.450001	0.085793	0.599998	124.763649
−2.400001	0.096305	0.649998	142.309143
−2.350001	0.108106	0.699998	162.485184

(continued)

x	$\phi_1\ (x)$	x	$\phi_1\ (x)$
−2.300001	0.121358	0.749998	185.718475
−2.250001	0.136238	0.799998	212.511246
−2.200001	0.152947	0.849998	243.455734
−2.150001	0.171711	0.899998	279.251709
−2.100001	0.192784	0.949998	320.727692
−2.050001	0.216451	0.999998	368.866669
−2.000001	0.243033	1.049998	424.837189
−1.950001	0.272892	1.099998	490.031219
−1.900001	0.306433	1.149998	566.110474
−1.850001	0.344112	1.199998	655.062378
−1.800001	0.386445	1.249998	759.268738
−1.750001	0.434007	1.299998	881.589783
−1.700001	0.487452	1.349998	1025.466919
−1.650001	0.547511	1.399998	1195.048706
−1.600001	0.615009	1.449998	1395.345581
−1.550001	0.690874	1.499998	1632.420044
−1.500001	0.776154	1.549998	1913.620728
−1.450001	0.872027	1.599998	2247.871338
−1.400002	0.979822	1.649998	2646.025635
−1.350002	1.101037	1.699998	3121.307861
−1.300002	1.237359	1.749998	3689.854980
−1.250002	1.390695	1.799998	4371.389160
−1.200002	1.563192	1.849998	5190.050781
−1.150002	1.757276	1.899998	6175.428223
−1.100002	1.975686	1.949998	7363.839844
−1.050002	2.221515	1.999997	8799.920898
−1.000002	2.498257	2.049998	10538.604192
−0.950002	2.809863	2.099998	12647.555664
−0.900002	3.160801	2.149997	15210.250977
−0.850002	3.556124	2.199997	18329.767578
−0.800002	4.001554	2.249997	22133.515625
−0.750002	4.503571	2.299997	26779.148438
−0.700002	5.069519	2.349997	32461.917969
−0.650002	5.707723	2.399997	39423.867188
−0.600002	6.427630	2.449997	47965.285156
−0.550002	7.239966	2.499997	58459.042969
−0.500002	8.156916	2.549997	71368.468750
−0.450002	9.192332	2.599997	87269.742188
−0.400002	10.361980	2.649997	106879.843750
−0.350002	11.683810	2.699997	131091.546875
−0.300002	13.178283	2.749997	161017.218750
−0.250002	14.868736	2.799997	198043.625000
−0.200002	16.781824	2.849997	243900.578125

(continued)

x	ϕ_1 (x)	x	ϕ_1 (x)
−0.150002	18.948009	2.899997	300746.906250
−0.100002	21.402149	2.949997	371278.218750
−0.050002	24.184177	2.999997	458861.812500
−0.000002	27.339884		

Appendix – 3

$$\phi_2(x) = 0.34 \times 10^{-x} + 3 \times 10^{-1.5x} + 24 \times 10^{-2x}$$

x	ϕ_2 (x)	x	ϕ_2 (x)
−3.000000	24095208.000000	0.049998	21.891256
−2.950000	19144006.000000	0.099998	17.537020
−2.900000	15210415.000000	0.149998	14.056290
−2.850000	12085252.000000	0.199998	11.272746
−2.800000	9602342.000000	0.249998	9.045821
−2.750000	7629671.500000	0.299998	7.263427
−2.700000	6062366.500000	0.349998	5.836161
−2.650000	4817110.500000	0.399998	4.692701
−2.600000	3827715.500000	0.449998	3.776132
−2.550000	3041598.000000	0.499998	3.041023
−2.500000	2416983.000000	0.549998	2.451102
−2.450001	1920682.750000	0.599998	1.977394
−2.400001	1526330.250000	0.649998	1.596753
−2.350001	1212977.875000	0.699998	1.290681
−2.300001	963983.125000	0.749998	1.044334
−2.250001	766123.750000	0.799998	0.846032
−2.200001	608894.562500	0.849998	0.686159
−2.150001	483949.187500	0.899998	0.557187
−2.100001	384656.312500	0.949998	0.453045
−2.050001	305747.031250	0.999998	0.368871
−2.000001	243035.062500	1.049998	0.300765
−1.950001	193194.156250	1.099998	0.245600
−1.900001	153581.343750	1.149998	0.200866
−1.850001	122096.609375	1.199998	0.164546
−1.800001	97071.242188	1.249998	0.135021
−1.750001	77179.296875	1.299998	0.110987
−1.700001	61367.105469	1.349998	0.091396
−1.650001	48797.386719	1.399998	0.075403
−1.600001	38804.777344	1.449998	0.062329

(continued)

x	$\phi_2\ (x)$	x	$\phi_2\ (x)$
−1.550001	30860.515625	1.499998	0.051622
−1.500001	24544.396484	1.549998	0.042841
−1.450001	19522.462891	1.599998	0.035627
−1.400002	15529.302734	1.649998	0.029689
−1.350002	12353.970703	1.699998	0.024794
−1.300002	9828.803711	1.74998	0.020750
−1.250002	7820.540527	1.799998	0.017403
−1.200002	6223.251955	1.849998	0.014628
−1.150002	4952.737305	1.899998	0.012322
−1.100002	3942.061768	1.949998	0.010402
−1.050002	3158.013672	1.999998	0.008800
−1.000002	2498.290039	2.049998	0.007461
−0.950002	1989.256958	2.099998	0.006339
−0.900002	1584.173584	2.149997	0.005397
−0.850002	1261.776611	2.199997	0.004604
−0.800002	1005.157776	2.249997	0.003936
−0.750002	800.870972	2.299997	0.003371
−0.700002	638.222656	2.349997	0.002893
−0.650002	508.707703	2.399997	0.002488
−0.600002	405.561096	2.449997	0.002143
−0.550002	323.401428	2.499997	0.001849
−0.500002	257.947510	2.549997	0.001598
−0.450002	205.793213	2.599997	0.001383
−0.400002	164.228333	2.649997	0.001199
−0.350002	131.096100	2.699997	0.001041
−0.300002	104.680077	2.749997	0.000905
−0.250002	83.614052	2.799997	0.000788
−0.200002	66.810440	2.849997	0.000687
−0.150002	53.403385	2.899997	0.000600
−0.100002	42.703415	2.949997	0.000524
−0.050002	34.161469	2.999997	0.000459
−0.000002	27.340212		

Notations

v_s Settling velocity
D_t Particle at depth D at time t
ρ_l Density of the liquid
ρ_s Density of the liquid
g Acceleration due to gravity at the place of observation
f Fraction

C_D Newton's drag coefficient
R Reynolds' number
ν Kinematic viscosity
d Diameter of the particle
s Specific gravity
μ Coefficient of viscosity

References

Camp TR (1946): Sedimentation and the design of settling tank. Trans ASCE 111:895–958
De A.: New methods of solutions of settling velocity problems. J IPHE India 2002(2)

Chapter 6
Flocculant Settling

Abstract Flocculation and flocculant settling are discussed. Expressions for contacts between particles due to differential settling and velocity gradients are deduced. Computation of removal of flocculant solids from their settling column test data is demonstrated.

Keywords Contacts in differential settling • Contacts in velocity gradient • Flocculants' removal • Settling Computation • Column design

6.1 Class-II Clarification or Flocculant Settling: Here the Particles Develop Flocs as They Settle and Fall with Accelerated Velocity

6.1.1 Discrete and Flocculant Settling

- Isoconcentration curve for discrete settling in depth-time coordinates

Figure 6.1a shows a column containing suspension. Let us imagine the suspension consisting of identical particles as regards their settling velocity $\boldsymbol{v}_s$. The particles maintain the same relative position with respect to each other as they settle and as such they are at concentration $\boldsymbol{C}_s$ wherever they are present.

Let us track the settling of the sphere (in Fig. 6.1a) as the particles settle. From its initial position at zero time, the sphere is located at D_1, D_2, D_3, at times t_1, t_2, t_3, respectively (say). We have

$$\frac{D_1}{t_1} = \frac{D_2}{t_2} = \frac{D_3}{t_3} = v_s$$

The concentration of solids C_S in the suspensions may be plotted in depth-time coordinates. The isoconcentration curve AB (in Fig. 6.1b) for discrete suspension is a straight line. *The settling and hence the removal of solids is independent of depth.*

Isoconcentration curve for flocculant suspension in depth-time coordinates:

Next consider the suspension in the column (in Fig. 6.1a) consisting of flocculating particles. As the particles start settling, flocs develop and the bigger mass accelerates.

A. De, *Sedimentation Process and Design of Settling Systems*,
Springer Transactions in Civil and Environmental Engineering,
DOI 10.1007/978-81-322-3634-4_6

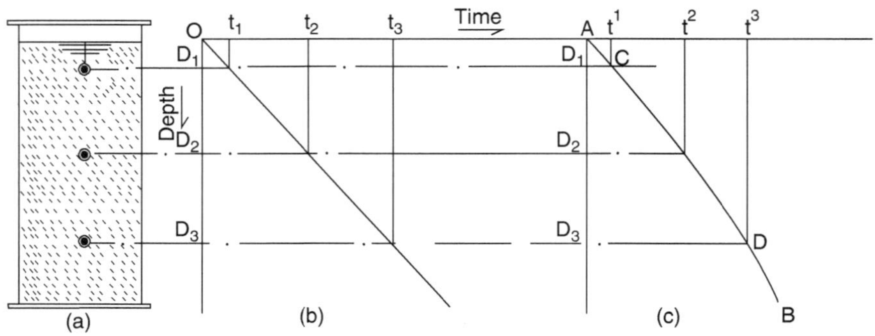

Fig. 6.1 (*a*) Settling of particles; (*b*) Discrete settling trajectory of particle; (*c*) Flocculant settling trajectory of particle

We track the movement of falling sphere changing its size but containing same mass at concentration C_S, leaving out the excess of size and mass formed from the developing flocs. From its initial position of zero time, the sphere located at depths $D_1, D_2, D_3 \ldots\ldots$ at time $t^1, t^2, t^3, \ldots$ can plot the isoconcentration curve in depth-time coordinates and we have

$$\frac{\mathbf{D}_1}{t^1} > \frac{\mathbf{D}_1}{\mathrm{t}_1}, \frac{\mathbf{D}_2}{t^2} > \frac{\mathbf{D}_2}{\mathrm{t}_2}, \frac{\mathbf{D}_3}{t^3} > \frac{\mathbf{D}_3}{\mathrm{t}_3} \ (\textbf{in Fig}.6.1\textbf{b})$$

The isoconcentration curve AB (in Fig. 6.1c) for flocculant suspension reveals three distinct characters.

Over smaller region AC, the curve is linear indicating no appreciable floc development or no accelerated movement.

Over the portion CD, it is curvilinear showing developing flocs and accelerating movements. The nature of curvature of curved portion depends upon the nature and concentration of particles. The increasing acceleration starts from C. The rate increases in the mid-region and then declines to almost nil at D to trace linear curve over the rest DB indicating no further development of floc.

Up to the point D, the settling and removal of flocculating solids is, of course, dependent on depth. But beyond D, the removal is independent of depth. Over AC and DB, settling is of discrete settling nature. The statement 'independent of depth' implies that the settling rate does not change with depth.

6.1.2 *Flocculation*

Water may contain solids. They vary in their characters. Settling or sedimentation is concerned with the removal of solids. Such removal of solids has to take into account the settleability of the solid particles. Here the size, shape and mass of the particles play their role.

Fines such as of the sizes of the order of 10^{-8} cm go into solution. Particles having sizes of the order of 10^{-4} cm are colloids. By virtue of their large surface area per unit volume, they acquire surface charges and repel each other. They are bombarded by the surrounding molecular kinetic heat motions. Due to their tiny masses, the unbalanced impacts of the kinetic molecular movement from the surrounding masses move them erratically in haphazard and random fashion. This helter and skelter movement of the particles is 'Brownian movement'.

Given a depth, whatever small it may be, to settle through in a given time, whatever large the time interval is, a colloidal particle will not settle because it cannot maintain its line of fall the same due to colloidal repulsion and 'Brownian movement'. Such particles are non-settleables.

Particles hundred times bigger than these have very small settling velocities. They require much more time to settle through the depth of the settling tank than the detention time conventionally provided. These are poorly settleable solids. Larger solids can readily be removed in conventional settling tank. They are settleable solids.

Flocculation, coagulation and coagulation-flocculation are the terms that are synonymously used for the process that aims at rendering the poorly settleables and non-settleables into settleable ones.

To render those solids settleable, it is required that the particles should be made to conglomerate to form bigger mass of particles when the settling velocity of the conglomerated mass of the particles will increase. For the conglomeration of the non-settleables such as colloids, the repulsive forces between them have to be reduced, and attractive forces between them are to be promoted.

The repulsive forces are reduced by neutralising the colloidal charges with counter ions, and attractive forces are promoted by making contact between them when short distance force, i.e. van der Waals' force comes into play. The more near the centres of masses of the particles are the more intense is the force to evade the repulsion and promote the growth of floc. Growth of floc also takes place due to surface adsorption and enmeshment on contact between the particles. The growth of floc is limited by the shearing from the surface eroding the further deposition on the same.

The reduction of repulsive forces is a chemical process. Making contacts between the particles is a physical process. If the term 'flocculation' is limited to making contacts between particles, it is a physical process. 'Flocculation', if the term is used in extended sense, is a physico-chemical process.

6.2 Contacts Between Particles

Contacts between the particles may be effected by (1) differential settling and (2) velocity gradient.

Contacts from differential settling: Due to the difference in the settling velocities, the faster moving particles will catch up the slower ones to make contacts between them.

Contacts from velocity gradient: In moving or agitated water, the liquid elements are in state of movement during the course of which they carry solid particles to impinge on to others bringing contacts between them.

6.2.1 Number of Contacts Between Particles Due to Differential Velocities

Let us consider a suspension containing particles of diameters $\boldsymbol{d}_1$ and $\boldsymbol{d}_2(\boldsymbol{d}_1, \boldsymbol{d}_2)$ (in Fig. 6.2) having settling velocities $\boldsymbol{v}_1$ and $\boldsymbol{v}_2(\boldsymbol{v}_1, \boldsymbol{v}_2)$, respectively.

The particles will come into contact when the distance of separation between their centres is $\frac{1}{2}(\boldsymbol{d}_1 + \boldsymbol{d}_2)$ (in Fig. 6.2). The particles are settling and the contacts between them takes place along vertical line when the particles lie on verticals separated by distance $\frac{1}{2}(\boldsymbol{d}_1 + \boldsymbol{d}_2)$.

The particles of diameter $\boldsymbol{d}_1$ are settling with respect to the particles of diameter $\boldsymbol{d}_2$ with velocity $(\boldsymbol{v}_1 - \boldsymbol{v}_2)$, and smaller particles may be visualised to be held stationary in the field of view. In 1 s particle of diameter $\boldsymbol{d}_1$ will move through $(\boldsymbol{v}_1 - \boldsymbol{v}_2)$ with respect to the particle of diameter $\boldsymbol{d}_2$.

If $\boldsymbol{n_1}$ = Number of particles of diameter $\boldsymbol{d_1}$ per unit volume,

$\boldsymbol{n_2}$ = Number of particles of diameter $\boldsymbol{d_2}$ per unit volume,

$$\boldsymbol{L} = (\boldsymbol{v}_1 - \boldsymbol{v}_2),$$

we concentrate on the number of the contacts that are taking place within the volume:

$$\frac{\pi}{4}(\boldsymbol{d}_1 + \boldsymbol{d}_2)^2(\boldsymbol{v}_1 - \boldsymbol{v}_2).$$

The total number of particles of diameter $\boldsymbol{d}_1$ within the volume

$$N_1 = \boldsymbol{n}_1\frac{\pi}{4}(\boldsymbol{d}_1 + \boldsymbol{d}_2)^2(\boldsymbol{v}_1 - \boldsymbol{v}_2).$$

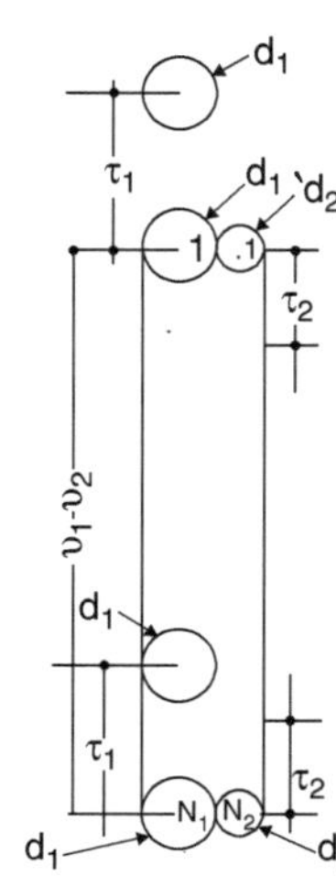

Fig. 6.2 Contacts from differential settling

Similarly, a total number of particles of diameter $\boldsymbol{d}_2$ within the volume

$$N_2 = n_2\frac{\pi}{4}(d_1 + d_2)^2(v_1 - v_2).$$

The distances of separation τ_1 and τ_2 between the particles of dia. d_1 and d_2, respectively, are

$$\tau_1 = \frac{L}{N_1}, \tau_2 = \frac{L}{N_2}$$

At any instant of time, the particles with dia, d_1 may be visualised to be in positions 1,2,3,. . . . . N_1

Over 1 s interval, particles from position marked '1' will move to the position marked N_1, and the particles from the positions 1,2,3,. . . . .N_1 will make contacts with particles of diameter d_2 within the volume which are

$$N_2, (L - \tau_1)\frac{N_2}{L}, (L - 2\tau_1)\frac{N_2}{L}, (L - 3\tau_1)\frac{N_2}{L} \cdots\cdots\cdots [L - (N_1 - 1)\tau_1]\frac{N_2}{L},$$

respectively.

Meanwhile, the particle of diameter d_1 from the position just above the position marked '1' will reach the position marked $(N_1 - 1)$. The others from above will follow to the subsequent positions. They will be making contacts within the volume per second:

$$[L - (N_1 - 1)\tau_1]\frac{N_2}{L}, [L - (N_1 - 2)\tau_1]\frac{N_2}{L}, \cdots\cdots\cdots \tau_1\frac{N_2}{L},$$

respectively.

Hence the total number of contacts between the particles within the volume per second

$$= N_2 + (L - \tau_1)\frac{N_2}{L} + (L - 2\tau_1)\frac{N_2}{L} + \cdots\cdots$$
$$+[L - (N_1 - 1)\tau_1]\frac{N_2}{L} + [L - (N_1 - 1)\tau_1]\frac{N_2}{L} + [L - (N_1 - 2)\tau_1]\frac{N_2}{L}$$
$$+\cdots\cdots + \tau_1\frac{N_2}{L}$$
$$= N_2 + N_2(N_1 - 1) - \frac{\tau_1 N_2}{L}[1 + 2 + 3 \ldots\ldots + (N_1 - 1)]$$
$$+N_2(N_1 - 1) - \frac{\tau_1 N_2}{L}[1 + 2 + 3 \ldots\ldots + (N_1 - 1)]$$

$$= \boldsymbol{N}_2 + \boldsymbol{N}_2(\boldsymbol{N}_1 - 1)$$

$$= \boldsymbol{N}_1\boldsymbol{N}_2 \,\textbf{i.e.}\, \boldsymbol{n}_1\boldsymbol{n}_2\left[\frac{\boldsymbol{\pi}}{4}(\boldsymbol{d}_1 + \boldsymbol{d}_2)^2(\boldsymbol{v}_1 - \boldsymbol{v}_2)\right]^2$$

So the total number of contacts per unit time per unit volume

$$= \boldsymbol{n}_1\boldsymbol{n}_2\frac{\pi}{4}(d_1 + d_2)^2(v_1 - v_2);$$

This is to be borne in mind that not all particles will suffer contacts. Depending upon the relative particle densities and distribution, some particles may avoid contacts. Even then the total number of contacts per unit volume per unit time

$$= \boldsymbol{k}\,\boldsymbol{n}_1\boldsymbol{n}_2\frac{\pi}{4}(d_1 + d_2)^2(v_1 - v_2)$$

where k is a fraction.

The observations that may be made with regard to the above expression are:

- Number of contacts per unit volume per unit time will increase with the numerical densities of the particles.
- Number of contacts per unit volume per unit time will increase with the increase in the sum of the diameters of the contacting particles.
- Number of contacts per unit volume per unit time will increase with the increase in the relative velocities of the contacting particles.
- No contact will take place between the particles of same settling velocities.

6.2.2 *Number of Contacts Between Particles Due to Velocity Gradients*

This is to find out the number of contacts between particles of diameters d_1 and d_2 per unit volume per unit time.

Consider a suspension in movement carrying solids. Because of their tiny sizes, the variation in diameters of a single solid particle across may be neglected, and it is very reasonable to consider them spherical with very high order of accuracy for our purpose. Contacts will take place between the spherical particles for the movement within the suspension.

For the number of contacts between two types of particles, we choose, say, the particles of diameters d_1 and d_2 (Fig. 6.3a, b) having their numerical densities n_1 and n_2 particles per unit volume within the suspension.

Contacts between them will take place when the distance between their centres will not exceed $\frac{1}{2}(d_1 + d_2)$. The sphere of radius $R = \frac{1}{2}(d_1 + d_2)$ is the sphere of influence (Fig. 6.3c).

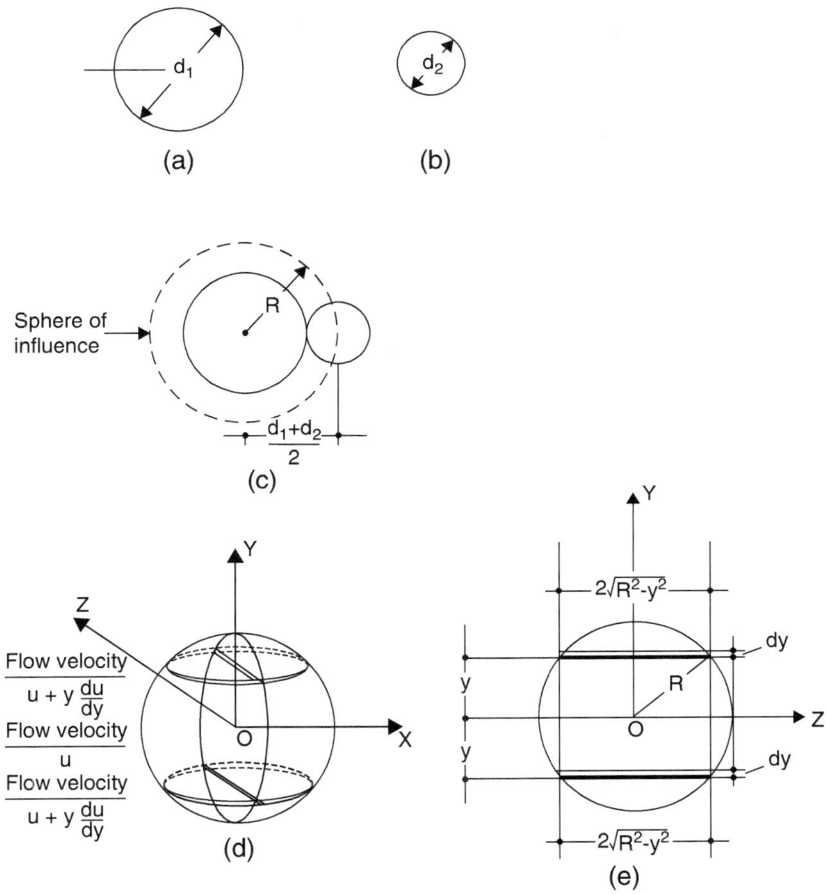

Fig. 6.3 (*a*) Particle of dia d; (*b*) Particle of dia d; (*c*) Sphere of influence for contact between particles of dia d and d; (*d*) Flow under velocity grdients around sphere of influence during contact; (*e*) Projected sphere of influence on Y-Z plane

Let us place the centre of the sphere of influence at the origin 'O' of the X,Y,Z rectangular coordinate system (Fig. 6.3d). If all particles were carried with same flow velocity, no contact between the particles could take place. For the contact between them to be possible, the flow carrying the particle has to move around a particle. This requires increasing flow velocity of the parallel flow vectors.

Let there be a particle of diameter d_1, say, with its centre at 'O'. Let 'u' be the flow velocity through its centre along the axis X with which the particle is being carried. For the contacts between the particles, flow velocity should increase as one moves along the Y-axis direction.

Whatever may the nature of variation of the point velocity gradient curve, the point velocity gradient may be taken to be constant across the tiny dimension of the sphere of influence. The flow velocity at distance y from X-axis is

$$u + y\frac{du}{dy} \text{ (Fig. 6.3d)}$$

It is the relative velocity with which the flow carrying solids will impinge on to the particle of diameter d_1.

We take two strips of thickness dy (Fig. 6.3d, e). The flow velocity with which the solids carried will impinge on the strips is

$$\left(u + y\frac{du}{dy}\right) - u \text{ i.e. } y\frac{du}{dy}.$$

The volume of suspension striking the projected strap areas per unit time is $y\frac{du}{dy}$. $2.2\sqrt{R^2 - y^2}\,dy$ (Fig. 6.3e).

The total volume of suspension striking the projected area of the sphere of influence/s is

$$\int_0^R y\frac{du}{dy}.4\sqrt{R^2 - y^2}dy$$

So the total number of contacts of a single particle of diameter d_1 with particles of diameter d_2/s

$$= n_2\int_0^R y\frac{du}{dy}.4\sqrt{R^2 - y^2}dy \tag{6.1}$$

The point velocity gradient varies from point to point and also from instant to instant. The totality of the above integration will not change if $\frac{du}{dy}$ in (Eq. 6.1.) is replaced by the statistical average $\frac{\bar{du}}{dy}$ over space and time. $\frac{\bar{du}}{dy}$ is known as 'mean temporal velocity gradient' and is represented by 'G'. This permits taking out of 'G' outside the sign of integration.

Hence the total number of contacts with a single particle of diameter d_1 by number of particles of diameter d_2 per sec

$$= n_2 \int_0^R y \frac{d\bar{u}}{dy} .4\sqrt{R^2 - y^2 dy}$$

$$= \boldsymbol{n}_2 \frac{d\bar{u}}{dy} . \frac{1}{6} (d_1 + d_2)^3$$

$\boldsymbol{n}_1$ being the number of particles of diameter d_1 per unit volume, the total number of contacts between the particles of diameter d_1 and the particles of diameter d_2 per unit time per unit volume of the suspension

$$= \frac{n_1 n_2}{6} (d_1 + d_2)^3 \frac{d\bar{u}}{dy} \tag{6.2}$$

Equation (6.2) is Smoluchowski's equation (Smoluchowski 1917).

The equation indicates that as the flocculation proceeds, $\boldsymbol{n}_1$ and $\boldsymbol{n}_2$ diminish and so diminish the rate of contacts between them.

6.2.3 Control on the Number of Contacts

The only factor that can be controlled externally is the 'mean temporal velocity gradient $\left(G = \frac{d\bar{u}}{dy}\right)$ This 'G' not being a measurable quantity has to be converted into a physically measurable one.

Camp and Stein (1943) deduced a parameter that can physically be measured and controlled to replace the 'mean temporal velocity gradient $\left(G = \frac{d\bar{u}}{dy}\right)$ in Smoluchowski's equation.

Water in movement or agitation has point velocity gradients everywhere within it. An elemental water cube is taken from such water, and its 'free body diagram' is presented in Fig. 6.4b.

The elemental cube (Fig. 6.4a) has dimensions $\Delta x, \Delta y, \Delta z$. The layer AB has velocity u under pressure gradient $\frac{dp}{dx}$. The velocity gradient being $\frac{d\bar{u}}{dy}$, the layer CD has velocity $u + \frac{d\bar{u}}{dy} \Delta y$.

Force acting on the surface $\Delta y, \Delta z$ is $p\Delta y, \Delta z$ and on the opposite surface $\left(p + \Delta x \frac{dp}{dx}\right) \Delta y . \Delta z$ where changed pressure is $\left(p + \Delta x \frac{dp}{dx}\right)$.

The upper contact surface is identified by (+) sign and lower one by (−) sign. Shear force opposes motion. It is $\tau . \Delta x . \Delta z$ where shear stress is τ. The shear stress on the surface $\Delta x . \Delta z$ is $\left(\tau . \Delta y . \frac{d\tau}{dy}\right) \Delta x . \Delta z$ where changed shear stress is $\left(\tau . \Delta y . \frac{d\tau}{dy}\right)$.

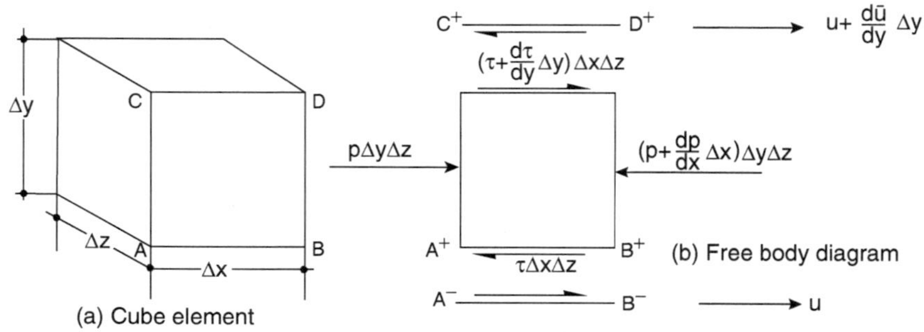

Fig. 6.4 Free body diagram of a cube element

Equating the forces for no acceleration, i.e. turbulence

$$p.\Delta y.\Delta z - \left(p + \Delta x\frac{dp}{dx}\right)\Delta y.\Delta z - \tau\Delta x.\Delta z + \left(\tau.\Delta y.\frac{d\tau}{dy}\right)\Delta x.\Delta z = 0$$

$$\text{i.e. } \frac{d\tau}{dy} = \frac{dp}{dx}$$

The torque acting on the element is $(\tau\Delta x.\Delta z)\Delta y$. This torque rotates the element with angular velocity $\frac{d\bar{u}}{dy}$.

Hence the rate of doing work P on the volume $(V = \Delta x.\Delta y.\Delta z)$

$$= \tau\Delta x.\Delta y.\Delta z.\frac{d\bar{u}}{dy}$$

i.e. the power input per unit volume of V

$\frac{P}{V} = \tau\frac{d\bar{u}}{dy} = \mu\left(\frac{d\bar{u}}{dy}\right)^2$, by definition of shear stress, where μ is the coefficient of viscosity,

$$\text{i.e.} \, G = \frac{d\bar{u}}{dy}$$

$$= \sqrt{\frac{P}{\mu V}}$$

Hence, the number of contacts between particles of diameters d_1 and d_2 per unit volume per unit time

$$= \frac{n_1 n_2}{6}(d_1 + d_2)\sqrt[3]{\frac{P}{\mu V}}$$

6.3 Computation of the Removal of Flocculant Solids

Eckenfelder and O'Connor (1957) advocated 'settling column analysis[1]' with flocculant suspension for the computation of total removal of flocculant solids through a settling tank.

1. *Settling column*: It is a cylinder provided with ports or openings at various depths for the collection of samples. In features it is same as that employed in the discrete suspension analysis (Fig. 5.7, Chap. 5). But this time there is one distinct necessity. In the Mode 1 analysis for discrete suspension, even a single port at the bottom of the cylinder could serve the purpose. But it cannot be the same with flocculant suspension analysis. This is so because, in this case, the concentrations of the samples obtained at depths at times are to be plotted in depth-time coordinates.
   This requires several ports along the depth of the column. The depth of the column depends on and should not be less than the depth of the settling tank for which the analysis is to be made. The number of ports is determined by the accuracy required for tracing the isoconcentration curves and also the concentration of solids.
   The total time over which the observations are to be made depends upon the concentration of solids and their settling characters as well.
   Diameter of the settling column should provide sufficient volume of suspension in the settling column such that drawing of the desired number of samples should not draw down the top surface of the suspension to such an extent so as to affect the accuracy of the observations.
2. *The settling test*: The test is to be carried out in following step-wise sequence.

   Step 1: With the designed settling column and the designed test for the suspension the settling column is filled up with the representative sample of the suspension.
   Step 2: It is vertically stirred very gently so as to make the suspension uniform throughout the depth of the column, while the openings remain closed. Care should be taken so that stirring does not change the character of the given suspension.
   Step 3: The instant when the stirring is discontinued marks the zero time for observations.
   Step 4: From all the depths, the samples are to be collected at different times, and the concentration of solids is determined.

[1]Settling column analysis for discrete suspension, advocated by Camp, and settling column analysis for flocculant suspension, advocated by Eckenfelder and O'Connor, differ in their modes of analysis. Both the methods have inadequacies in their own terms. Elimination of these inadequacies ends up in a single method. To avoid repetition, the inadequacies in both of the methods and a single mode of analysis for both are presented in Chap. 8.

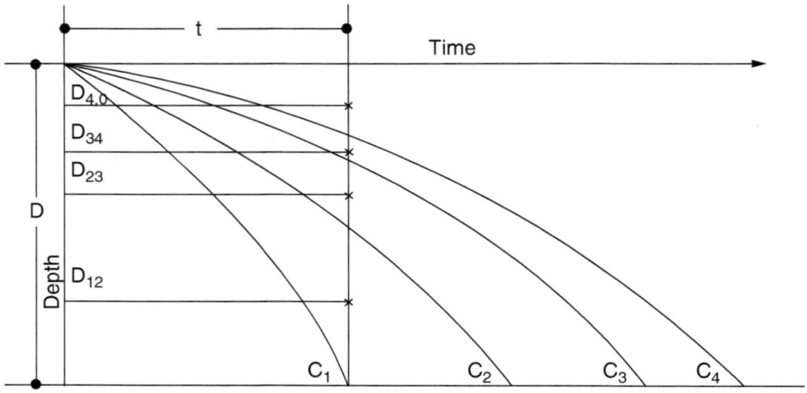

Fig. 6.5 Concentrations in depth-time coordinates

3. *The analysis*: The coordinates (depth-time) for concentrations being observed from the test the concentrations are plotted in depth-time coordinates (Fig. 6.5). The following steps are to be followed.

Step 1: Isoconcentration lines $C_1, C_2, C_3 \ldots\ldots\ldots\ldots\ldots\ldots C_n$ are drawn through them.

Step 2: Lines parallel to time and depth axes are drawn at depth D (depth of the settling tank) and at time t (theoretical detention time of the settling tank).

Step 3: Isoconcentration line C_1 passes through their point of intersection. If C_1 (the isoconcentration curve through the point of intersection) were not drawn initially through the point of intersection, such a curve through the same has to be traced out by interpolation.

Step 4: The midpoints of interceptions of the time coordinate t between C_1 and C_2, C_2 and $C_3 \ldots\ldots\ldots$, C_{n-1} and C_n are identified at depths $D_{12}, D_{23}, D_{34} \ldots$ $..D_{n-1}$, respectively.

Step 5: From Fig. 6.5, the settling velocities of particles constituting concentrations $C_1, (C_1 - C_2), (C_2 - C_3) \ldots\ldots (C_{n-1} - C_n)$ are $\frac{D}{t}$ or less, (average) $\frac{D_{12}}{t}$, (average)$\frac{D_{23}}{t} \ldots\ldots \frac{D_{n-1},\, n}{t}$, respectively.

Step 6: If the initial concentration of solids in suspension was C_0, according to Eckenfelder and O'Connor, the computed total fractional removal of solids through the settling tank

$$= \frac{\text{Total solids removed in mg/l}}{\text{Initial concentration of solids in mg/l}}$$

$$= \frac{1}{C_0}\left[(C_0 - C_1) + \frac{D_{12}/t}{D/t}(C_1 - C_2) + \frac{D_{23}/t}{D/t}(C_2 - C_3) + \ldots \frac{D_{n-1},n/t}{D/t}(C_{n-1} - C_n)\right] \tag{6.3}$$

$$= \frac{1}{C_0}\left[(C_0 - C_1) + \frac{D_{12}}{D}(C_1 - C_2) + \frac{D_{23}}{D}(C_2 - C_3) + \ldots \frac{D_{n-1,n}}{D}(C_{n-1} - C_n)\right]$$

$$= \left(\frac{(C_0 - C_1)}{C_0}\right) + \frac{D_{12}}{D}\left(\frac{(C_0 - C_2)}{C_0} - \frac{(C_0 - C_1)}{C_0}\right)$$

$$+ \frac{D_{23}}{D}\left(\frac{(C_0 - C_3)}{C_0} - \frac{(C_0 - C_2)}{C_0}\right) + \ldots \frac{D_{n-1,\mathrm{n}}}{D}\left(\frac{(C_0 - C_n)}{C_0} - \frac{(C_0 - C_{n-1})}{C_0}\right) \quad (6.4)$$

$$= X_1 + \frac{D_{12}}{D}(X_2 - X_1) + \cdots\cdots\cdots + \frac{D_{n-1,n}}{D}(X_n - X_{n-1}) \quad (6.5)$$

$C_1, C_2, \ldots\ldots\ldots\ldots\ldots\ldots\ldots C_n$ may also be designated as iso-removal curves:

$$X_1 = \frac{(C_0 - C_1)}{C_0}, X_2 = \frac{(C_0 - C_2)}{C_0}, X_n = \frac{(C_0 - C_n)}{C_0} \text{ respectively.}$$

In writing Eq. (6.5), Eckenfelder and O'Connor *utilised a conclusion deduced by T.R. Camp for discrete settling in ideal settling tank. This conclusion is not true for flocculant settling.*

Problem 6.1 A domestic sewage was subjected to settling column analysis and the following observations were tabulated.

Find out the expected removal of solids through a settling tank of depth 1.8 m and detention time 30 min.

Solution:

The tabulated observations from Table 6.1 were plotted in depth-time coordinates in Fig. 6.6. The lines parallel to the time axis and depth axis are drawn at depth 1.8 m and 30 min, respectively. The isoconcentration curves of 211 mg/l, 187 mg/l, 174 mg/l, 143 mg/l, 90 mg/l and 77 mg/l are drawn through the interpolated points. The laid depths of the intercepted portions are marked and noted.

Table. 6.1 Concentrations in mg/l at observed depths and times

	Time in min						
Depth, m	0	10	20	30	40	50	60
0.3	275	218	167	107	83	57	34
0.6	275	248	198	164	120	88	77
0.9	275	253	220	179	152	121	90
1.2	275	259	227	193	174	142	118
1.5	275	264	231	206	184	151	132
1.8	275	267	235	211	187	174	143

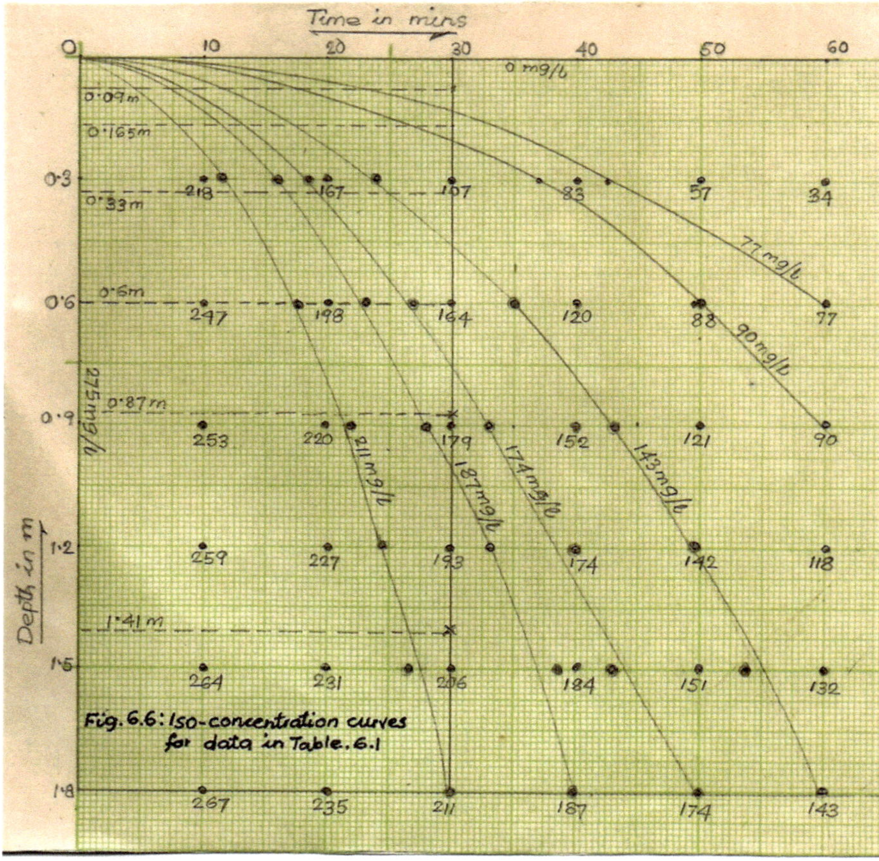

Fig 6.6 Isoconcentration curves for data in Table 6.1

Total removal of solids

$$
\begin{aligned}
&= \frac{1}{275}(275-211)+\frac{1.41}{1.8}(211-187)+\frac{0.87}{1.8}(187-174)\\
&\quad+\frac{0.6}{1.8}(174-143)+\frac{0.33}{1.8}(143-90)+\frac{0.165}{1.8}(90-77)\\
&\quad+\frac{0.09}{1.8}(77-0)\\
&= 0.415 \text{ i.e. } 42\,\%
\end{aligned}
$$

Notations

D Depth
t Time
d Diameter of particle
v Settling velocity of particle
n Numerical density of particles
N Total number of particles
τ Shear stress and also the vertical distance between two particles
u Flow velocity
V Volume of the element
μ Coeff. of viscosity
G Mean temporal velocity gradient
C_n Concentration of the nth isoconcentration curve

References

Camp TR, Stein PC (1943): Velocity gradients and internal work in fluid motion. J Boston Soc Civil Eng 30:219
Eckenfelder WW Jr, O'Connor DJ (1957): Biological waste treatment. Pergamon Press, New York
Smoluchowski M (1917): Versuch einer mathe matischen Theory der Koagulationskinetik Kolloider Losungen, Z. phys.chem92(129):pl55

Chapter 7
Zone Settling and Compression

Abstract Sludge settling characteristics are discussed. Theories of zone settling and compression are presented. The theory is applied to the design of thickener in continuous operation. Design problem is solved in the way of illustration.

Keywords Settling of sludge • Zone settling • Zone settling thickener • Compression • Continuous thickening in compression

7.1 Settling of Sludge

Be it water and waste water treatment, metallurgical, chemical or mining industrial processes, wherever sludge is produced, the bulk of sludge has to be reduced prior to its suitable disposal or recycling. Zone settling and compression take care of these sludges.

The onset of zone settling may or may not follow a certain sequence, always depending upon the settling characteristics of the sludge. Thus settling characteristics of that sludge can be studied with the settling of the sludge in a transparent glass cylinder.

7.1.1 Characteristic Zones in Batch Settling of Sludge

Figure 7.1a shows a transparent glass cylinder filled up with sludge. The sludge is gently stirred vertically to make the sludge distribution uniform throughout the depth, taking care not to change the character of the sludge. The point of discontinuation of stirring marks the zero time for observations.

At an instant, after some time, zonal difference in settling character of the settling sludge may be exhibited as shown in Fig. 7.1b.

Between AB and CD is clear water zone free from settleable solids with a conspicuous solid-liquid surface of separation CD marked (1).

A. De, *Sedimentation Process and Design of Settling Systems*,
Springer Transactions in Civil and Environmental Engineering,
DOI 10.1007/978-81-322-3634-4_7

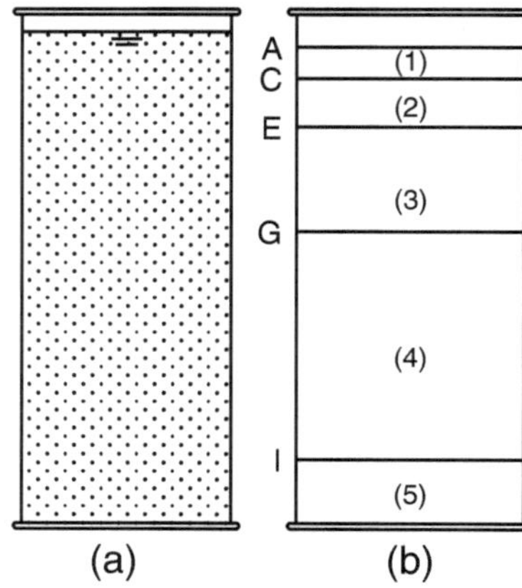

Fig. 7.1 Batch settling of sludge: (**a**) sludge settling; (**b**) zonal characteristics

Between CD and EF is shown a zone of flocculating particles. The concentration of solids in different layers between CD and EF is different, the concentration of solids in the layers being increasing with their distances from the layer CD.

The layer of separation EF of zone (2) may or may not be visually identifiable depending upon the dilution and nature of the sludge in the cylinder.

A particle falling through a fluid has to make its way through it.

This sends out disturbance to its surrounding medium with its movement. The distance to which this disturbance is transmitted may be referred to as its 'velocity field'.

When the particles in their suspension come closer, the velocity fields interfere. This interference leads to the sharing of momentum between particles. At high concentration of solids, the sharing of momentum produces equitable distribution of the same among the particles, and the particles march downwards with the same velocity. This is 'hindered settling'.

Between EF and GH in Fig. 7.1b, shown as marked (3) is the 'zone of hindered settling'. The interface EF sinks with constant velocity. The concentration of the solids in the layers between EF and GH does not vary.

The concentration of solids between GH and IJ in Fig. 7.1b increases further to exhibit settling of different characteristics. The zone marked (4) is a zone identified as 'zone settling'.

In the zone exhibiting 'zone settling', the particles, because of their closeness, form a latticed structure where no faster moving particle can cross past the slower ones. Any layer receives solids from its adjoining top and releases the same to its adjoining bottom. Thus the solids do not settle with their particle identity.

Between IJ and the bottom of the cylinder marked (5) are the particles more closer than those in zone settling such that the bottom lying particles share the weight of the upper ones, and thus the layers in the zone are compressed. This is the 'zone of compression'.

The instantaneous picture depicted in Fig. 7.1b changes from instant to instant. Whether or not all the characteristic zones marked (1), (2), (3), (4) and (5) will appear at any instant of time is determined by the concentration of solids and the degree of their flocculant nature. The relative comparison of the extent of the zones shown in Fig. 7.1b at that instant of time would also depend on the two factors mentioned.

For illustration, increasing either the concentration of the sludge or with increased degree of its flocculant nature or both zone (2) might be reduced or eliminated. Similarly increasing any or both of the factors further might reduce or eliminate zone (3).

All the zones in batch settling of sludge may be difficult to be identified by visual inspection even in the case of dilute sludge. The settling of interface with time can reveal the settling characteristics of the sludge. The concentration of solids in the interface at zero time was C_0, the initial concentration of sludge and the final concentration of the same when it reaches the bottom was the concentration of the compressed sludge surface C_u. At any intermediate position of the interface, its concentration lies in between.

7.1.2 Interface Settling Characteristics

Figure 7.2 shows the movement of interface with time for the batch settling of sludge shown in Fig. 7.1a.

A very small horizontal portion AB may appear initially. This may be due to the disturbance from the stirring discontinued.

Curved portion BC of interface settling may appear showing flocculant settling. CD is a straight portion revealing uniform settling rate of the interface and shows hindered settling of sludge.

From D to E, the curve shows the gradual slowing down of the interface settling in the phase of zone settling of sludge.

From E onwards, the curve shows the very slow exponentially decaying rate of the interface settling under compression.

7.2 Zone Settling [(Kynch 1952); (Talmadge and Fitch 1955)]

Depending upon high relative values of the concentration of solids and their degree of flocculant nature, the particles come very close to form into a sort of open framework.

Settling of solids simply shrinks the framework depthwise where no particle can cross past the bottom lying particles and the particles lying just above move on to

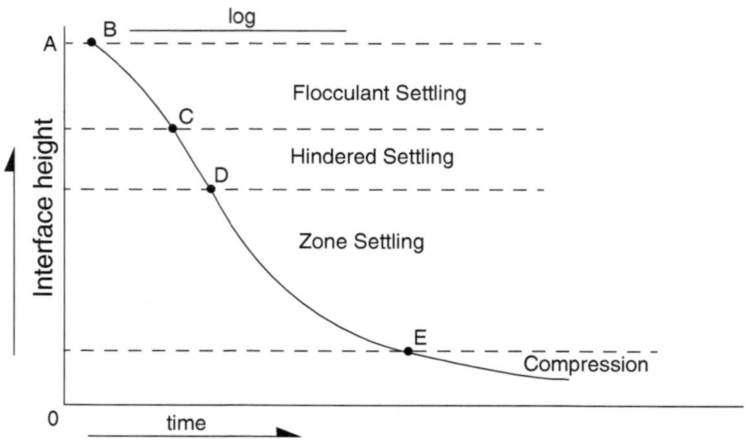

Fig. 7.2 Characteristic zones in interface versus time curve of settling sludge

the particles just lying below. This is just what may be visualised while the framework is shrinking depthwise from the top.

Thus for any layer considered, the layer receives to it particles from the layer lying above and releases its particles to the layer lying just under.

The concentration of the solids in interface at the onset of zone settling was the initial concentration of the solid sludge. Finally when the interface reaches the bottom, its concentration reaches maximum. For any intermediate position of the interface, its concentration of solids lies in between.

The interface concentration of solids, thus, always changes being increased from the initial concentration to the final concentration when zone settling is over.

Any layer within the framework undergoes the similar changes that the interface follows. Any particular concentration of solids that interface assumes at any point of time during zone settling must have been present at every depth at different other points of time before. This is the same as saying that the particular concentration value travelled all through from the bottom to end up in the interface. This is true for all concentration values between the initial and final values in the interface.

The more is the depth of the position of the interface, the more is the concentration of solids in it and, from the graph in Fig. 7.2, the lesser is the interface settling rate, i.e. the settling rate of the layer zone of the particles in the interface.

7.2.1 Theory of Zone Settling

h_0 is the height of a transparent glass cylinder of cross-sectional area A, which is filled up with sludge at concentration C_0 and, thus, contains total solids C_0h_0A.

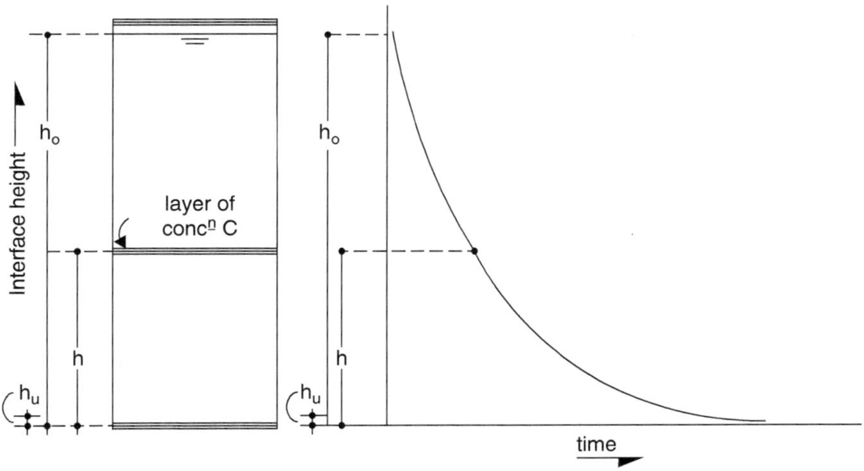

Fig. 7.3 Zone settling

Fig. 7.4 A zone layer

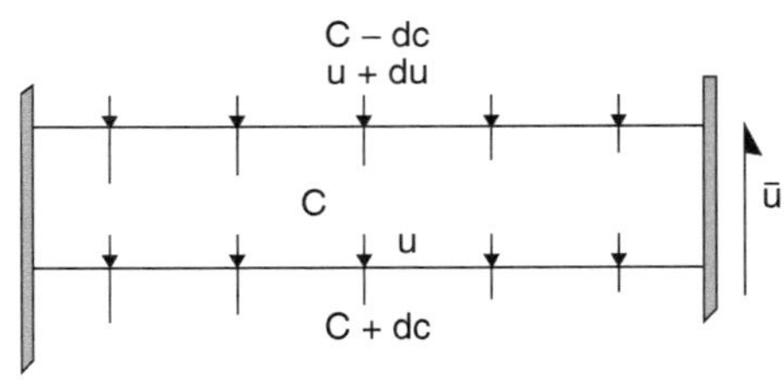

It is in the phase of zone settling. The interface heights at different times are plotted and shown by curve in Fig. 7.3.

The concentration of solids in the interface at its initial height h_0 is C_0, and it is C_u when the interface height is h_u. The lesser is the height of the interface, the more is the concentration of solids in it and the lesser is its settling rate (Fig. 7.3). The interface settling rate, i.e. the settling velocity 'u' of the zone layer of particles in the concentration 'c' solids in the interface, may be put down as

$$u = \phi(c) \tag{7.1}$$

A layer of concentration 'c' is being considered in Fig. 7.4. Particles from this layer are settling with velocity 'u' into the, just underlying, layer of concentration $c + dc$. It is receiving solids from the layer of concentration $c - dc$, just lying above, with particle settling velocity $u + du$. The velocity of the upward movement of the concentration value 'c' is $\overline{u}$.

The settling velocity of the particles from the layer of concentration $(c - dc)$ with respect to the layer with concentration value 'c' being $(u + du + \overline{u})$ the layer of solid concentration 'c' will be receiving solids from the layer above per unit time:

$$= (c - dc)A(u + du + \overline{u}) \tag{7.2}$$

The settling velocity of the particles of this layer being 'u' the particles is being released with respect to the layer itself per unit time:

$$= cA(u + \overline{u}) \tag{7.3}$$

Since the upward moving concentration value 'c' of the layer remains unchanged, incoming and outgoing solids from the layer are equated:

$$(c - dc)A(u + du + \overline{u}) = cA(u + \overline{u}) \tag{7.4}$$

Neglecting the higher order of smallness from the simplification of the above equation:

$$\overline{u} = c\frac{du}{dc} - u \tag{7.5}$$

From Eq. (7.1)

$$\overline{u} = c\phi'(c) - \phi(c) \tag{7.6}$$

i.e. the concentration value moves upwards with constant velocity characteristic of its own.

If the interface concentration is observed at a height 'h' at time 't' (at the point P in the Fig. 7.5), it has traversed to that height 'h' in time 't' with constant upward moving velocity:

$$\overline{u} = \frac{h}{t}, \tag{7.7}$$

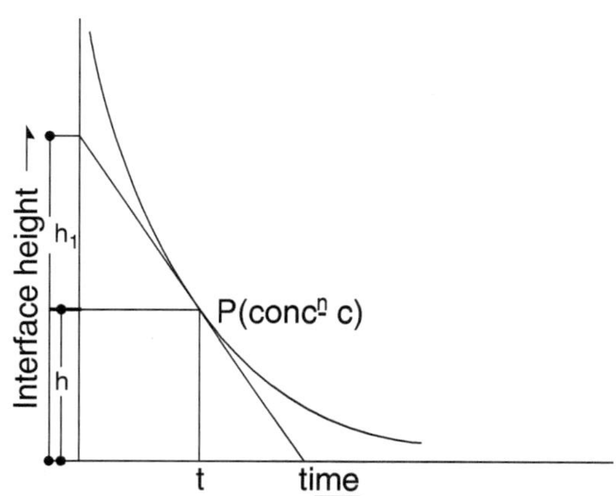

Fig. 7.5 Evaluation of '$\overline{u}$' and 'u'

while the particles from the layer will settle with velocity 'u' given by the slope of the tangent at P:

$$u = \frac{h_1 - h}{t} \tag{7.8}$$

This implies that the concentration 'c' has travelled all through the distance 'h' releasing from it the solids over the time interval 't':

$$\begin{aligned} &= cA(u + \overline{u})t \\ &= cA\left(\frac{h_1 - h}{t} + \frac{h}{t}\right)t \\ &= cAh_1 \end{aligned}$$

This should include all the solids of the column $= C_0 h_0 A$:

$$\text{i.e.}\, cAh_1 = C_0 h_0 A,\ \text{i.e.}\, h_1 = \frac{C_0 h_0}{c} \tag{7.9}$$

Equation (7.9) interprets the height h_1 to be the height to which all the solids of the column would occupy if distributed throughout at uniform concentration 'c', the interface concentration at height 'h'.

7.2.2 Application of Zone Settling Theory to the Thickener in Continuous Operation

What is happening in batch settling of sludge may be visualised to be continually repeating in continuous operation of thickener. In Fig. 7.6a, the thickener in

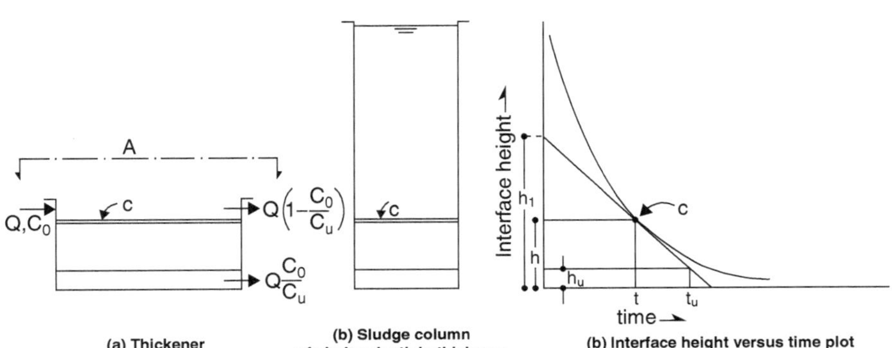

Fig. 7.6 Zone settling in thickener in continuous operation

continuous operation is fed with Q (Volume) $\times$ C_0 (concentration of sludge) sludge per unit time. The rate of withdrawal of underflowing sludge is $Q_u C_u$.

Batch settling of sludge is conducted in a transparent glass cylinder (Fig. 7.6b). The interface versus time curve is plotted (Fig. 7.6c) for the solids filling the cylinder to the height h_0.

To utilise the data of the batch settling of the sludge through the interface concentration c in a continuous thickener of base area A to release thickened sludge at concentration C_u, Ah_1 volume of sludge has to be reduced to Ah_u volume such that

$$Ah_1 c = Ah_u C_u.$$

This requires $(Ah_1 - Ah_u)$ volume of water to be removed through the layer of interface concentration c. Interface settling rate of this layer being $\frac{h_1 - h}{t}$ the rate of overflowing water through the interface concentration 'c' of the surface layer in the thickener is

$$A\left(\frac{h_1 - h}{t}\right).$$

To remove water volume $A(h_1 - h_\mathrm{u})$, settling time that is required to be provided to the incoming sludge is

$$t_u = \frac{A(h_1 - h_u)}{A\left(\frac{h_1 - h}{t}\right)}$$

$$\text{i.e}\ \frac{h_1 - h}{t} = \frac{h_1 - h_u}{t_u}$$

This provides the geometry for the determination of t_u from the graph as shown in Fig. 7.6c.

$C_0 h_0 A$ sludge having required the settling time t_u, the solid handling capacity of the thickener per unit time is

$$\frac{C_0 h_0 A}{t_u}$$

and this should be equal to the solid input to the thickener QC_0 per unit time

$$\text{i.e}\ \frac{C_0 h_0 A}{t_u} = QC_0$$

i.e. the thickener area $A = \frac{Qt_u}{h_0}$; This allows water overflow rate of $A\left(\frac{h_1 - h}{t}\right)$ and sludge underflowing rate $\frac{Ah_u C_u}{t_u}$ for the solid-input rate of QC_0. In other terms, overflow rate of water is $Q\left(1 - \frac{C_0}{C_u}\right)$ and volume rate of underflowing sludge is $\frac{QC_0}{C_u}$.

Question may arise as to what should be the stable interface concentration of solids on the surface of the thickener. The answer to this question may have three solutions as follows:

1. To find out a point by trial with every point on the steep curvature of the interface versus time curve from batch settling of sludge that gives the maximum value of t_u. This works out the interface concentration of solids on the thickener surface.
2. To find out the interface concentration of sludge at which compression starts.
3. To find the interface concentration at the point on the interface versus time curve where the bisector of the angle between the two arms of the curve intersects (Eckenfelder and Melbinger 1957). This concentration is usually very near to the concentration sought in solution no.2.

The third solution is preferred.

7.3 Compression (Coulson and Richardson 1955)

Compression in settling begins when the underlying particles start sharing the weights of the overlying ones. Settling proceeds with more and more water coming out through the interstices between the particles with compression nearing completion.

Figure 7.7 shows a typical interface height versus time plot in compression phase of batch settling of sludge.

The interface height decreases with the increase of time, i.e. $\frac{dh}{dt}$ is − ve.

The negative values of the slopes of the tangent at different interface heights are decreasing with decreasing interface heights.

The negative values, i.e. $-\frac{dh}{dt}$ values, may be plotted against interface heights from Fig. 7.7 to give a straight line passing through the origin as shown in Fig. 7.8. This provides the equation of the curve:

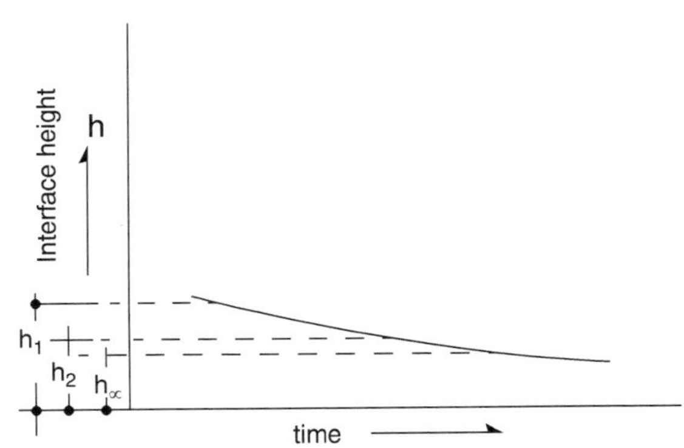

Fig. 7.7 Interface height versus time curve in zone compression

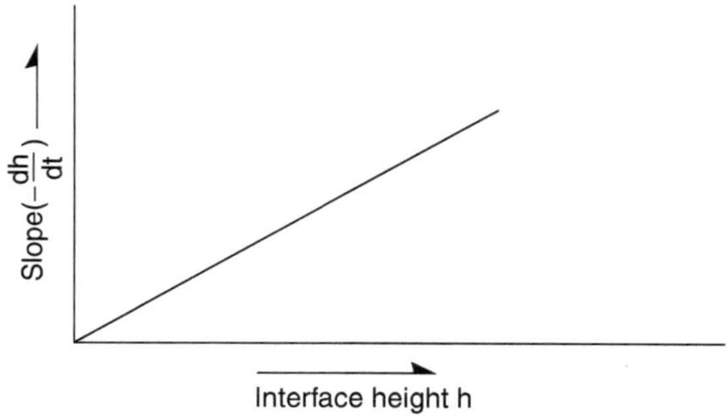

Fig. 7.8 Slope versus height plot from the curve in Fig. 7.7

$$-\frac{dh}{dt} = kh$$

describing the progressing compression in sludge settling. 'k' is constant, characteristic of the compression settling of the sludge.

Let h_∞ be the interface height of the sludge after very long time of settling when further reduction in height is not perceptible. Then reckoning the height of the sludge measured from h_∞, the heights of the sludge $(h_1 - h_\infty)$ at $t = t_1$ and $(h_2 - h_\infty)$ at $t = t_2$ are related as

$$\int_{h_1-h_\infty}^{h_2-h_\infty} (-)\frac{dh}{h} = \int_{t_1}^{t_2} kdt$$

$$\text{i.e} \log_e \frac{h_1 - h_\infty}{h_2 - h_\infty} = k(t_2 - t_1)$$

$$\text{i.e}\, h_1 - h_2 = (h_1 - h_\infty)\left(1 - e^{-k(t_2-t_1)}\right)$$

Problem 1 A sludge having concentration of solids 3500 mg/l is to be thickened @30 l/s in a thickener in continuous operation to produce thickened sludge of solid concentration of 14,000 mg/l.

Batch settling of sludge was carried out in a transparent glass cylinder with 50 cm height of sludge column. The interface crossing different depths at different times was noted and tabulated as under:

Interface height(cm)	50	45	40	35	30	25	20	15	14	13	13
Time (min)	0	1.0	2.5	4.5	7	10	13.5	24	33.5	42	45

1. Plot the interface height versus time curve.
2. Find the point on the curve where settling enters into compression phase.

3. Find the interface concentration of solids in the thickener.
4. Find the sludge settling time (t_u) through the interface solid concentration to thicken the sludge to the desired underflow concentration of solids.
5. Find the interface height versus interface settling velocity plot till the sludge settling enters into compression phase.
6. Find out the thickener area A.

Solution

1. Figure 7.9 shows the interface versus time curve for the batch settling observations for the sludge. The variation of the slope of the curve indicates that the sludge settling is in the zone settling phase from the zero time of observations.
2. Two points on two arms of the curve in Fig. 7.9 are selected for sharp turning, and tangents are drawn to the points. The angle between the tangents shown in the figure is bisected. The bisector intersects the curve at the point 'P'. 'P' is the point where sludge settling enters into compression phase.
3. The tangent at 'P' intersects the height axis at 29.5 cm. Then the interface concentration of solids at 'P' is –
$c = \frac{50 \times 3500}{29.5}$, i.e. 5932, i.e. 5900 mg/l.

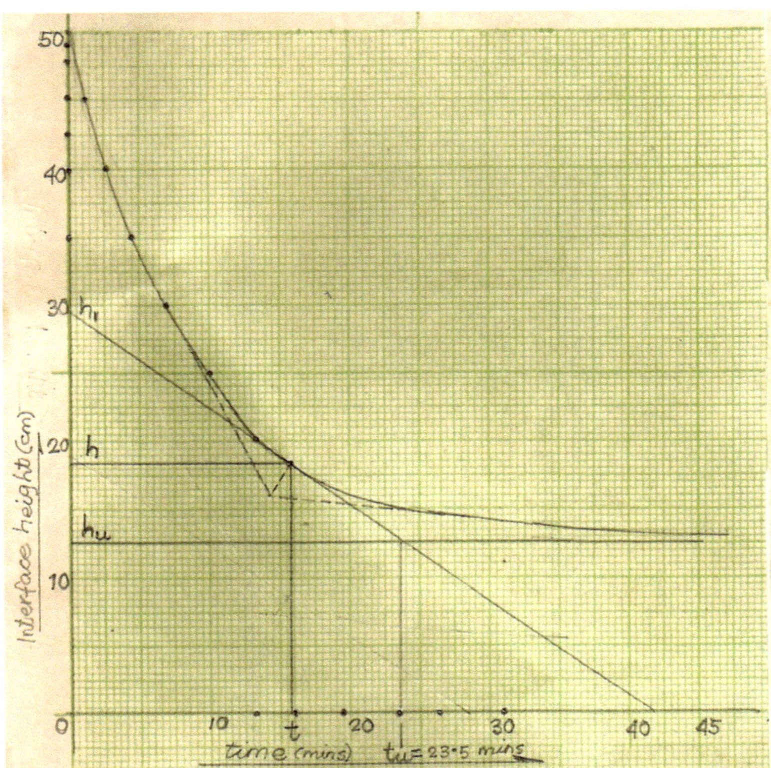

Fig. 7.9 Interface height versus time plot

Fig. 7.10 Interface height versus interface settling velocity plot

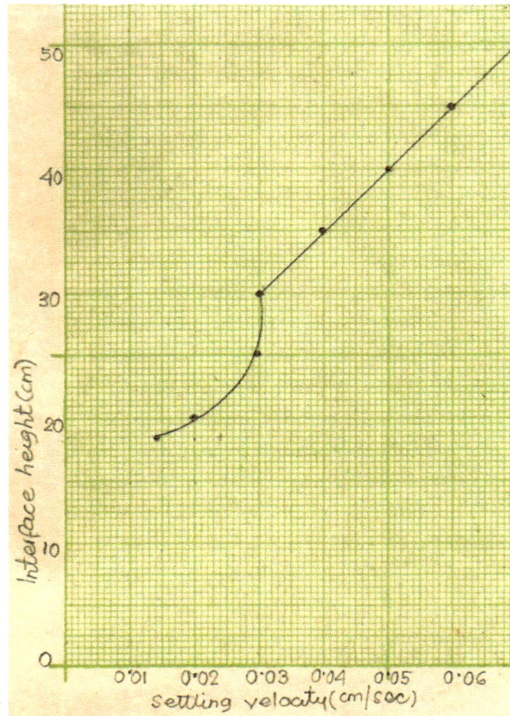

4. To produce a thickened sludge of concentration 14,000 mg/l, 50 cm sludge column is to be reduced to sludge column of height h_u, given by $h_u = \frac{50\times3500}{14,000}$ cm, i.e. 12.5 cm.
   Tangent is drawn at the point 'P', and a straight line is drawn parallel to the time axis through the height 12.5 cm. The point of intersection of the two lines corresponds to the time – $t_u = 23.5\,\text{min}$.
5. Several interface heights are chosen. The corresponding points on the curve are located. Tangents at the points are drawn. The tangents intersect the interface height axis and the time axis. The heights and the corresponding times are noted and are tabulated. From these interface settling, velocities are calculated as shown in the table and are plotted against interface height as in Fig. 7.10.

Interface	Height(cm)	45	40	35	30	25	30	18.5
Tangents intersect at	Height	49	48	45	42.5	40	35	29.5
	Time	13.25	16	19.5	23.5	26.25	31	41.75
Settling velocity	(cm/sec)	0.06	0.05	0.04	0.03	0.03	0.02	0.012

6. Thickener area may be calculated as required for the thickening of sludge:

$$A = \frac{Qt_u}{h_0}$$

$$= \frac{30 \times 23.5 \times 60 \times 100}{1000 \times 50} \text{ i.e.} 84.6\text{m}^2;$$

30 l of sludge having solid concentration of 3500 mg/l is to be reduced to sludge volume with solid concentration of 14,000 mg/l per second.
The volume of water to be removed per second is
$= 30\left(1 - \frac{3500}{14,000}\right)$, i.e. 22.5 l/s.
Interface settling rate at interface concentration of solids of 5900 mg/l in the thickener being 0.012 cm/s thickener area required to release the above water at the above rate is

$$A = \frac{22.5}{1000} \times \frac{100}{0.012} \text{m}^2 \text{ i.e.} 187\text{m}^2;$$

Hence to serve both the purpose of thickening the sludge and releasing the volume of water as necessary, thickener area that has to be provided is 187 m^2. With 15 cm of free board, the depth of the thickener is 65 cm.

Notations

h_0	Sludge height in batch settling test of sludge
C_0	Initial solid concentration in the sludge
A	Cross-sectional area of batch settling cylinder and also the thickener area
h	Interface height
h_u	Sludge height desired for thickening of sludge
c	Concentration of any sludge layer in settling cylinder, also interface concentration at which compression phase of the settling of sludge begins
u	Interface settling velocity
$\overline{u}$	Upward velocity of the concentration value
h_1	Interface height axis intersected by the tangent at any point on the batch settling curve of the sludge *j*
t	Time
C_u	Desired solid concentration in the thickened sludge
t_u	Settling time to reach the desired thickened solid concentration in the underflowing sludge
h_1, h_2	Also used to indicate sludge height at times t^ and tg respectively in compression phase
h_∞	Sludge height after a long time when no perceptible change of height may be observed

References

Coulson JM, Richardson JB (1955): Chemical Engineering, vol 2. Mc-Graw Hil Book Co., New York, p 515

Eckenfelder WW Jr, Melbinger N (1957): Settling and compaction characteristics of biological sludges. Sewage Ind Waste 29:1114–1122

Kynch GJ (1952): A theory of sedimentation. Trans Faraday Soc 48:166

Talmadge WP, Fitch EB (1955): Determining thickener unit areas. Ind Eng Chem 47:38

Chapter 8
New Mode of Column Settling Data Analysis

Abstract The modes of conventional analysis of 'column settling data' differ with nature of suspension. The methods are inadequate. The inadequacies are pointed out. A single method without any consideration to the nature of suspension and without any assumption is presented. The applications are illustrated with actual analysis of laboratory settling data.

Keywords Discrete analysis • Inadequacies • Flocculant analysis • Revised analysis

8.1 Introduction

The process of sedimentation consumes a large portion of investment in water and waste water treatment. Development of the theory on the above subject aims at the understanding of the operation, maintenance and economic design of settling tanks. Hazen 1904 deduced the removal of discrete particles in ideal settling. It depended on the surface area of the settling tank. Fitch 1957 pointed out the removal of flocculant particles to depend on the overflow rate as well as the depth of the tank. Camp (1946) stated that in quiescent settling in a test cylinder, the flocculation is due to differential settling. Settling in a cylinder with proper stirring might simulate flocculation due to velocity gradient, the effect of turbulence and bottom scour and could predict the performance in plant scale settling tanks after proper corrections for hydraulic short circuiting as reflected in their geometrical model studies. Flocculation in a relatively deep tank with low flow-through velocity was shown to be due to differential settling only.

For the efficient operation, maintenance and economic design of settling tank, the characteristics of settleable solids in the raw water suspension should be related to their removal by a settling tank in its plant scale performance. The characteristics of settleable solids are studied by analysing the column settling data collected, in laboratory.

To predict the removal in a plant scale settling tank, Camp advocated 'settling column analysis' and described an analysis for discrete suspension.

The suspended solids encountered in domestic and industrial waste waters are usually flocculant in nature. (O'Connor and Eckenfelder, Jr) employed a different

A. De, *Sedimentation Process and Design of Settling Systems*,
Springer Transactions in Civil and Environmental Engineering,
DOI 10.1007/978-81-322-3634-4_8

mode of analysis for flocculant suspension. *They based their method on a conclusion that is valid for discrete particles only*. Since then all the standard textbooks on the subject, world over, describe two different modes of analysis—one for discrete suspension and the other for flocculant one.

Zanoni et al. (1975), Krishnan (1976); Berthouex and Stevens (1982); Ong (1985); Hasan Ali (1989); Bhaskar et al. (1992); Overcamp (2006) and Pise and Halkude (2011) worked with the analysis of column settling data.

Zanoni et al., Bhaskar et al. and Overcamp worked with flocculant solids and claimed the reproducibility of their methods by comparing with the results with that produced by conventional methods.

Berthouex and Stevens, Ong and Hasan Ali attempted mathematical description model and worked with it in their own ways.

In dealing with flocculant solids, no one mentioned any drawback of the conventional methods of analysis. They did not attempt to resolve the analysis for discrete suspension and also left out of their consideration the suspension with flocculant and discrete solids in their heterogeneous distribution. Although the discrete settling can be contained in mathematically described model, such models cannot be tried for flocculant solids and are futile. Krishnan's method is simple in utilising solid removals at different depths of column for computation of solids in the same.

Pise and Halkude (2011) remarked:

1. Analysis by conventional method of analysis results in variation in overflow rates, settling velocity, detention time and suspended solid removal.
2. Krishnan's method of analysis compared well with the conventional method in their end results. But finding suspended solid concentration at every sample depth at constant time interval is difficult, tedious and time consuming.

In their modified method, solid removals at different port depths were summed up and averaged to even out the deviations from the constant time interval of sample collection. From this average value, solids in the column were determined.

They tried to establish their method comparing their result with the results of the conventional methods, the reproducibility of which they themselves had questioned.

For the critical appraisal of the method, let us consider a settling column provided with, say, three ports at quarter point depth of column. The column is filled up with uniform suspension of similar particles with regard to their settling velocities. Let us imagine that the samples are collected from the ports at a time when the solid-liquid surface of separation just has not crossed the port at first quarter point. The average of removals of solids will record 0 %, although the top quarter depth is free from solids and is settled, i.e. the removal is 25 %.

Again if the samples are collected from the ports at a time when the surface of solid-liquid separation has just crossed the port at last quarter point, the average removal will calculate 100 %, although only 75 % solids are then settled. Similar observations will be reflected when the solid-liquid surface of separation is at any intermediate point. The method therefore appears to be of questionable reproducibility. In spite of the forgoing attempts, the conventional methods are the methods that are mostly talked about but without any much needed evaluation anywhere.

8.2 Need for Revision of the Method of Analysis

The critical evaluation of the conventional modes of analysis is presented here below to reveal the need for revision of the same.

8.2.1 *Conventional Analysis for 'Discrete Suspension'*

The mode of analysis has been presented in Chap. 5. The same is being presented here in a form that is encountered in the textbooks on the subject. This is for lucid discussion that follows on the mode of analysis.

If $'c_{D,t}'$ is the concentration of solids at depth 'D' at time 't' from the start of settling, then $'c_{D,t}'$ will consist of particles having settling velocity $v_s \leq D/t$. The ratio $c_{D,t}/c_0$ (c_0 being the initial uniform concentration of solids in the settling column) will indicate the weight fraction of particles in the suspension having settling velocity $v_s \leq D/t$. This will be so provided the drawing of samples does not cause appreciable lowering of the surface in the cylinder to affect the result.

The ratio $c_{D,t}/c_0$ can be plotted against D/t to draw what is known as 'cumulative frequency distribution diagram' for the settling velocities of particles in the suspension as shown in Fig. 8.1.

From the curve the total fractional removal of the solids in the suspension corresponding to the overflow velocity v_0 can be put down as

X_T (total fractional removal of solids)

$$= (1 - X_0) + \int_0^{X_0} \frac{v_s}{v_0} dX$$

$$= \frac{\text{OABCDO area}}{v_0} \tag{8.1}$$

Camp (1946) Bertheoux and Stevens (1982) was right to remark that the foregoing analysis was true for discrete settling only and that in case of flocculant settling, since flocculation takes place as the particles settle, the distribution of settling velocities among the particles at any time will vary with depth. The above analysis cannot be employed in case of flocculant suspension as such.

Fig. 8.1 Cumulative frequency distribution diagram

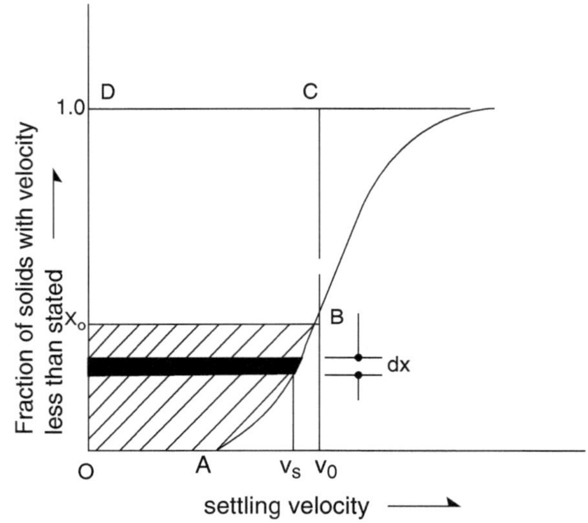

8.2.2 *Inadequacies in the Analysis of Discrete Suspension (De 1998)*

A hypothetical composition of discrete suspension is presented in Table 8.1. This is designated as SUSPENSION.I for the purpose of our discussion.

Let the suspension be now subjected to settling column analysis in a column provided with ports at 60 cm, 120 cm and 180 cm depths (say) from the surface of the suspension.

The concentration of particles in any sample that may be drawn from any port at any time is calculated from Table 8.1. They are presented in Table 8.2.

If the samples are collected from the port at 120 cm depth (say) at times 19 min 55 s, 39 min 55 s, 66 min 35 s, 99 min 55 s and 199 min 55 s, the concentrations in respective samples from Table 8.2 are 600 mg/l, 450 mg/l, 250 mg/l, 150 mg/l and 50 mg/l.

The cumulative frequency distribution of settling velocities of particles in the suspension according to the conventional method is presented in Table 8.3.

This is represented in Fig. 8.2.

If the samples are collected from the same port at times 10 min, 20 min 5 s, 50 min, 75 min and 150 min, then from Table 8.2 the concentrations of the samples are 600 mg/l, 450 mg/l, 250 mg/l, 150 mg/l and 50 mg/l, respectively. According to the conventional method of analysis, the cumulative frequency distribution of settling velocities of particles in the suspension is presented in Table 8.4.

This is represented by a curve different from Fig. 8.2. Thus depending upon the times of collection of samples at different ports, an infinite number of such curves may be obtained for the same suspension.

Table 8.1 A discrete suspension of hypothetical composition

SUSPENSION I	
Concn. in mg/l	Settling velocity v_s of the particles in cm/s
50	0.01
100	0.02
100	0.03
200	0.05
150	0.10

Table 8.2 Concentration of particles in a sample collected from any port at indicated depth at any time

Port at depth	At time[a]	Concentration in mg/l
60 cm	0–10 min	600
	10–20 min	450
	20–33 mins 20 s	250
	33 min 20 s–50 min	150
	50 min–100 rains	50
120 cm	0–20 min	600
	20–40 min	450
	40–66 min 40 s	250
	66 min 40 s–100 min	150
	100 mins–200 min	50
180 cm	0–30 mins	600
	30 min–60 min	450
	60 min–100 min	250
	100 min–150 min	150
	150 mins–300 min	50

[a]Time measurement commences from the start of settling

Table 8.3 Cumulative frequency distribution of settling velocities of particles in the suspension

Concentration in mg/l	Consists of particles having settling velocity v_s less than the stated in cm/s
600	0.1
450	0.05
250	0.03
150	0.02
50	0.01

Again we can consider a suspension of any hypothetical composition by dividing the ordinate 600 mg/l in Fig. 8.2, into any number of parts and assuming each part concentration consists of particles having settling velocity equal to the velocity corresponding to the upper limit of that part concentration in the graph of Fig. 8.2. Any such suspension like one presented in Table 8.5 may be described by the same curve in Fig. 8.2.

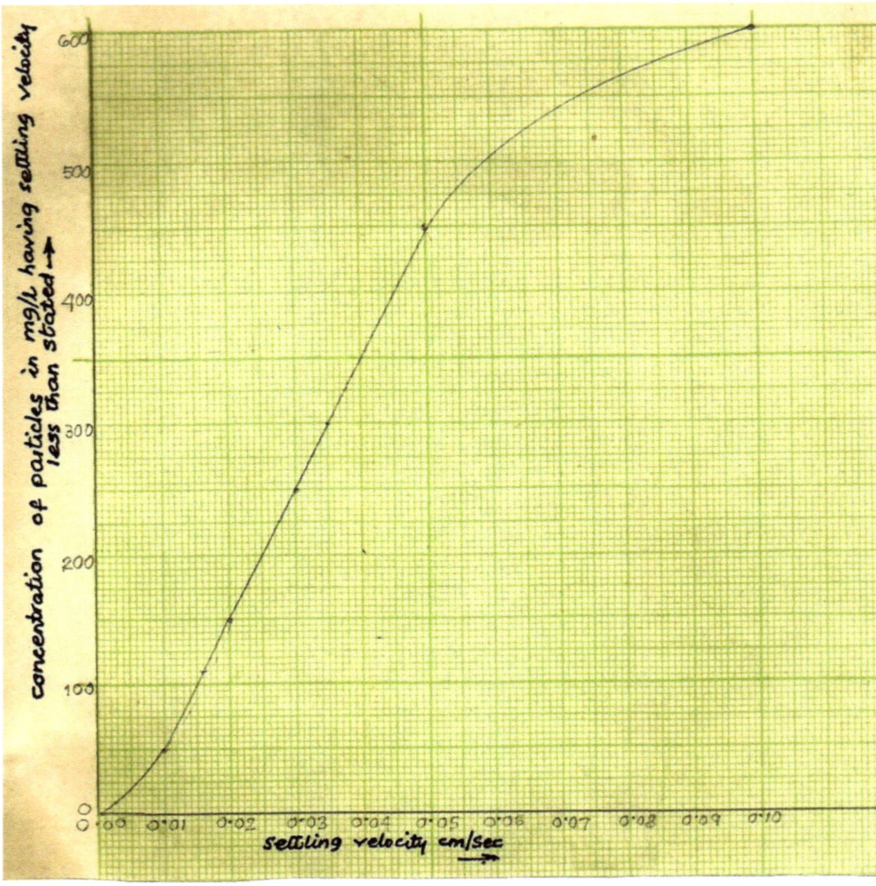

Fig. 8.2 Cumulative frequency distribution of settling velocities of particles in the suspension presented in Table 8.3

Table 8.4 Cumulative frequency distribution of settling velocities of particles in the suspension

Concentration in mg/l	Consists of particles having settling velocity $v_s$cm/s less than stated
600	0.2
450	0.1
250	0.04
150	0.027
50	0.013

An infinite number of such suspensions are possible. The same curve in Fig. 8.2 may claim to have represented the settling characteristics of all of them according to the conventional method of analysis.

Table 8.5 A suspension of the settling characteristics of which may also be claimed to have been represented by the curve in Fig. 8.2

SUSPENSION II	
Concn. in mg/l	Settling velocity v_s cm/s
50	0.01
60	0.016
40	0.2
150	0.035
150	0.05
150	0.10

Table 8.6 Comparison of arithmatically and graphically computed removal values for SUSPENSION I and SUSPENSION II corresponding to different overflow rates

1	2	3	4	5
	Arithmatically computed removal values in mg/l		Graphically computed removal values in mg/l	
Overflow velocity in cm/s	SUSPENSION I	SUSPENSION II	SUSPENSION I	SUSPENSION II
0.02	575	563	538	same as in col.4
0.03	534	526	492	
0.04	488	488	444	
0.05	460	450	395	

For different overflow velocities, the removal values for SUSPENSION I in Table 8.1 and SUSPENSION II in Table 8.5 are computed arithmetically. They are also found out graphically from the graph in Fig. 8.2 according to Eq. 8.1. They are presented in Table 8.6 for comparison.

In Table 8.6, it is seen that the graphically computed removal values for SUSPENSION I and SUSPENSION II are the same corresponding to an overflow velocity.

It is so because the diagram in Fig. 8.2 claims to have represented both the suspensions. In case of SUSPENSION II where more variations of velocities of settling particles are present than that in SUSPENSION I, arithmatically computed removal values are closer to the graphically computed values.

The discrepancy between the results in column 2 and that in column 4 may be due to the assumption that all variations of velocities of settling particles are present in between any two settling velocity values.

Foregoing discussion shows that the cumulative frequency distribution curve of settling velocities plotted in accordance with the conventional method may not represent the settling characteristics of the suspension. The computation of removal values from such curve, therefore, appears to be erroneous and misleading.

8.2.3 *Need for the Revision of the Mode of Analysis (De 1998)*

The inadequacies and anomalies in the conventional mode of analysis appear only because the method ignores one fact as illustrated in the following.

In Fig. 8.3 a settling cylinder is filled up with discrete suspension at concentration 'c'. All particles contained therein are identical as regards their settling velocity v_s.

The interface (the surface of solid-liquid separation) depths are plotted at different settling times. The drawn straight line through the points, i.e. the straight line 'OC', traces the interface positions of the concentrations 'c' at varying settling times,

Let the interface settle through the depth $D = v_s t_i$ at time t_i measured from the begining of settling at 'O'. Sample may be collected through a port at depth $D = v_s$ t_i of the column at any time over the interval of time $0 - t_i$.

For all the points on AB (over the interval of time $(0 - t \leq t_i)$, samples collected through the port located at depth $D = v_s t_i$ will give the concentration of solids 'c' and settling velocity $(v_s t_i / t > v_s)$ excluding only one point at B where the sample will give the concentration of solids 'c' as well as the settling velocity $(v_s t_i / t_i) = v_s$.

This resolves that the characteristic settling velocity of the solids composing the concentration 'c' cannot be obtained unless the sample is collected from the port while the interface of the concentration of solids is just crossing it. Interface settling rates are the settling velocities of the fastest moving particles composing the concentrations, and this is unique characteristic feature of the individual concentration so far as settling is concerned.

Sample should be collected from the interface of the concentration separating it from the other.

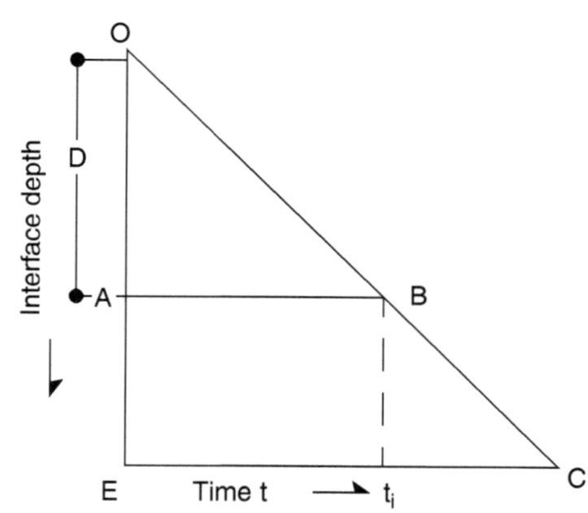

Fig. 8.3 Interface trajectory of uniform discrete suspension

8.2.4 *In Quest of a Revised Mode of Analysis (De 1998)*

A revised mode of column settling analysis should find out the unique characteristic of each concentration of solids that differentiates one concentration of solids from the other. In other words revised mode of analysis should enable one to draw the characteristic interface depth versus time diagram for the concentration 'c' of the suspension in Sect. 8.2.3 as shown in Fig. 8.3 from the column settling observation.

The above suspension, in Sect. 8.2.3, is settling in a cylinder in Fig. 8.4a provided with ports at depths d, $2d$, $3d$ from the surface of the suspension (Fig. 8.4a).

Samples are collected from the ports as time progresses, and the concentration of solids in the samples is plotted to obtain concentration versus time curves for the observations obtained at each of the above ports in Fig. 8.4b.

In case of port no. 1, the concentration values are repeated along $AB = t$ where $d = v_s t$, B is a point on the interface. Similarly from port no. 2, the concentration values are repeated over $DE = 2t$, where $2d = v_s 2t$ and E is a point on the interface.

Similarly H is also a point of the interface at the port at depth $3d$ at time $3t$. The interface depth versus time graph obtained in Fig. 8.4c is the unique settling characteristic curve for the concentration 'c' in Fig. 8.3.

The slope of the straight line is $v_s = \frac{d}{t}$.

Problem 8.1 A settling column test of a suspension was performed in a cylinder provided with ports at depths of (1) 0.6 m, (2) 1.2 m and (3) 1.8 m. Samples were collected from the ports in quick succession at noted times. The concentration of solids in the samples was determined. The concentration versus time graphs at the ports are plotted as shown in Fig. 8.5a–c.

1. Find the characteristic settling curves (i.e. interface depth versus time curve) of the suspension.

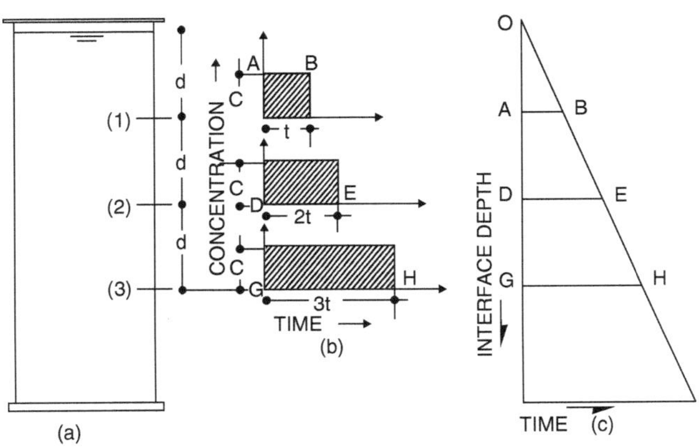

Fig. 8.4 In quest of unique characteristics of a concentration

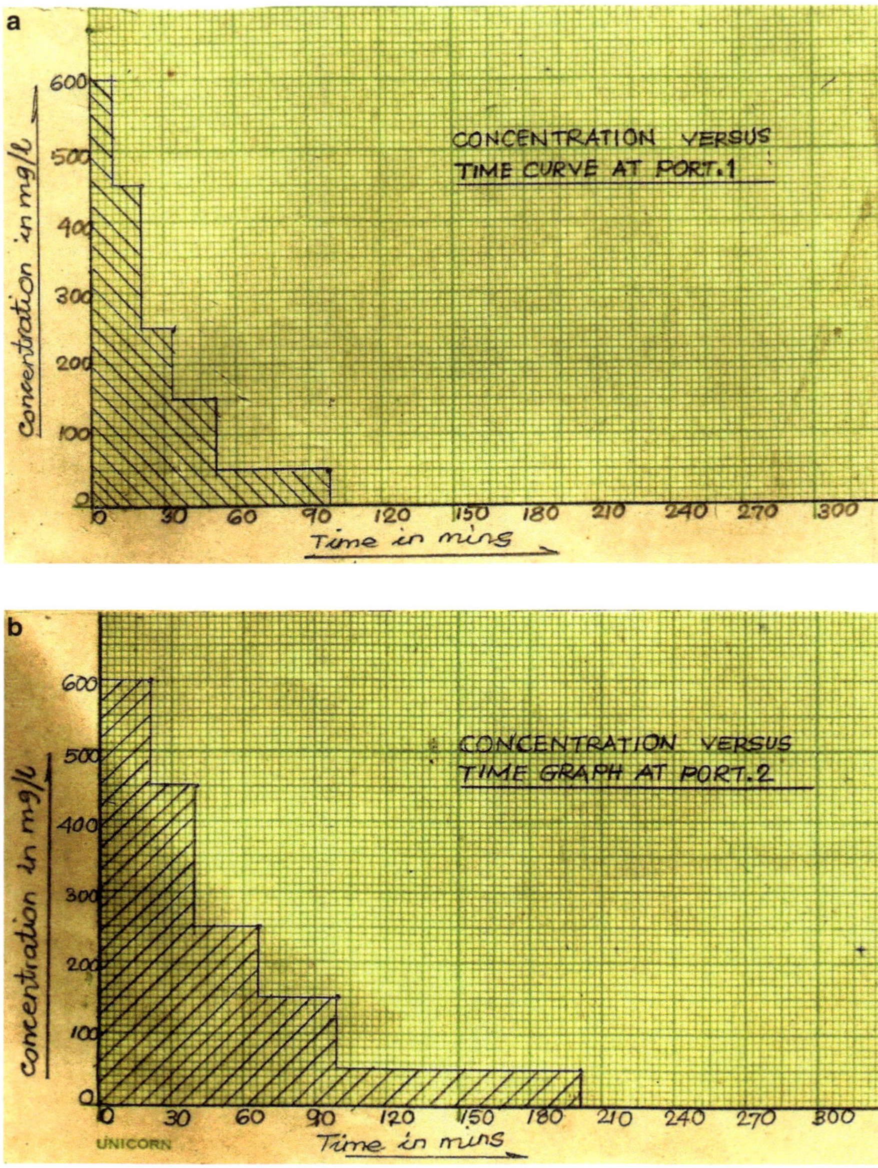

Fig. 8.5 Column settling analysis

2. Find the composition of solids in the suspension.
3. Find the amount of solids that will be settled at time $t = 60$ min.

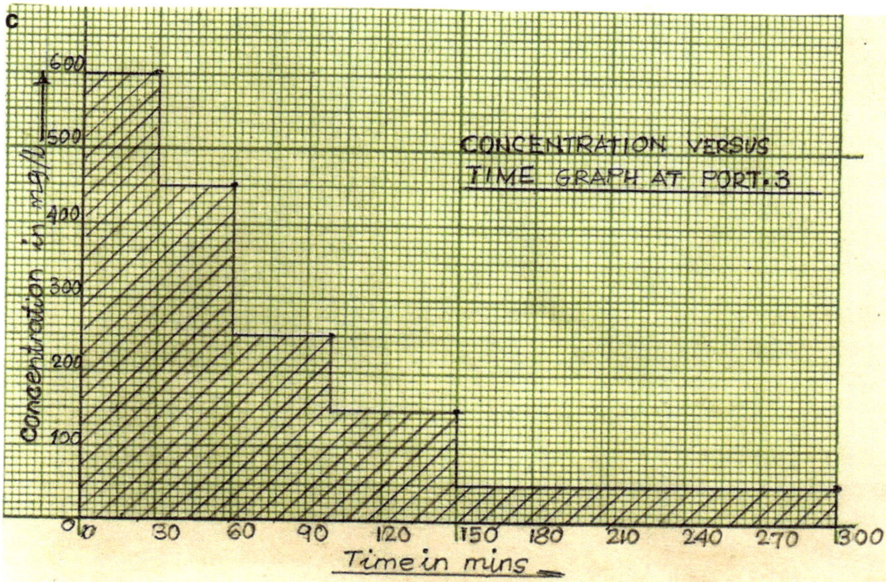

Fig. 8.5 (continued)

Table 8.7 Interface depth versus time observations for different concentrations

	Time of crossing the interface depths (in mins)		
Concentration in mg/l	0.6 m	1.2 m	1.8 m
600	9	21	30
450	21	39	60
250	53	66	99
150	51	99	150
50	99	201	300

Solution

1. Scaled from Fig. 8.5a–c are the interface depths of the concentrations at different times as tabulated hereunder (Table 8.7).
   The points are plotted in Fig. 8.6.
   Through the plotted points, interface depth versus time curves for the concentrations are drawn. These curves are the characteristic settling curves of the suspension.
2. In between two curves, in Fig. 8.6 there cannot be any other curve. This is so because if any concentration is chosen in between, the interface depth versus time curve for that concentration coincides with the curve for the upper concentration. This suggests that all the particles between the two concentrations are identical as regards their settling velocity and the settling velocity is the slope of

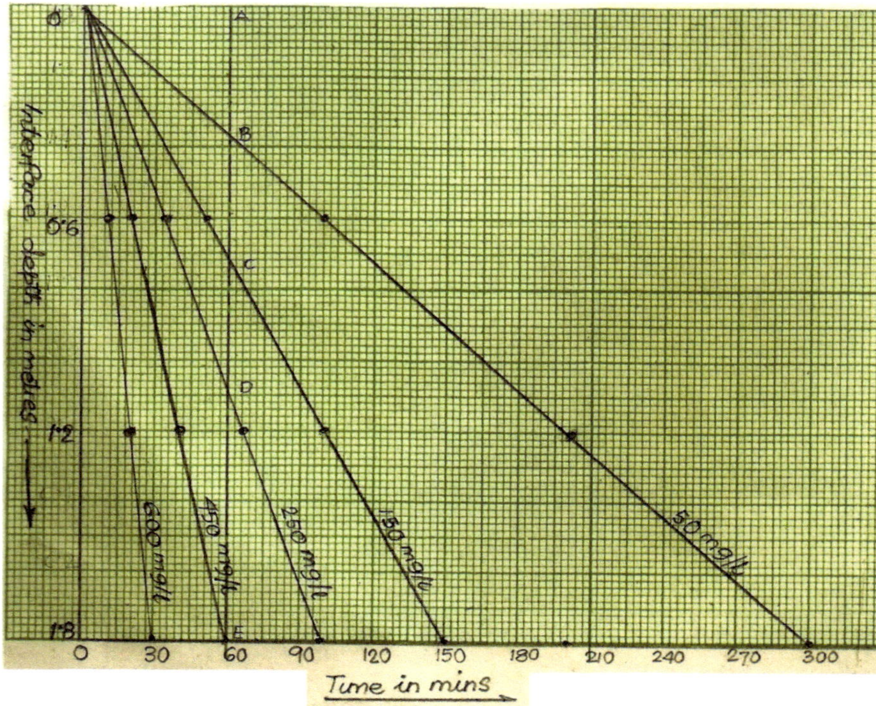

Fig. 8.6 Interface concentration trajectories

the curve for higher concentration. Hence the composition of the solids in the suspension can be worked out as follows:

(600–450) i.e. 150 mg/l consists of solid particles of settling velocity $\frac{180}{30\times60}$ cm/s, i.e. 0.1 cm/s
(450–250) i.e. 200 mg/l consists of solid particles of settling velocity $\frac{180}{60\times60}$ cm/s, i.e. 0.05 cm/s
(250–150) i.e. 100 mg/l consists of solid particles of settling velocity $\frac{180}{99\times60}$ cm/s, i.e. 0.03 cm/s
(150–50) i.e. 100 mg/l consists of solid particles of settling velocity $\frac{180}{150\times60}$ cm/s, i.e. 0.02 cm/s
(50–0) i.e. 50 mg/l consists of solid particles of settling velocity $\frac{180}{300\times60}$ cm/s, i.e. 0.01 cm/s

The composition of the solid particles in the suspension can be presented in Table. 8.8.

Table 8.8 Composition of solid particles in the suspension

Concentration in mg/l	Settling velocity of particles v_s in cm/s
150	0.1
200	0.05
100	0.03
100	0.02
50	0.01

3. Since in between two interface settling curves there is no other particles of different settling velocity, the concentration of solids is between two curves, i.e. at all points defined by depth and time, the concentration of solids remains at concentration of the upper curve.
   The solids remaining over the depth of 1.8 m of the cylinder after 60 mins. of settling can be calculated by and from placing a vertical line at $t = 60$ min. as shown in Fig. 8.6.
   $AE = 1.8$ m is such a line. From Fig. 8.6 the length of $AB = \frac{18.5 \times 1.8 \times 100}{90}$, i.e. 37 cm, is free from solids.
   The length BC contains solids $= \frac{17.5 \times 1.8 \times 100\text{cm}}{90} \times \frac{50\,\text{mg}}{1000\,\text{cm}^3}$
   $= 35\,\text{cm} \times 0.05\,\text{mg/cm}^3$, i.e. $1.75\,\text{mg/cm}^2$.
   The length CD contains solids $= \frac{19 \times 1.8 \times 100\text{cm}}{90} \times \frac{150\,\text{mg}}{1000\,\text{cm}^3}$
   $= 38\,\text{cm} \times 0.15\,\text{mg/cm}^3$, i.e. $5.7\,\text{mg/cm}^2$;.
   The length DE contains solids $= \frac{35 \times 1.8 \times 100\text{cm}}{90} \times \frac{250\,\text{mg}}{1000\,\text{cm}^3}$, i.e. $17.5\,\text{mg/cm}^2$;.
   Hence after 60 min, the solids in suspension over the length of 1.8 m are $(17.5 + 5.70 + 17.5)\text{mg/cm}^2$, i.e. $24.95\,\text{mg/cm}^2$.
   The solids initially present in the cylinder over the length $1.80\,\text{m} = 180\,\text{cm} \times 0.6\,\text{mg/cm}^2$, i.e. $108\,\text{mg/cm}^2$.
   The solids settled from the length of $1.8\,\text{m} = \frac{(108 - 24.95) \times 100}{108}$
   $= 76.9\%$.

8.2.5 *Need for the Critical Evaluation Mode of Analysis for Flocculant Suspension*

The critical evaluation of the method of analysis for flocculant suspension is required to reveal the drawbacks of the method of analysis. The critical evaluation should follow after presenting the conventional method of analysis, although the same has already been presented in Chap. 6.

8.2.6 Conventional Analysis for Flocculant Suspension

The mode of analysis for flocculant suspension was reported by O'Connor and Eckenfelder (1957).

The solid concentrations obtained at different times at different depths in the settling column test with flocculant suspension are expressed as

$$\frac{C_{D,t}}{C_0} = \text{fraction of initial concentration,}$$

$1 - \frac{C_{D,t}}{C_0} = X_{D,t}$ fraction of particles which settled past the point of tap in the test cylinder at depth D and at time t from the start of the test. $X_{D,t}$ fraction of particles, therefore, has had average velocity D/t or more. These $X_{D,t}$ values are plotted in depth-time coordinates. The so-called isoconcentration curves or smooth curves identifying the same fractional removal are drawn through with the help of plotted values as shown in Fig. 8.7. (In the presentation that follows, it will be shown that isoconcentration curve is not characteristic to any concentration. Many such isoconcentration curves may result for the same concentration. It is actually an isoconcentration area and not an isoconcentration curve that one obtains in depth-time coordinates). The curvilinear nature of the curves reflects the flocculating nature of the particles. The overall removal X_T in an ideal basin of depth D corresponding to an overflow velocity v_0 and theoretical detention time $t_0 = D/v_0$ (by def^{n}) may be computed as follows.

In time t_0, x_B fraction of particles has had settling velocity D/t_0 or greater and hence will be removed completely. $(x_C - x_B)$ fraction of particles has had average velocity D_1/t_0 and hence will be removed in the ratio $(D_1/t_0)(D/t_0)$, i.e. (D_1/D). (This relation cannot be permitted to be used for it is true for discrete suspension only.) In similar note $(x_D - x_C)$ will be removed in the ratio D_2/D, and the overall removal would be written approximately:

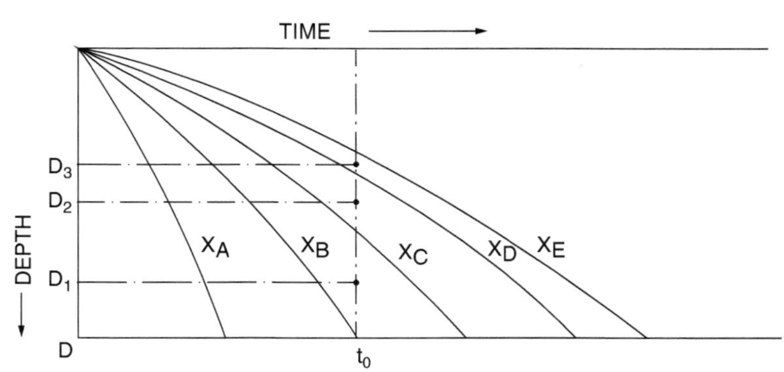

Fig. 8.7 Fractional removal trajectories for flocculant solids

$$X_{\mathrm{T}} = x_B + \frac{D_1}{D}(x_C - x_B) + \frac{D_2}{D}(x_D - x_C) + \frac{D_3}{D}(x_E - x_D) \tag{8.2}$$

The summation is extended till $(1 - x_E)$ is insignificantly small to affect the overall removal.

8.2.7 *Inadequacies in the Analysis for Flocculant Suspension (De 1998)*

Let us redraw Fig. 8.3 with a few salient changes as in Fig. 8.8.

We consider a suspension in a settling column of height '*h*'. Identical settling particles with settling velocity v_s are uniformly distributed at concentration *C* within it. At any time *t* from start of settling, the solid-liquid surface of separation will sink to a depth $D = v_s t$. The particles maintain invariable position with respect to each other as they settle identically. They remain at concentration *C* wherever they are present.

The height $(h - v_s t)$ of the settling column will be filled up with suspension at uniform concentration *C*. If we plot the concentrations with time at depth *D*, we get a straight line AB parallel to abscissa, which will extend to B, i.e. to the time t, as shown in Fig. 8.8. The slope of OB is $v_s t/t$ settling velocity of each particle, which is also surface settling rate. The points on the line OBC, therefore, will indicate the position of the surface (of separation) at different times so that any point within the space ODC indicates the presence of concentration C at coordinate depth D at time t, and the ordinate intercepted between lines OC and DC will indicate the height to which the settling column will remain filled up with suspension at concentration *C* at time *t* from the start of settling.

Now let us imagine that the particles just considered are identically flocculant. Each particle settles identically suffering identical collisions with identical

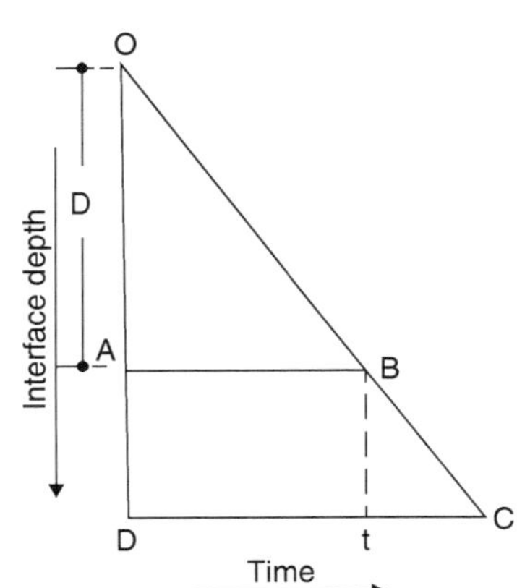

Fig. 8.8 Discrete suspension

distribution of other particles having different other settling velocities. This may be so if all particles in the suspension are uniformly distributed at start and quiescent settling of them is assured. Each of the particles will grow in size as a floc identically during the identical movements. Each floc will, therefore, accelerate identically. We confine our attention to particles we are considering only, leaving aside other particles forming floc with them. We are attentive also to their accelerated movements. Since all the particles accelerate identically, they maintain their invariable position with respect to each other as they settle. As such they are at concentration C wherever they are present.

Under this situation instead of getting OC a straight line in Fig. 8.8, we get OC curved as shown in Fig. 8.9. The curvature indicates that the particles on the surface (of separation) hence within the body of the settling column settle with an accelerated velocity.

Next we consider that there are three types of particles, say, in the settling column. In each type the particles are identical. Concentrations of three types are C_1, C_2, C_3. The foregoing arguments appear to be valid individually for each type. The total composite picture may be presented in Fig. 8.10. Samples collected at depth D and time t will indicate concentration of particles $(C_1 + C_2 + C_3)$ if D,t coordinates lie within the space ODC in Fig. 8.10, $(C_1 + C_2)$ if it is within the space OCE and C_3 if it is within OEF. This is going to show, it appears, that it is actually an isoconcentration area and not an isoconcentration line that we obtain in depth-time coordinates.

Depending upon variations of types of particles in suspension, the space may increase or decrease. Theoretically, an infinite number isoconcentration lines may be possible. Drawing of isoconcentration line appears to be misleading.

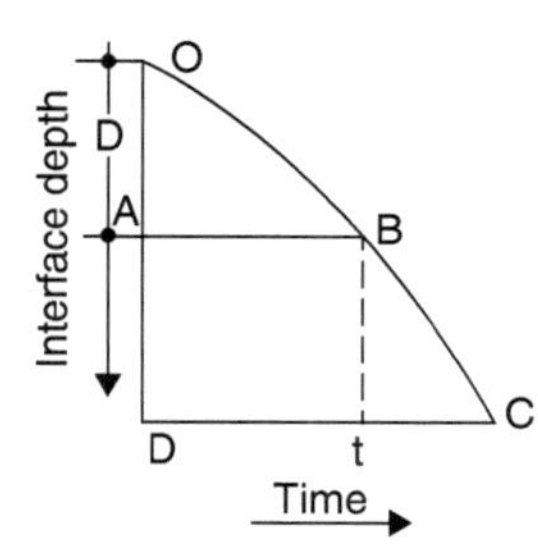

Fig. 8.9 Flocculant suspension

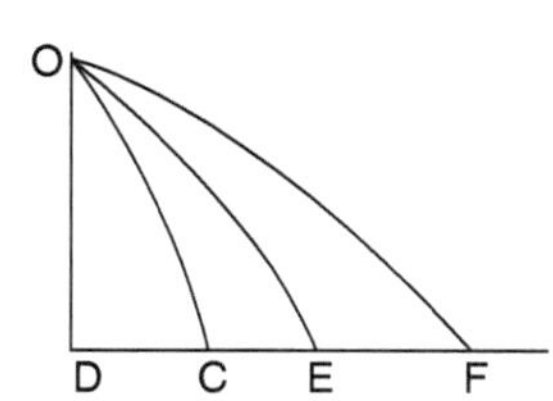

Fig. 8.10 Flocculant suspension (composite)

8.2.8 Need for Revision of the Settling Analysis for Flocculant Suspension (De 1998)

The inadequacies of the conventional method of analysis have been pointed out. They appear because *conventional method of analysis fails to plot the interface settling curve which is the only unique characteristic feature separating one concentration from the other so far as settling is concerned.*

The method of analysis may be *loudly questioned for having based the method of analysis on a conclusion that is strictly valid for discrete settling only.*

Revised method should:

1. Draw the interface settling curves
2. Not use any assumption that is not valid for flocculant suspension

The method of analysis outlined in Sect. 8.2.4 satisfies both of them.

8.3 Revised Mode of Analysis of Column Settling Data (De 1998)

The revised mode of analysis aims at assessing the total amount of solids that are present in the suspension of the settling column at any time t.

8.3.1 Test Procedure and Analysis

The following steps are to be followed:

1. The settling test is performed with suspension in a cylinder for the collection of samples and determination of concentrations of solids in them at various depths at different times.
2. The concentration versus time curves for each of the depths are plotted as shown in Fig. 8.11.
3. The times at which the particular concentration crosses the different depths are to be found out from the above curves in Fig. 8.11. For example, the surface of separation (i.e. interface) of the concentration C_0, the uniform concentration at start of the test, crosses the depths D_1, D_2, D_3, D_4 at times t^1, t^2, t^3, t^4, respectively.
   The surface of separation of any other concentration C_1 can be located to be at depths D_1, D_2, D_3, D_4 at times t_1, t_2, t_3, t_4, respectively, as shown in Fig. 8.11.
4. The positions of the surface of separation of different concentrations C_1, C_2, C_3, C_4, etc. can now be plotted as shown, and smooth curves are drawn through them as shown in Fig. 8.12. These are showing the surface of separation of each concentration crossing different depths with time. The points on the

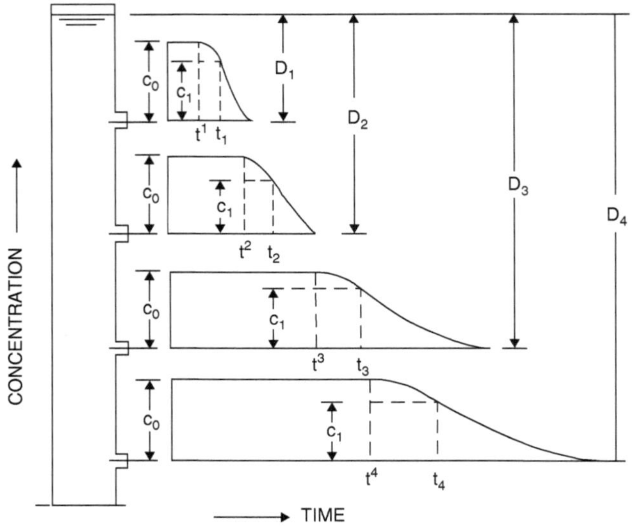

Fig. 8.11 Concentration versus time curve at different depths

Fig. 8.12 Position of surface of separation (interface)

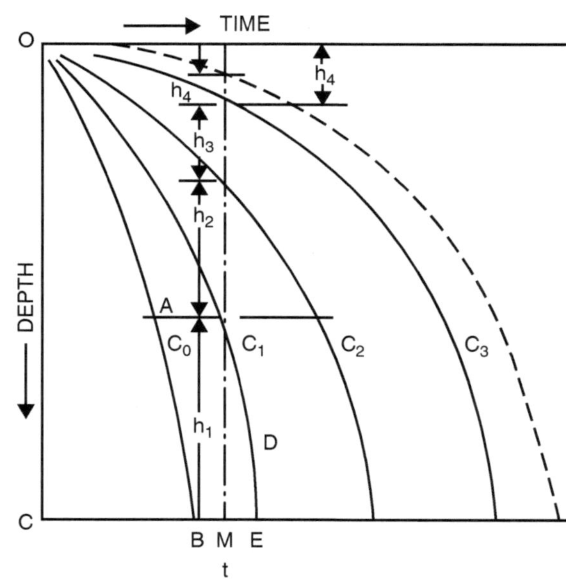

curve OAB give the positions of the surface of separation of concentration C_0 at different times. Any point (D,t) within the space OABCO will give the concentration C_0 present at depth D at time t. Similarly other curves for concentrations C_1, C_2, C_3 are drawn in the same way. At any point (D,t) in space OABEDO, it will indicate the presence of concentrations the values of which lie between C_0 and C_1 at depth D at time t.

The concentrations $C_1, C_2, C_3 \ldots\ldots$ are to be chosen in such a way that the curve in between two concentrations in concentration versus time curve in Fig. 8.11 at a particular depth will be straight line.

5. At any time t amount of solids present in suspension in settling column may be computed. We draw a vertical line at time t.

In between C_1 and C_2, the concentration intercept is h_2 (Fig. 8.12). Since we have chosen the variation between C_1 and C_2 to be linear, the amount of solids in h_2 of the settling column per unit cross-sectional area is

$$\frac{h_2(C_1 + C_2)}{2} \tag{8.3}$$

The height h_3 will contain the amount of solids per unit cross-sectional area of column:

$$\frac{h_3(C_2 + C_3)}{2} \tag{8.4}$$

Now the curve of concentration C passing through the point M on BE can be found as

$$C = C_0 - (C_0 - C_1)\frac{\text{BM}}{\text{BE}} \tag{8.5}$$

The height h_1 will contain the amount of solids per unit cross-sectional area of the column

$$= h_1\left(\frac{C_0 + C_1}{2} - \frac{C_0 - C_1}{2}\cdot\frac{\text{BM}}{\text{BE}}\right) \tag{8.6}$$

Similarly the amount of solids contained in the height h_4 of the settling column per unit cross-sectional area is

$$\frac{h_4C_3}{2} \tag{8.7}$$

So the amount of solids removed per unit cross-sectional area at time t is

$$\begin{aligned} &= hC_0 - \left(h_1\left(\frac{C_0 + C_1}{2} - \frac{C_0 - C_1}{2}\cdot\frac{\text{BM}}{\text{BE}}\right) + h_2\frac{C_1 + C_2}{2} + h_3\frac{C_2 + C_3}{2} + h_4\frac{C_3}{2}\right. \\ &= h_1\left(C_0 - \left(\frac{C_0 + C_1}{2} - \frac{C_0 - C_1}{2}\cdot\frac{\text{BM}}{\text{BE}}\right) + h_2\left(C_0 - \frac{C_1 + C_2}{2}\right)\right. \\ &\quad + h_3\left(C_0 - \frac{C_2 + C_3}{2}\right) + h_4\left(C_0 - \frac{C_3}{2}\right) \end{aligned} \tag{8.8}$$

8.3.2 Discussion

The last term for summation in Eq. 8.8 is an approximation. The term will give a fairly accurate result if the concentration versus time curves at different depths have long tails and if the tails in those curves are approximately straight lines subsequent to the last concentration for which the positions of the surface of separation have been plotted in depth-time coordinates. Such a concentration as shown in Fig. 8.12 is C_3.

For accurate estimation, the position of the solid-liquid line of separation should be plotted in depth-time coordinates as shown by dotted curve in Fig. 8.12, from the concentration versus time curves shown in Fig. 8.11. In that case h_4 in Eq. 8.8 should be replaced by h_4' shown in Fig. 8.12. During the computation, it must be borne in mind that if there is a vertical drop in the concentration versus time curve drawn at any depth, it will mean the absence of all concentrations in between the upper value and lower value and necessary changes should be incorporated during the computation for fractional removal.

8.3.3 Conclusion

Revised mode of settling column analysis *draws isoconcentration curve of any concentration from the determination of this concentration from the samples collected from its interface* for its different positions in depth-time coordinates. *This is unique characteristic* for its settling *that no other so-called isoconcentration curve used by conventional method can entrap.*

This method of analysis assesses the settled solids after the settling for the stipulated period of time. *The computation is direct and is not based on any assumption like the one on which conventional method of analysis is based.*

Problem 8.2 Laboratory settling data showing the concentration of suspended solids in mg/l at different depths in metres at different times in min is shown in Table 8.9.

Calculate the removal through an ideal settling tank of 1.8 m depth at an overflow velocity of 0.0015 $m^3/s/m^2$.

Solution

From the data presented, in Table 8.9 concentration versus time curves are plotted for each of the ports at 60 cm, 120 cm and 180 cm depth as presented in Fig. 8.13

From Fig. 8.13, the times at which the interfaces of different selected concentrations pass through the ports at depths 60 cm, 120 cm and 180 cm are found out. These are tabulated in Table 8.10.

The interface settling curves for the selected concentrations are drawn in Fig. 8.14 through the points plotted from the data in Table 8.10.

Table 8.9 Settling data

Time in min	0.6 m	1.2 m	1.8 m	Time in min	0.6 m	1.2 m	1.8 m
0	900	900	900	0	900	900	900
10	660	770	790	45	290	330	350
20	380	490	550	60	260	300	320
30	330	370	450	120	210	240	270

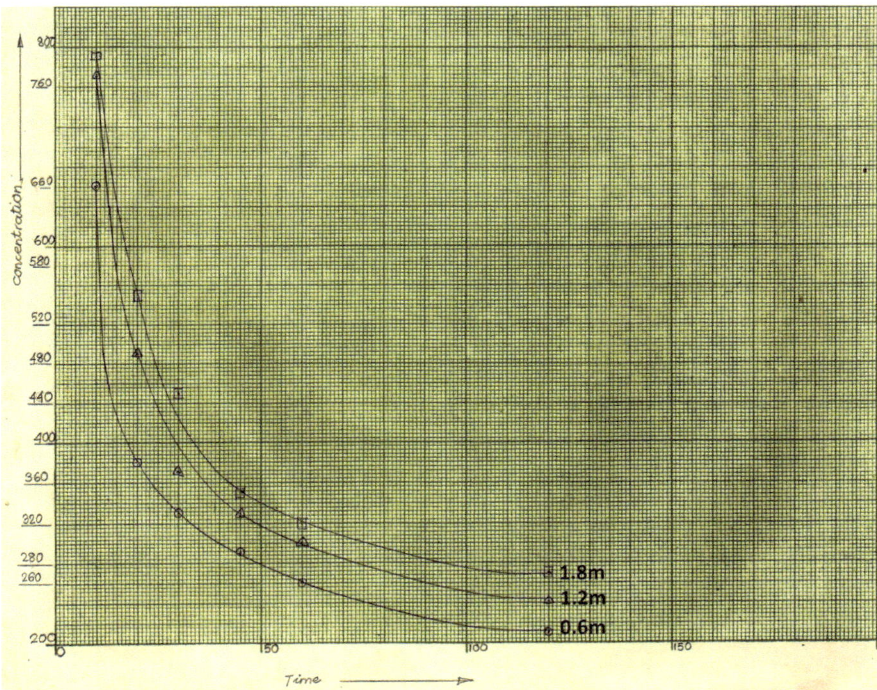

Fig. 8.13 Concentration versus time curve for the solids in settling suspension

Table 8.10 Interface of different concentrations crossing the different depths at different times in min

Cone^{n} in mg/l	Interface crossing the depth		
	60 cm	120 cm	180 cm
760	–	10	11
660	10	12.5	14
580	11	15	18
520	11.5	18	22
480	12	21	24.5
400	17.5	29	33.5
360	23	36.5	42
280	49	72.5	93

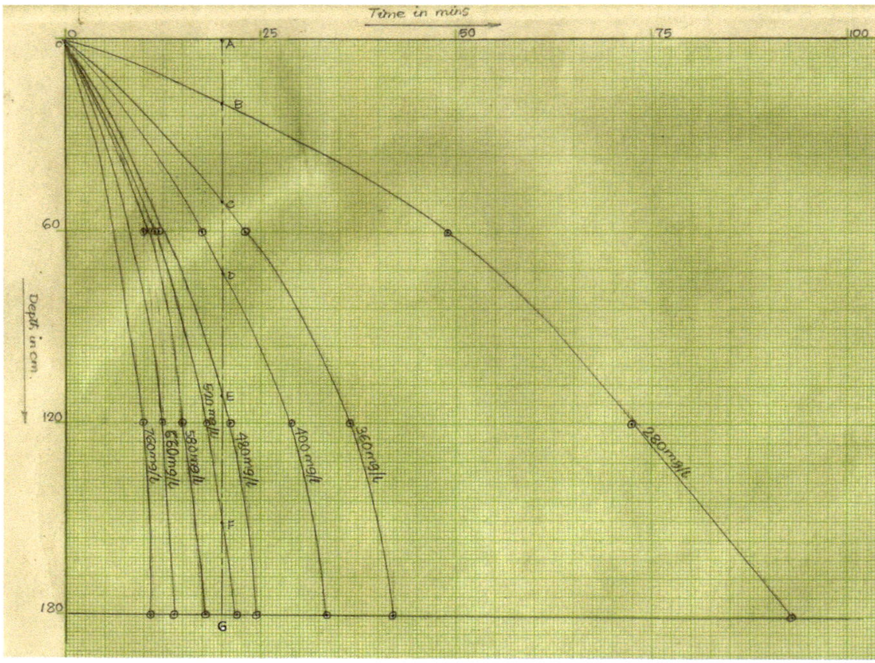

Fig. 8.14 Interface settling curves for the solids in the settling suspension in depth-time coordinates

Table 8.11 Solid distribution in cylinder after $t = 20$ min settling

Portions	length in cm	Concentration range of solids in mg/l	Average concentration in mg/l
AB	20.4	0–280	140
BC	30	280–360	320
CD	22.8	360–400	380
DE	38.4	400–480	440
EF	39.6	480–520	500
FG	28.8	520–550[a]	535

[a]550 mg/l is the concentration of the interface settling curve interpolated through the point G

The clarification rate of 0.0015 $m^3/s/m^2$ in an ideal settling tank provides settling time (theoretical detention time):

$$t = \frac{1 \cdot 8}{0.0015 \times 60} \text{ min i.e. } 20 \text{ min}$$

A vertical line AG is drawn at $t = 20$ min. The line shows that the portions of the height of the column contain solids between the concentrations as tabulated in Table 8.11.

Settling solids per cm^2 of the settling base can be computed from Table 8.11 or Eq. 8.8:

$$
\begin{aligned}
&= [(900-140)20.4+(900-320)30+(900-380)22.8 \\
&\quad +(900-440)38.4+(900-500)39.4+(900-535)28.8]0.001 \\
&= 15.504+17.400+11.856+17.664+15.760+10.512 \\
&= 88.696\,\mathrm{mg}.
\end{aligned}
$$

Initial solids present in 180 cm length of the cylinder/cm^2 of base area

$$= 900 \times 180 \times 0.001, \text{ i.e.} 162\,\mathrm{mg}.$$

Hence the percentage removal is 88.696 × 100/162, i.e. 55 %.

Notations

$C_{D,t}$	Concentration of solids at depth D at time t
C_0	Initial concentration of solids
X_T, x_T	Total fractional removal of solids
X_0, x_0	Fraction of solids having settling velocity $v_s \leq v_0$, the overflow velocity
c, C	Concentration of solids
t_i	Any particular time t_i
t	Any time
d, D	Depth
$X_{D,t}$	Fraction of particles having settling velocities D/t or more
h	Height of the suspension
h1 etc.	Intercept of the cylinder In between two interface surfaces

References

Bertheoux PM, Stevens DK (1982) Computer analysis of settling data. J Env Eng Div ASCE 108 (5):1065–1069

Bhaska PU, Chaudhuri S, Jawed M (1992) Type II sedimentation – Removal efficiency from column settling tests. J Env Eng Div ASCE 118(3):757–760

Camp TR (1946) Sedimentation and the design of settling tank. Trans ASCE 111:895

De A (1998) Revised mode of analysis of column settling data. Indian Chem Eng Section B 40(4)

Fitch EB. (1957): Sedimentation process fundamentals. Biological treatment of sewage and Industrial wastes, vol 2. Reinhold PublishingCorporation, New York

Hasan A (1989): Analytical approach for evaluation of settling data. J Env Eng Div ASCE 115 (2):455–461

Hazen A (1904): On sedimentation. Trans ASCE LIII:63

Krishnan P (1976): Column settling test for flocculant suspension. J Eng Div EEI:227–229

O'Connor, Eckenfelder WW Jr. (1957): Evaluation of laboratory settling data for process design. Biological treatment of sewage and industrial wastes, vol 2. Reinhold Publishing Corporation, New York

Ong SL (1985): Least square analysis of settling data under discrete settling conditions. Water SA 11(4)

Overcamp TJ (2006) Type II settling data analysis. J Environ Eng ASCE 123:137–139

Pise CP, Halkude SA (2011): A modified method of settling column data analysis. Int J Eng Sci Technol 3(4)

Zanoni AE et al (1975): Column settling test for flocculant suspension. J Env Eng Div ASCE 101 (3):109–118

Chapter 9
Analysis of Short Circuiting Phenomena

Abstract Hydraulic short circuiting is an important factor that impairs settling performance of settling tank. It has been shown that depth-wise variation of velocities produces short circuiting that does not impair settling, but the short circuiting resulting from widthwise variation of velocities does. Interestingly, this analysis resolves an age-old dilemma.

Keywords Short circuiting • Eliassen's demonstration • Depth-wise flow variation • Widthwise flow variation • Short circuiting and Velocity Profile Theorem

9.1 Introduction

Unequal times of passage of liquid elements through a tank give rise to the phenomenon of short circuiting. A settling particle entering into a settler must spend time, within the settler, that is required by it to fall through a vertical distance from its point of entry to the bottom of the settler before it is removed from the flow. Short circuiting, therefore, affects settling.

An analysis leading to the understanding of the phenomenon of short circuiting and its effect on settling particles will help to control the same in the process of designing an efficient settling system. Non-uniform distribution of velocities over the cross section results in unequal times of passage of the fluid elements through their length of travel. Dead space, density currents and wind blowing over the surface of settling tank also contribute to it further.

The effect of this non-uniform distribution of velocities on settling of solids may be studied under:

- Variation of velocities along the width of the cross section
- Variation of velocities along the depth of the same, for the sake of generality

A. De, *Sedimentation Process and Design of Settling Systems*,
Springer Transactions in Civil and Environmental Engineering,
DOI 10.1007/978-81-322-3634-4_9

9.2 Background of the Present Study

9.2.1 A 'Thought Evoking Debate'

In 1946 Eliassen (Eliassen 1946), while discussing Camp, demonstrated that short circuiting did not impair settling according to the ideal basin concept. Eliassen was countered by Camp with the statements 'The literature is full of experimental evidence that the short circuiting impairs the removal in settling tanks' and that 'Short circuiting affects the overflow rate in precisely the same manner as it does the detention time'.

'This answer', Fitch (Eliassen 1946) correctly pointed out, 'does not in any way resolve the dilemma presented by Eliassen's demonstration'. He stated further that the two assertions made by Camp such as 'the removal is governed by overflow rate and not by detention' and 'short circuiting decreases removal' are not compatible to each other in explaining settling phenomenon in a basin. If one is valid the other must be invalid. The search of literature reveals that since Eliassen presented the so-called dilemma, no attempt has yet been made to resolve the problem. It is also true at the same time that in order to base the study of the settling phenomenon on a sound theoretical background for the design of an efficient settling tank, the theoretical analysis on the effect of hydraulic short circuiting on the overall removal by the basin is very much needed.

This present chapter will resolve the dilemma already stated and will provide a theoretical analysis on the effect of short-circuiting phenomenon on the removal efficiency of the basin. This may give direction to the control of velocity distribution over the cross section of the settling tank by providing properly designed inlet to the tank.

9.2.2 Eliassen's Demonstration

Eliassen considered the tanks A, B, C and D as shown in Fig. 9.1. All of the tanks have the same surface area. Flow into each of the tanks is Q per unit width. Tanks A, C and D have the same depth $3d$ and the depth of the tank B is d. The top one third of the tank A is active and the bottom one third of the tank D is active. The inactive portion of the above tanks is stagnant with only exception to the tank C in which there is no such stagnant zone.

The velocity through the active portion of the tanks has been assumed to be uniform. The trajectories of a particle entering at the top of the tanks have been shown in Fig. 9.1. In all tanks A, B, C and D, a particle entering at the top of the settling zone will reach the bottom at a distance $QT/3d$ measured from the beginning of the zone.

T is the time required by the particle to fall through distance $3d$.

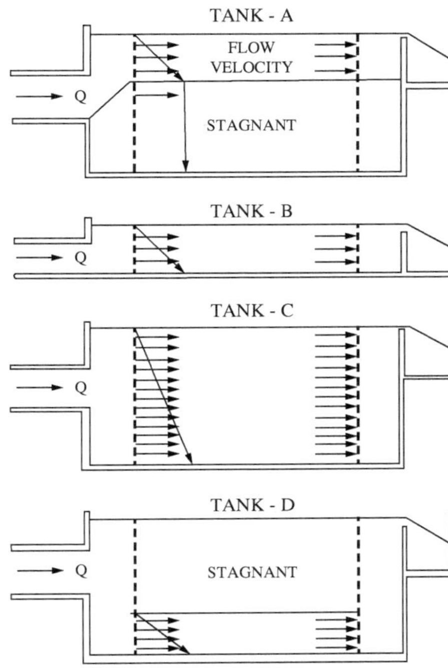

Fig. 9.1 Eliassen's demonstration

Any particle entering identically into the active portion of any of the above tanks will settle identically. For identical flow and suspension characteristics, therefore, each of the above tanks will accomplish identical removals. Short-circuiting tanks A or D of the same dimensions as the ideal tank C would make the same removal as the shallower and otherwise identical tank B.

9.3 Analysis (De 1990)

In an ideal rectangular or circular basin, each fluid element is detained inside the basin for the same interval of time. In actual basin, all the fluid elements do not spend the same interval of time inside the basin. Some pass out in less than the theoretical detention time and some spend more than that.

This is due to the uneven distribution of velocities over the cross section of the tank. For the purpose of our discussion, we shall consider a rectangular basin, the settling zone of which has length L, width B and depth D. It is fed with a flow rate Q containing identical discrete particles each having settling velocity v_s. When the basin functions ideally, all particles having settling velocity $v_s \geq$ the overflow velocity $v_0 = \frac{Q}{BL}$ will be removed completely, and the particles having $v_s < v_0$ will be removed in the ratio v_s/v_0.

9.3.1 *Effect of Short Circuiting with the Velocities Varying Along the Width*

Here we consider that in the above tank short circuiting is present. The flow rate remains at Q. The influent contains identical discrete particles each having settling velocity v_s. Velocities change along the width only. At any point on the same vertical, the velocity remains the same. At any two points separated by a horizontal distance, the velocities may vary. We want to evaluate the effect of short circuiting, thus resulting in the removal efficiency of the settling tank.

For definite evaluation for the purpose, a definite flow pattern should be assumed. We assume a parabolic distribution arbitrarily as shown in Fig. 9.2.

The variation of velocities has been assumed in accordance with the following equation (Appendix):

$$y = \frac{6L}{D} v_0 \left(\frac{x}{B} - \frac{x^2}{B^2} \right) \tag{9.1}$$

where y is the velocity at distance x measured along the width on the basin and $v_0 = \frac{Q}{BL}$.

The basin under this condition can be looked upon as an assembly of an infinite number of elementary ideal basins each of length L and depth D all connected in parallel.

Let us consider an elementary ideal basin at a distance x_i of width dx_i and of length L and depth D, respectively. The flow-through velocity of this elementary basin is

$$y_i = \frac{6L}{D} v_0 \left(\frac{x_i}{B} - \frac{x_i^2}{B^2} \right) \tag{9.2}$$

The overflow velocity of this basin is

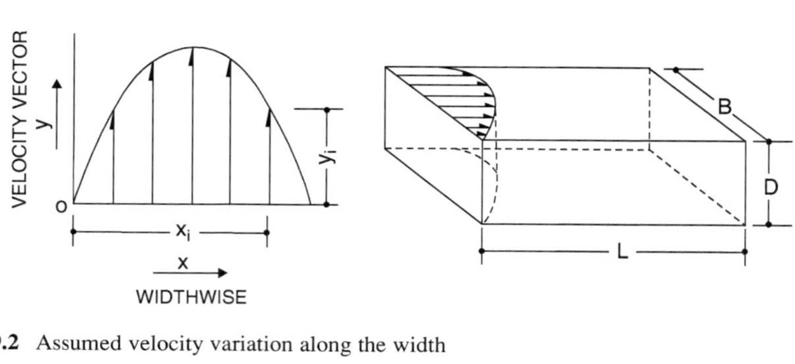

Fig. 9.2 Assumed velocity variation along the width

$$v_{ox_i} = \frac{y_i D}{L} \text{ i.e. } 6v_0\left(\frac{x_i}{B} - \frac{x_i^2}{B^2}\right) \tag{9.3}$$

If the concentration of discrete particles in the influent is C_s, total solids entering into the tank over very small interval of time $\tau = Q\tau C_s$. They are distributed uniformly over the cross sections of the elementary basins in proportion to the flow through them.

The basins in which $v_{ox_i} \leq v_s$ of the particles will be completely removed. Considering the limiting case,

$$v_s = v_{ox_i} \text{ i.e. } = 6v_0\left(\frac{x_i}{B} - \frac{x_i^2}{B^2}\right) \tag{9.4}$$

$$\text{From Eq. 9.4 the value of } x_i \text{ for such basin is } = \frac{B}{2} \pm \frac{B}{2}\left(1 - \frac{2v_s}{3v_0}\right)^{\frac{1}{2}} \tag{9.5}$$

This means that from the flow entering from $x_i = 0$ to

$x_i = \frac{B}{2} - \frac{B}{2}\left(1 - \frac{2v_s}{3v_0}\right)^{\frac{1}{2}}$ and from

$x_i = \frac{B}{2} + \frac{B}{2}\left(1 - \frac{2v_s}{3v_0}\right)^{\frac{1}{2}}$ to $x_i = B$, all particles will be removed.

The particles entering with the flow from

$$x_i = \frac{B}{2} - \frac{B}{2}\left(1 - \frac{2v_s}{3v_0}\right)^{\frac{1}{2}} \text{to} x_i = \frac{B}{2} + \frac{B}{2}\left(1 - \frac{2v_s}{3v_0}\right)^{\frac{1}{2}}$$

into different elementary ideal basins will be removed in the ratio v_s/v_{ox_i}.

Accordingly, the fractional removal by the entire basin can be written as

$$= \frac{1}{Q\tau c_s}\left[2\int_0^{\frac{B}{2}-\frac{B}{2}\left(1-\frac{2v_s}{3v_0}\right)^{\frac{1}{2}}} y_i D \mathrm{dx}_i \tau c_s + \int_{\frac{B}{2}-\frac{B}{2}\left(1-\frac{2v_s}{3v_0}\right)^{\frac{1}{2}}}^{\frac{B}{2}+\frac{B}{2}\left(1-\frac{2v_s}{3v_0}\right)^{\frac{1}{2}}} y_i D \mathrm{dx}_i \tau c_s \frac{v_s}{v_{ox_i}}\right]$$

$$= \frac{1}{Q\tau c_s}\left[2\int_0^{\frac{B}{2}-\frac{B}{2}\left(1-\frac{2v_s}{3v_0}\right)^{\frac{1}{2}}} y_i D \mathrm{dx}_i \tau c_s + \int_{\frac{B}{2}-\frac{B}{2}\left(1-\frac{2v_s}{3v_0}\right)^{\frac{1}{2}}}^{\frac{B}{2}+\frac{B}{2}\left(1-\frac{2v_s}{3v_0}\right)^{\frac{1}{2}}} y_i D \mathrm{dx}_i \tau c_s \frac{v_s}{\frac{y_i D}{L}}\right]$$

$$= 1 - \left(1 - \frac{2v_s}{3v_0}\right)^{\frac{3}{2}} \ldots\ldots\ldots\ldots\ldots\ldots\ldots (\text{Appendix}) \tag{9.6}$$

$$= \left(\frac{v_s}{v_0}\right) - \left\{\frac{1}{6}\left(\frac{v_s}{v_0}\right)^2 + \frac{1}{54}\left(\frac{v_s}{v_0}\right)^3 + \frac{1}{216}\left(\frac{v_s}{v_0}\right)^4 + \ldots\ldots\ldots\right\}$$

$$= \frac{v_s}{v_0} - \text{a positive quantity less than} \frac{v_s}{v_0} \tag{9.7}$$

i.e. the removal in this case is lesser than that accomplished by the basin when it functions ideally. This result has been derived for short circuiting resulting from the widthwise parabolic variation of velocities and can also be derived for any other variation.

9.3.2 *Effect of Short Circuiting with the Velocities Varying Along the Depth*

Here we consider that in the above tank short circuiting is present. Flow rate remains at Q. The influent contains identical discrete particles each having settling velocity v_s. Velocity of the fluid elements changes with depth only. At any point on the same horizontal level, the velocity remains the same. At any two points separated by a vertical distance, the velocities may vary. We want to evaluate the effect of short circuiting, thus resulting in the removal efficiency of the basin.

For definite evaluation for the purpose, a definite flow pattern should be assumed. We assume a parabolic distribution arbitrarily as shown in Fig. 9.3. The variation has been assumed in accordance with the following equation (Appendix):

$$y = \frac{3L}{2D} v_0 \left(1 - \frac{x^2}{D^2}\right) \tag{9.8}$$

where y is the velocity at depth x and $v_0 = \frac{Q}{BL}$.

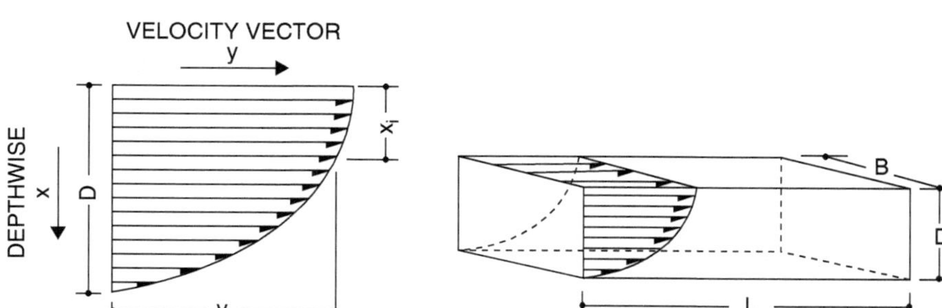

Fig. 9.3 Assumed velocity variation along the depth

We consider a particle entering at a particular depth x_i. It will start moving forward with velocity

$y_i = \frac{3L}{2D} v_0 \left(1 - \frac{x_i^2}{D^2}\right)$ and falls through a small vertical distance $\mathrm{d}x_i$ in time $\frac{\mathrm{d}x_i}{v_s}$ during which it will move forward through a distance:

$$\frac{\mathrm{d}x_i}{v_s} \cdot \frac{3L}{2D} v_0 \left(1 - \frac{x_i^2}{D^2}\right).$$

The condition that the particle should be settled requires that

$$\int_{x_i}^{D} \frac{\mathrm{d}x_i}{v_s} \cdot \frac{3L}{2D} v_0 \left(1 - \frac{x_i^2}{D^2}\right) \leq L$$

$$\text{i.e.} \ \frac{3L}{2D} \cdot \frac{v_0}{v_s} \cdot \frac{(D - x_i)^2 (2D + x_i)}{3D^2} \leq L \tag{9.9}$$

Here it is assumed that the particles that cannot touch the sludge zone before reaching the end of the settling zone will not be settled and the particles that touch the sludge zone will be removed. This means all the particles entering with the flow through the depth from x_i to D will be settled where x_i is given by the equation:

$$\frac{3L}{2D} \cdot \frac{v_0}{v_s} \cdot \frac{(D - x_i)^2 (2D + x_i)}{3D^2} = L \tag{9.10}$$

So if the concentration of particles in the influent is C_s, total amount of particles entering in very small interval of time $\tau = Q\tau C_s$ of which the amount that is settled

$$= C_s \tau \int_{x_i}^{D} B. \frac{3L}{2D} v_0 \left(1 - \frac{x_i^2}{D^2}\right) \mathrm{d}x_i$$

$$= C_s \tau B. \frac{3L}{2D} v_0 \frac{(D - x_i)^2 (2D + x_i)}{3D^2}$$

$$= C_s \tau B. \frac{3L}{2D} v_0 L. \frac{2D}{3L} \cdot \frac{v_s}{v_0} \ \text{from Eq. (9.10)}$$

$$= C_s \tau B L v_s$$

So the fractional removal $= \frac{C_s \tau B L v_s}{C_s Q \tau}$

$$= \frac{v_s}{\frac{Q}{BL}} \ \text{i.e.} \ \frac{v_s}{v_0} \tag{9.11}$$

i.e. the removal is the same as that accomplished by an ideal basin.

The above result has been deduced for the parabolic variation. The same result will be arrived at for any other variation. This shows that the short circuiting resulting from the depth-wise variation of velocities does not change the removal efficiency of the basin.

9.4 General Treatment of the Foregoing Analyses

The foregoing analyses are based upon the assumed parabolic distribution of velocities. In the following, the same analyses have been carried out with widthwise and depth-wise variation of velocities without assuming any particular pattern of velocity variation. These have been analysed with the application of 'Velocity Profile Theorem' (Chap. 3)

9.4.1 Velocity Profile Theorem (De 2009)

Let us recall to enunciate the theorem prior to its application to the present analyses. It is a new concept. The theorem has been deduced and established (De 2009). It can be applied to solve any settling problem analysis. The theorem states:

In a settler inclined at an angle θ with the horizontal if a settling particle with settling velocity v_s moves from (x_1, y_1, α_1) to (x_2, y_2, α_1), then $(x_2 - x_1)$

$$= \frac{(\text{Area of flow diagram} - \text{area of particle velocity diagram})\text{ between } y_1 \text{ and } y_2}{v_s \cos\theta} \tag{9.12}$$

$$= \frac{\text{Area of velocity profile diagram between } y_1 \text{ and } y_2}{v_s \cos\theta} \tag{9.13}$$

and also area of flow diagram between y_1 and y_2

$$= (x_2 - x_1)\, v_s \cos\theta + (y_1 - y_2)\, v_s \sin\theta \tag{9.14}$$

9.4.2 Analysis of the Effect of Short Circuiting on Settling

9.4.2.1 Short Circuiting from Widthwise Variation of Velocity

Water containing settleable solids of concentration C_S consisting of identical particles as regards their settling velocity v_s enters into ideal settling zone

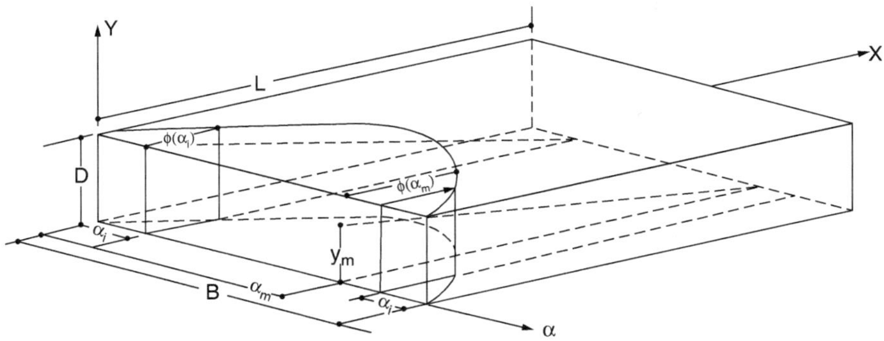

Fig. 9.4 Widthwise variation of flow velocity

$L \times B \times D$. Over an infinitely small interval of the time τ, solids entering into the zone are $Q\tau C_S$. They distribute themselves uniformly over the entire cross section.

The flow velocity is not uniform over the cross section. Let the flow velocity vary along the width. At $\alpha = \alpha$ the flow velocity is $\phi(\alpha)$. The flow velocity distribution is symmetrical about central longitudinal section, i.e.

$$\phi(\alpha) = \phi(B - \alpha) \tag{9.15}$$

This is so because the flow condition maintains symmetry about the central longitudinal section shown in Fig. 9.4.

A particle entering at $\alpha = \alpha_i$ at its top travels through the length L to reach just the bottom. By Velocity Profile Theorem,

$$\frac{D\phi(\alpha_i)}{v_s} = L \text{ i.e.} \, \phi(\alpha_i) = \frac{Lv_s}{D} \tag{9.16}$$

This implies that the particles entering through the areas $D\alpha_i$ at either ends of the cross section will be completely removed.

Hence the solids entering through the areas over the interval of time τ and removed completely

$$= 2C_s\tau \int_0^{\alpha_i} \phi(\alpha)D\,d\alpha$$

$$= 2C_s\tau D \int_0^{\alpha_i} \phi(\alpha)d\alpha \tag{9.17}$$

$$= 2C_s\tau Dk\alpha_i\phi(\alpha_i), \text{ since } \int_0^{\alpha_i}\phi(\alpha)d\alpha = k\alpha_i\phi(\alpha_i) \text{ where } k < l$$

$$= 2C_s\tau Dk\alpha_i\frac{Lv_s}{D} \text{ from Eq. 9.16}$$

$$= 2C_s\tau k\alpha_i Lv_s \quad (9.18)$$

For $\alpha_i < \alpha < (B - \alpha_i)$, let at any section $\alpha = \alpha_m$ the flow velocity be $\phi(\alpha_m)$,and a particle entering at a height y_m above the bottom will move through the length L to reach just the bottom, and by Velocity Profile Theorem,

$$y_m\phi(\alpha_m) = Lv_s \quad (9.19)$$

All particles entering through the area $y_m d\alpha$ will be completely removed. The solids entering through the area $y_m d\alpha$ over the interval of time τ

$$= C_s\tau\phi(\alpha_m)y_m d\alpha$$

$$= C_s\tau Lv_s\, d\alpha \text{ using Eq.9.19} \quad (9.20)$$

Hence the solids entering through the area $D(B - 2\alpha_i)$ of the cross section over the interval of time τ that will be removed

$$= \int_{\alpha_i}^{B-\alpha_i} C_s\tau Lv_s\, d\alpha$$

$$= C_s\tau Lv_s(B - 2\alpha_i) \quad (9.21)$$

The rest of the solids entering into the zone will be carried with the effluent. Then, solids $Q\tau C_s$ entering into the ideal zone that will be settled

$$= 2C_s\tau k\alpha_i Lv_s + C_s\tau Lv_s(B - 2\alpha_i) \text{ from Eqs.9.18 and 9.21}$$

$$= C_s\tau Lv_s[B - 2\alpha_i(1 - k)] \quad (9.22)$$

Hence the fraction of solids settled

$$= \frac{C_s\tau Lv_s[B - 2\alpha_i(1 - k)]}{Q\tau C_s}$$

$$= \frac{BLv_s[B - 2\alpha_i(1 - k)]}{QB}$$

$$= \frac{v_s}{v_0} \times \text{Fraction (Less than unity), where } v_0 = \frac{Q}{BL};$$

Here the removal is less than ideal removal. This implies that short circuiting resulting from widthwise variation of flow velocity deteriorates the settling of particles.

9.4.2.2 Short Circuiting Resulting from Depth-Wise Variation of Velocity

Water containing settleable solids at concentration C_s, consisting of ideal particles as regards their settling velocity v_s, enters into an ideal settling zone $L \times B \times D$ (Fig. 9.5). The flow rate is Q.

Flow velocity varies along the depth. At height y from the bottom, the flow velocity is $\phi(y)$.

This implies that at all points on a horizontal plane, the flow velocity is the same and it varies with depth only. Over a very small interval of time τ, water that enters into the tank is Q carrying with it solids $Q\tau C_s$. On entering into the settling zone, they distribute themselves uniformly over the cross section.

Let at $y = y_i$ a particle starts moving forward as it settles with velocity v_s and just reach the bottom travelling through a distance L. By Velocity Profile Theorem,

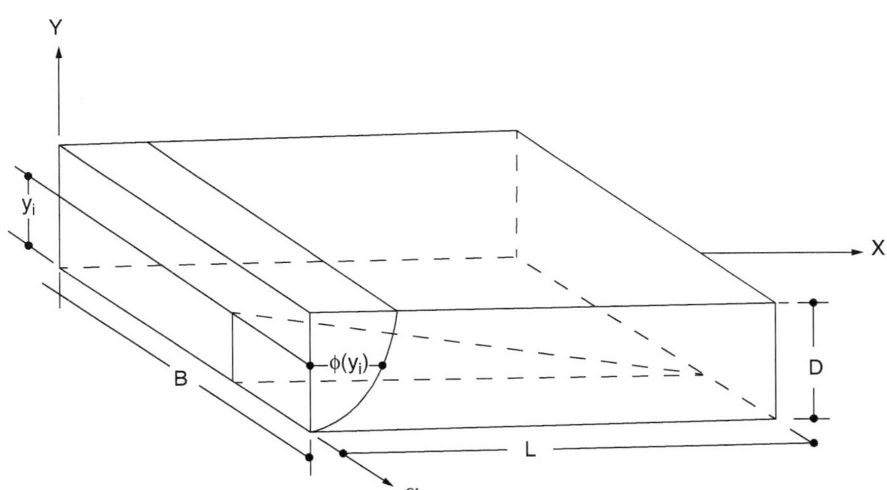

Fig. 9.5 Depth-wise variation of flow velocity

$$L = \frac{\text{Area of flow diagram between}\, y = 0\, \text{to}\, y = y_i}{v_s}$$

$$\text{i.e.}\, L = \frac{\int_0^{y_i} \phi(y)\mathrm{d}y}{v_s} \quad \text{i.e.} \int_0^{y_i} \phi(y) = Lv_s \tag{9.23}$$

Solids entering into the settling zone through the area By_i will be settled, and the particles entering through the area $B(D - y_i)$ are carried with the effluent. Hence the solids carried into the tank by the flow over a very small interval of time and settled into it are

$$= \int_0^{y_i} C_s \tau \phi(y) B\, \mathrm{d}y$$

$$= C_s B \tau \int_0^{y_i} \phi(y)\, \mathrm{d}y$$

$$= \tau C_s B L v_s \quad \text{from Eq.9.23} \tag{9.24}$$

Hence the fraction of solids settled into the tank

$$= \frac{\tau C_s B L v_s}{Q \tau C_s}\ \text{i.e.} \frac{v_s}{v_0}\ \text{where}\ v_0 = \frac{Q}{BL}$$

This shows that the removal in this case is ideal. Short circuiting resulting from the depth-wise variation of velocity does not affect settling.

9.4.3 Discussion

The foregoing analysis could conclude:

1. Widthwise variation of flow velocity in a settling tank deteriorates the settling of particles.
2. Depth-wise variation of flow velocity in a tank does not affect settling of particle.

Conclusion 1 suggests that the inlet width of settling tank should be made as narrow as possible for an efficient settling system. Narrowing the width implies the making of length to width ratio larger. This explains the age-old experience that a long narrow channel is an efficient settler.

In the context of conclusion 1, let us consider a flow rate through a vertical section of an infinitesimally small width. The flow rate, in terms of Reynolds' number, is in the zone of turbulence.

In turbulence the velocity vectors are varying and also randomly distributed. The vectors may be resolved into components in the direction of flow and perpendicular to it. The summation of the component vectors in the direction of flow at any cross section amounts to the flow rate Q. The summation of the perpendicular components amounts to zero since there is no flow in that direction. This implies that the perpendicular components move the settling particle as much up as to the downward direction resulting in no net vertical movement under their influence. The particles settle under their settling velocity only.

This is true for every path followed by every particle entering through every point of the cross section. This points to the fact that the component vectors parallel to the direction of flow will carry the settling particle forward during which time the particle drops from the point of its entry to the bottom of the settler under its settling velocity only. This condition for settling of the particles remains invariant irrespective of the conditions of flow, laminar or turbulent. Turbulence, therefore, appears not to affect settling so long flow rate remains the same and scour does not occur.

Conclusion 2 that depth-wise variation of flow velocity does not affect settling suggests that whatever the conditions of flow (turbulent or laminar) may be, the flow velocity component vectors, in the direction of flow that carry the particles forward, may conveniently be redistributed depth-wise to the advantage of calculating the actual settling of particles provided the rate of flow remains the same and no scour occurs.

Eliassen's demonstration considered the short circuiting resulting from the depth-wise variation of velocities only.

One should not wonder, therefore, that his demonstration came out with the conclusion that 'short circuiting does not affect removal', but facts reveal otherwise. In fact short circuiting decreases the performance of an actual basin.

In actual basin short circuiting results from both the depth-wise and widthwise variation of velocities. Though the former variation does not affect the removal, the basin performance is decreased due to the widthwise variation of velocities. Theoretical analysis on the effect of short-circuiting phenomenon on the removal efficiency of the basin carried herein shows that the general conclusion derived from partial observation made Eliassen's demonstration appear as a dilemma.

His analysis cannot be referred to have contradicted facts as done by Fitch (1957).

9.4.4 Conclusions

1. Short circuiting resulting from the widthwise variation of flow velocity deteriorates the settling of particles in a settler.

2. The inlet width should be made as narrow as possible in the design of an efficient settler.
3. Short circuiting resulting from the depth-wise variation of flow velocity does not affect settling.
4. Whatever the conditions of flow (laminar or turbulent) may be, the flow velocity components in the direction of flow may be redistributed depth-wise to the convenience of calculating the actual removal of particles through the settler. This is true so long flow rate remains the same and scour does not occur.

Notations

Q	Rate of flow
$D, (3d)$	Depth of the tank
L	Length of the tank
B	Width of the tank
$\frac{Q}{BL}, v_0$	Overflow velocity
y	Velocity at a distance x
y_i	Any particular velocity at particular distance X_i;
C_s	Concentration of particles having settling velocity v_s;
τ	Infinitesimally small interval of time
v_{Ox_i}	Overflow velocity of an elementary ideal basin at a particular distance x_i and of width dx_i, length L and depth D
x, y, α	Coordinates
$\phi(y)$	Flow velocity at y

Appendix

Derivation of Eqs. 9.9, 9.2, and 9.6

(a) We assume the distribution of velocities parabolic as shown in Fig. 9.3. The distribution is subjected to the following conditions:

$$\text{at } x = 0, \qquad y = m \tag{i}$$

$$\frac{dy}{dx} = 0 \tag{ii}$$

$$\text{at } x = D, \qquad y = 0 \tag{iii}$$

$$\text{and} \int_0^D B y dx = Q$$

$$= \frac{Q}{BL}.BL \text{ i.e } v_0 BL \tag{iv}$$

$$\text{The general equation } y = ax^2 + bx + c \tag{9.25}$$

From Eq. 9.25 and (i), $m = c$

$$\text{and Eq. 9.25 becomes } y = ax^2 + bx + m \tag{9.26}$$

$$\frac{dy}{dx} = 2ax + b \tag{9.27}$$

From Eq. 9.27 and (ii), $0 = b$

$$\text{and Eq. 9.26 becomes } y = ax^2 + m \tag{9.28}$$

From Eq. 9.28 and (iii), $0 = aD^2 + m$

$$\text{i.e } a = -m/D^2$$

and Eq. 9.28 becomes $y = -\frac{m}{D^2} x^2 + m$

$$= m\left(1 - \frac{x^2}{D^2}\right) \tag{9.29}$$

From Eq. 9.29 and (iv),

$$\int_0^D Bm\left(1 - \frac{x^2}{D^2}\right)\mathrm{d}x = v_0 BL$$

$$\text{i.e } BmD\left(1 - \frac{1}{3}\right) = v_0 BL$$

$$\text{i.e m} = \frac{3L}{2D} v_0$$

$$\text{and Eq. 9.29 becomes } y = \frac{3L}{2D} v_0 \left(1 - \frac{x^2}{D^2}\right) \tag{9.30}$$

(b) We assume the distribution of velocities parabolic as shown in Fig. 9.2. The distribution is subjected to the following conditions at

$$x = 0, \qquad y = 0 \tag{i}$$

$$x = \frac{B}{2}, \qquad y = m, \tag{ii}$$

$$x = B, \qquad y = 0, \tag{iii}$$

$$\int_0^B Dydx = Q$$

$$= \frac{Q}{BL}.BL\,\text{i.e}\,v_0BL \tag{iv}$$

$$\text{The general Eq. } y = ax^2 + bx + c \tag{9.31}$$

From Eq. 9.31 and (i), $0 = c$, and Eq. 9.31 becomes

$$y = ax^2 + bx \tag{9.32}$$

$$\text{From Eq. 9.32 and (ii) } 4m = aB^2 + 2bB \tag{9.33}$$

$$\text{and Eq. 9.32 and (iii) } 0 = aB^2 + bB \tag{9.34}$$

From Eqs. 9.33 and 9.34,

$$4m = bB\,\text{i.e.}\,b = \frac{4m}{B}$$

$$a = -\frac{bB}{B^2}$$

$$= -\frac{4m}{B^2}$$

and Eq. 9.32 becomes

$$y = -\frac{4m}{B^2}\mathrm{x}^2 + \frac{4m}{B}x$$

$$= 4m\left(\frac{x}{B} - \frac{x^2}{B^2}\right) \tag{9.35}$$

From condition (iv), $\int_0^B Dy\,dx = Q$

$$= \int_0^B D4m\left(\frac{x}{B} - \frac{x^2}{B^2}\right)\mathrm{d}x$$

$$= 4Dm\frac{B}{6}$$
$$= v_0 BL$$

i.e. $m = \frac{3L}{2D} v_0$, and Eq. 9.35 becomes

$$y = 4.\frac{3L}{2D} v_0 \left(\frac{x}{B} - \frac{x^2}{B^2}\right)$$
$$= \frac{6L}{D} v_0 \left(\frac{x}{B} - \frac{x^2}{B^2}\right) \tag{9.36}$$

At a distance x_i, the flow-through velocity in the elementary basin of thickness $\mathrm{d}x_i$ is

$$y_i = \frac{6L}{D} v_0 \left(\frac{x_i}{B} - \frac{x_i^2}{B^2}\right) \tag{9.37}$$

The overflow velocity v_{0x_i} of this basin

$$= \frac{y_i D}{L}$$
$$= 6v_0 \left(\frac{x_i}{B} - \frac{x_i^2}{B^2}\right) \tag{9.38}$$

The distance x_i at which the overflow velocity v_{0x_i} is equal to the v_s settling velocity of the particles can be obtained from

$$v_s = v_{0x_i}$$
$$= 6v_0 \left(\frac{x_i}{B} - \frac{x_i^2}{B^2}\right) \tag{9.39}$$
$$\text{i.e. } \left(\frac{x_i}{B}\right)^2 - \left(\frac{x_i}{B}\right) + \frac{v_s}{6v_0} = 0$$
$$\frac{x_i}{B} = \frac{1 \pm \sqrt{1 - 4 \cdot 1 \cdot v_s/6v_0}}{2}$$
$$\text{i.e.} x_i = \frac{B}{2} \pm \frac{B}{2}\left(1 - \frac{2v_s}{3v_0}\right)^{\frac{1}{2}} \text{ i.e.} \alpha \pm \beta \tag{9.40}$$

The fractional removal by the entire basin

$$= \frac{1}{Q\tau c_s}\left[2\int_0^{\frac{B}{2}-\frac{B}{2}\left(1-\frac{2v_s}{3v_0}\right)^{\frac{1}{2}}} y_i D \mathrm{d}x_i \tau c_s + \int_{\frac{B}{2}-\frac{B}{2}\left(1-\frac{2v_s}{3v_0}\right)^{\frac{1}{2}}}^{\frac{B}{2}+\frac{B}{2}\left(1-\frac{2v_s}{3v_0}\right)^{\frac{1}{2}}} y_i D \mathrm{d}x_i \tau c_s v_s / v_{0x_i}\right]$$

$$= \frac{1}{Q\tau c_s}\left[2\int_0^{\alpha-\beta} y_i D \mathrm{d}x_i \tau c_s + \int_{\alpha-\beta}^{\alpha+\beta} y_i D \mathrm{d}x_i \tau c_s \frac{v_s}{\frac{y_i D}{L}}\right]$$

$$= \frac{1}{Q}\left[2D\int_0^{\alpha-\beta} 6v_0 \frac{L}{D}\left(\frac{x_i}{B}-\frac{x_i^2}{B^2}\right)\mathrm{d}x_i + Lv_s 2\beta\right]$$

$$= \frac{2D}{Q}.6v_0\frac{L}{D}\left[\frac{(\alpha-\beta)^2}{2B}-\frac{(\alpha-\beta)^3}{3B^2}\right] + \frac{2Lv_s}{Q}.\beta \tag{9.41}$$

$$\text{Now}\,(\alpha-\beta)^2 = \frac{B^2}{4}+\frac{B^2}{4}\left(1-\frac{2v_s}{3v_0}\right) - 2.\frac{B}{2}.\frac{B}{2}\left(1-\frac{2v_s}{3v_0}\right)^{\frac{1}{2}}$$

$$= \frac{B^2}{4}+\frac{B^2}{4}\left(1-\frac{2v_s}{3v_0}\right) - \frac{B^2}{2}\left(1-\frac{2v_s}{3v_0}\right)^{\frac{1}{2}}$$

$$\frac{(\alpha-\beta)^2}{2B} = \frac{B}{8}+\frac{B}{8}\left(1-\frac{2v_s}{3v_0}\right) - \frac{B}{4}\left(1-\frac{2v_s}{3v_0}\right)^{\frac{1}{2}}$$

$$(\alpha-\beta)^3 = \left(\frac{B}{2}\right)^3$$

$$- 3\left(\frac{B}{2}\right)^2.\frac{B}{2}\left(1-\frac{2v_s}{3v_0}\right)^{\frac{1}{2}} + 3\frac{B}{2}.\left(\frac{B}{2}\right)^2\left(1-\frac{2v_s}{3v_0}\right) - \left(\frac{B}{2}\right)^3\left(1-\frac{2v_s}{3v_0}\right)^{\frac{3}{2}}$$

$$= \frac{B^3}{8} - \frac{3B^3}{8}\left(1-\frac{2v_s}{3v_0}\right)^{\frac{1}{2}} + \frac{3B^3}{8}\left(1-\frac{2v_s}{3v_0}\right) - \frac{B^3}{8}\left(1-\frac{2v_s}{3v_0}\right)^{\frac{3}{2}}$$

$$\frac{(\alpha-\beta)^3}{3B^2} = \frac{B}{24} - \frac{B}{8}\left(1-\frac{2v_s}{3v_0}\right)^{\frac{1}{2}} + \frac{B}{8}\left(1-\frac{2v_s}{3v_0}\right) - \frac{B}{24}\left(1-\frac{2v_s}{3v_0}\right)^{\frac{3}{2}}$$

$$\frac{(\alpha-\beta)^2}{2B} - \frac{(\alpha-\beta)^3}{3B^2}$$

$$= \frac{B}{8}+\frac{B}{8}\left(1-\frac{2v_s}{3v_0}\right) - \frac{B}{4}\left(1-\frac{2v_s}{3v_0}\right)^{\frac{1}{2}} - \frac{B}{24} - \frac{B}{8}\left(1-\frac{2v_s}{3v_0}\right)$$
$$+\frac{B}{8}\left(1-\frac{2v_s}{3v_0}\right)^{\frac{1}{2}} + \frac{B}{24}\left(1-\frac{2v_s}{3v_0}\right)^{\frac{3}{2}}$$

$$= \frac{B}{12} - \frac{B}{8}\left(1 - \frac{2v_s}{3v_0}\right)^{\frac{1}{2}} + \frac{B}{24}\left(1 - \frac{2v_s}{3v_0}\right)^{\frac{3}{2}}$$

Equation 9.41 becomes

$$\frac{2D}{Q}.6v_0\frac{L}{D}\left[\frac{B}{12} - \frac{B}{8}\left(1 - \frac{2v_s}{3v_0}\right)^{\frac{1}{2}} + \frac{B}{24}\left(1 - \frac{2v_s}{3v_0}\right)^{\frac{3}{2}}\right] + \frac{2Lv_s}{Q}.\frac{B}{2}\left(1 - \frac{2v_s}{3v_0}\right)^{\frac{1}{2}}$$

$$= \frac{2D}{v_0BL}6v_0\frac{L}{D}\frac{B}{24}\left[2 - 3\left(1 - \frac{2v_s}{3v_0}\right)^{\frac{1}{2}} + \left(1 - \frac{2v_s}{3v_0}\right)^{\frac{3}{2}}\right] + \frac{BLv_s}{BLv_0}\left(1 - \frac{2v_s}{3v_0}\right)^{\frac{1}{2}}$$

$$= \frac{v_s}{v_0}\left(1 - \frac{2v_s}{3v_0}\right)^{\frac{1}{2}} + 1 - \frac{3}{2}\left(1 - \frac{2v_s}{3v_0}\right)^{\frac{1}{2}} + \frac{1}{2}\left(1 - \frac{2v_s}{3v_0}\right)^{\frac{3}{2}}$$

$$= 1 + \left(1 - \frac{2v_s}{3v_0}\right)^{\frac{1}{2}}\left(\frac{v_s}{v_0} - \frac{3}{2} + \frac{1}{2} - \frac{1v_s}{3v_0}\right)$$

$$= 1 - \left(1 - \frac{2v_s}{3v_0}\right)^{\frac{1}{2}}\left(1 - \frac{2v_s}{3v_0}\right)$$

$$= 1 - \left(1 - \frac{2v_s}{3v_0}\right)^{\frac{3}{2}} \tag{9.42}$$

$$\begin{aligned}\left(1 - \tfrac{2v_s}{3v_0}\right)^{\frac{3}{2}} = \; & 1 - \frac{3}{2}\left(\frac{2v_s}{3v_0}\right) + \frac{\frac{3}{2}\left(\frac{3}{2} - 1\right)}{1.2}\left(\frac{2v_s}{3v_0}\right)^2 \\ & - \frac{\frac{3}{2}\left(\frac{3}{2} - 1\right)\left(\frac{3}{2} - 2\right)}{1.2.3}\left(\frac{2v_s}{3v_0}\right)^3 + \frac{\frac{3}{2}\left(\frac{3}{2} - 1\right)\left(\frac{3}{2} - 2\right)\left(\frac{3}{2} - 3\right)}{1.2.3.4}\left(\frac{2v_s}{3v_0}\right)^4 \\ & - \ldots \\ = \; & 1 - \frac{v_s}{v_0} + \frac{3}{2}.\frac{1}{2}.\frac{1}{1.2}.\frac{4}{9}\left(\frac{v_s}{v_0}\right)^2 + \frac{3}{2}.\frac{1}{2}.\frac{1}{2}.\frac{1}{1.2.3}.\frac{8}{27}\left(\frac{v_s}{v_0}\right)^3 \\ & + \frac{3}{2}.\frac{1}{2}.\frac{1}{2}.\frac{3}{2}.\frac{1}{1.2.3.4}.\frac{16}{81}\left(\frac{v_s}{v_0}\right)^4 + \ldots \\ = 1 - & \left(\frac{v_s}{v_0}\right) + \frac{1}{6}\left(\frac{v_s}{v_0}\right)^2 + \frac{1}{54}\left(\frac{v_s}{v_0}\right)^3 + \frac{1}{216}\left(\frac{v_s}{v_0}\right)^4 + \ldots.\end{aligned}$$

Equation 9.42 becomes

$$= 1 - \left(1 - \frac{2v_s}{3v_0}\right)^{\frac{3}{2}}$$

$$= 1 - 1 + \left(\frac{v_s}{v_0}\right) - \left[\frac{1}{6}\left(\frac{v_s}{v_0}\right)^2 + \frac{1}{54}\left(\frac{v_s}{v_0}\right)^3 + \frac{1}{216}\left(\frac{v_s}{v_0}\right)^4 + \ldots\right]$$

$$= \frac{v_s}{v_0} - \left[\frac{1}{6}\left(\frac{v_s}{v_0}\right)^2 + \frac{1}{54}\left(\frac{v_s}{v_0}\right)^3 + \frac{1}{216}\left(\frac{v_s}{v_0}\right)^4 + \ldots\right] \text{i.e.} \frac{v_s}{v_0} - \text{a positive quantity less than} \frac{v_s}{v_0} \quad (9.43)$$

References

De A (1990) Effect of short circuiting on the basin efficiency. J IPHE 2:37

De A (2009) Velocity profile theorem – concept for solving settling problem analysis. J IPHE 1:25

De A (2010) Analysis of the effect of short circuiting on settling -An application of velocity profile theorem, J IPHE India 2009–2010 No.(4)

Eliassen R (1946) Discussion of "Sedimentation and the design of settling tanks" by T.R. Camp. Trans ASCE 111:950

Fitch FB (1957) The significance of detention in sedimentation. Sewage Ind Waste 1.29:1123

Chapter 10
In Quest of Parameter for Settling Comparison

Abstract In any settling performance study settling performance may have to be compared. This calls for a parameter reflective of settling tank characteristics only corresponding to overflow velocity. 'Exponential efficiency' is such a parameter. It has been defined and its simple determination is presented.

Keywords Ideal efficiency • Operational efficiency • Overflow residual efficiency • Exponential efficiency • Determination of parameters

10.1 Introduction

The removal of solids through a settling tank depends on the settling characteristics of solids carried by the influent, the flow-through velocity or overflow velocity through the tank and the tank geometry.

The flow-through velocity and the tank geometry together are responsible for eddies, dead space, etc. that give rise to short circuiting that in turn impedes settling of solids through the tank. Temperature gradient promoting density currents may also contribute to the impairment of settling of solids.

The proper design for an efficient settling tank looks for proper design of tank geometry and its inlet and outlet devices.

Whether or not a particular design is more effective in its performance requires the understanding of the comparison of performances.

This calls for a parameter, for observation, that will be a measure of performance independent of the settling characteristics of the influent solids and will only reflect the influence of tank geometry together with the overflow velocity or critical fall velocity on the solid removal.

10.2 Concerned 'Parameters' Under Review

The following parameters for settling performance comparison have come up in the literature on the concerned subject.

A. De, *Sedimentation Process and Design of Settling Systems*,
Springer Transactions in Civil and Environmental Engineering,
DOI 10.1007/978-81-322-3634-4_10

10.2.1 Ideal Efficiency (Camp 1946)

Figure 10.1 shows the cumulative frequency distribution curve OABC for the settling velocities of solids in the influent of a settling tank operated at overflow velocity $v_0 (= \text{OF})$.

The 'ideal efficiency' of the tank

$$= \frac{\text{Area OABEDO}}{\text{OF}}$$
$$= \text{Average ordinate of the shadded diagram} \times 100\,\% \tag{10.1}$$

10.2.2 Operational Efficiency

A settling tank operated with influent solid concentration C_i and effluent solid concentration

C_e at overflow velocity v_0 has operational efficiency

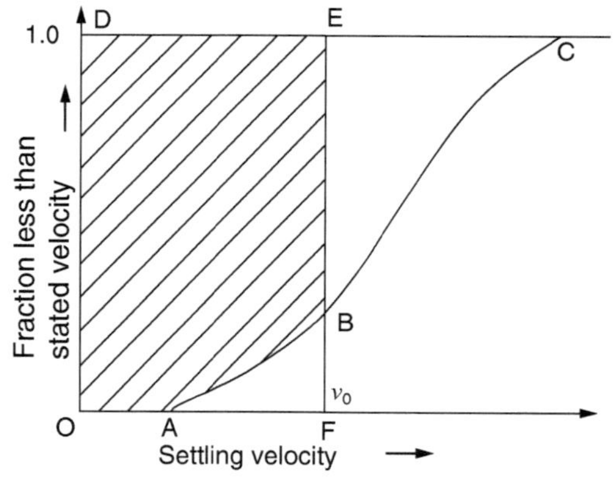

Fig. 10.1 Ideal efficiency

$$= \frac{C_i - C_e}{C_i} \times 100\% \tag{10.2}$$

10.2.3 *Overflow Residual Efficiency (Ingersoll et al. 1956)*

A settling tank operated at an overflow rate v_0 with influent concentration of all solids having settling velocity $v_s \geq v_0$, C_{i0}, and effluent concentration of solids having settling velocity $v_s \geq v_0$, C_{e0}, has 'overflow residual efficiency'

$$= \frac{C_{i0} - C_{e0}}{C_{i0}} \times 100\% \tag{10.3}$$

10.2.4 *Exponential Efficiency (De 1976, 1983)*

A settling tank operated at overflow velocity v_0 releasing into its effluent the particles with maximum settling velocity v_{max} has 'exponential efficiency'

$$E = \text{Exp}(-)\left[\frac{v_{max} - v_0}{v_0}\right] \times 100\% \tag{10.4}$$

10.3 Desirable Characters of a Suitable 'Parameter' for Settling Performance Comparison

The parameter should reflect the deterioration in settling through the tank due to tank geometry and overflow velocity only. This should have the following characters:

(i) The maximum value of the parameter should not go beyond $E = 1$, for reason that is obvious. As regards its lower limit whatever poor the performance of settling may be, the parametric value E may tend to zero but can never be equal to the same. This is so because, for a set of solid particles in the influent to a settling tank operated at a particular overflow velocity may not remove a single particle through settling but still the same tank operated at the same overflow velocity may remove all the solids when they constitute set of much heavier particles with regard to that overflow velocity.

(ii) Higher value of the parameter for the settling tank operated at an overflow velocity v_0 should not have any relation with the distribution of solids in its influent, i.e. the value of the parameter *E* for the tank at a overflow rate should be independent of its influent solids.
This requires that the value of the parameter for the tank at an overflow rate is to be determined without considering the distribution of influent solids.

(iii) The determination procedure for the parameter *E* is required to be a simple one.

10.4 Review of the Parameters

The parameters in 10.2 are to be reviewed in the light of desirability of the characters mentioned in 10.3.

10.4.1 Ideal Efficiency (1946, Camp)

Ideal efficiency is calculated from the cumulative frequency distribution of settling velocities in the influent of the tank.

In Fig. 10.2 three distribution curves are for same concentration of influent solids. Their distribution of settling velocities equal to and greater than the overflow velocity v_0 differs. But they calculate the same value of E.

In Fig. 10.3 three distribution curves marked (1), (2) and (3) are for same concentration of solids in the influent. The distribution of settling velocities of influent solids having settling velocities less than and equal to the overflow velocity v_0 differs in them. They calculate different values of 'ideal efficiency'.

Under the cover of ideal assumptions, ideal efficiency dispenses away with all the effects of overflow velocity and the tank geometry on the impairment of settling.

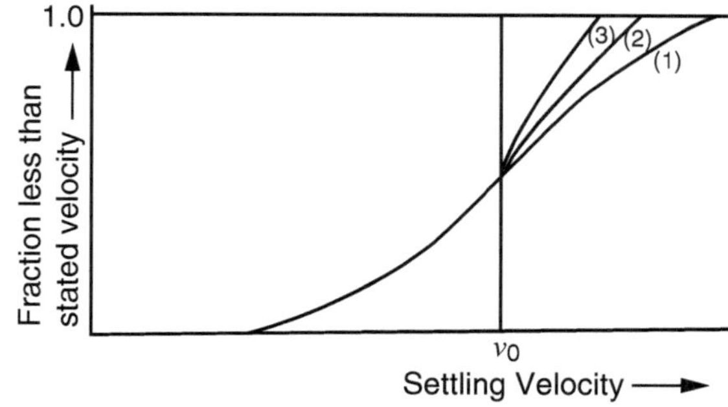

Fig. 10.2 Same ideal efficiency of three equal solid concentrations having different settling velocity distributions $\geq v_0$

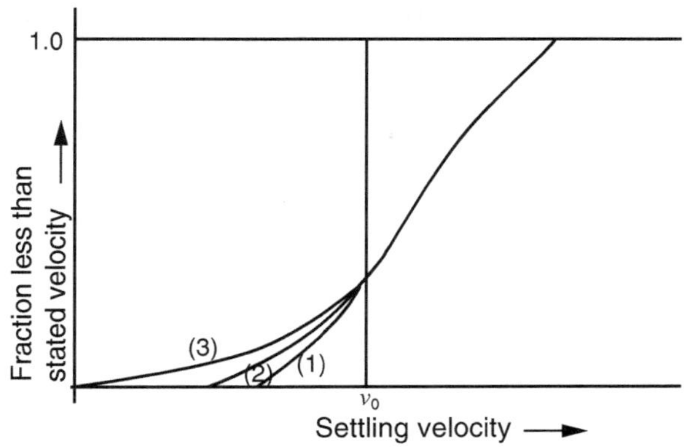

Fig. 10.3 Different ideal efficiencies for same concentrations of solids having same settling velocity distributions $\geq v_0$

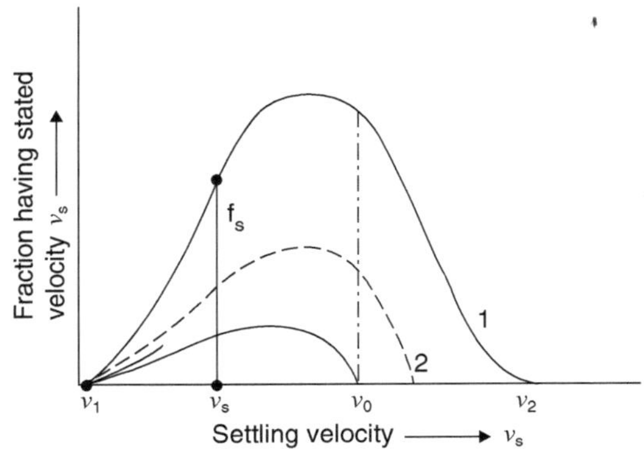

Fig. 10.4 Operational efficiency

10.4.2 Operational Efficiency

Let the concentration of solids in the influent be C_0.f_s is the fraction of solids having settling velocity v_s. Then from Fig. 10.4,

$$\int_{v_1}^{v_2} C_0 f_s \mathrm{d}v_s = C_0 \text{ i.e. } \int_{v_1}^{v_2} f_s \mathrm{d}v_s = 1 \tag{10.5}$$

For an ideal tank, f_s will be reduced to $\left(1 - \frac{v_S}{v_0}\right) f_s$ in the effluent. Then the concentration of solids in the effluent through the ideal settling tank

$$C_{ei} = \int_{v_1}^{v_2} \left(1 - \frac{v_S}{v_0}\right) f_s C_0 \mathrm{d}v_s \tag{10.6}$$

Wherever the ratio v_S/v_0 appears, its maximum value should be limited to 1, i.e. where $v_S \geq v_0$, the ratio is =1.

In actual settling tank in operation, f_s will be reduced by $k_s f_s \frac{v_S}{v_0}$ in its effluent where k_s is less than unity.

Its value reflects the deterioration in settling due to overflow velocity and tank geometry. The concentration of solids in the effluent

$$C_{ea} = \int_{v_1}^{v_2} \left(1 - k_s \frac{v_S}{v_0}\right) f_s C_0 \mathrm{d}v_s, \quad \text{For } v_S \geq v_0 \quad \frac{v_S}{v_0} = 1 \tag{10.7}$$

$$\begin{aligned} \text{Operational Efficiency} &= \left[\int_{v_1}^{v_2} C_0 f_s \mathrm{d}v_s - \int_{v_1}^{v_2} \left(1 - k_s \frac{v_S}{v_0}\right) f_s C_0 \mathrm{d}v_s\right] / C_0 \\ &= 1 - \int_{v_1}^{v_2} \left(1 - k_s \frac{v_S}{v_0}\right) f_s \mathrm{d}v_s \\ &= 1 - \int_{v_1}^{v_2} f_s \mathrm{d}v_s + \int_{v_1}^{v_2} k_s \frac{v_S}{v_0} f_s \mathrm{d}v_s \\ &= \int_{v_1}^{v_2} k_s \frac{v_S}{v_0} f_s \mathrm{d}v_s \end{aligned} \tag{10.8}$$

Operational efficiency does take care of the factors that impairs settling. But being dependent on the influent solids and their settling, velocity distribution cannot serve as a parameter for settling performance comparison.

10.4.3 Overflow Residual Efficiency (1956, Ingersoll et al.)

Ingersoll et al. (1956) considered the removal of solids from the influent of an actual tank that had settling velocities $v_S \geq v_0$.

The solids in the influent having settling velocity $v_S \geq v_0$ (curve marked (1) in Fig. 10.5)

$$C_{i0} = \int_{v_0}^{v_2} f_s C_0 \mathrm{d}v_s \tag{10.9}$$

From the curve marked (2), the solids in the effluent of actual tank in operation, having settling velocity $v_S \geq v_0$,

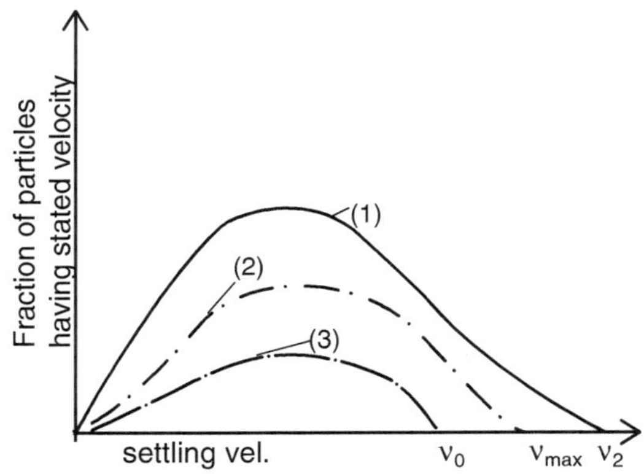

Fig. 10.5 Exponential efficiency

$$C_{e0} = \int_{v_0}^{v_2} \left(1 - k_s \frac{v_S}{v_0}\right) f_s C_0 \mathrm{d}v_s,$$

$$For\ v_S \geq v_0$$

$$\frac{v_S}{v_0} = 1$$

$$= C_0 \int_{v_0}^{v_2} f_s \mathrm{d}v_s - C_0 \int_{v_0}^{v_2} k_s \frac{v_S}{v_0} f_s \mathrm{d}v_s \tag{10.10}$$

Overflow residual efficiency (ORE):

$$= \frac{C_{i0} - C_{e0}}{C_{i0}}$$

$$= \frac{\int_{v_0}^{v_2} f_s C_0 \mathrm{d}v_s - C_0 \int_{v_0}^{v_2} f_s \mathrm{d}v_s + C_0 \int_{v_0}^{v_2} k_s \frac{v_S}{v_0} f_s \mathrm{d}v_s}{C_0 \int_{v_0}^{v_2} f_s \mathrm{d}v_s}$$

$$= \frac{\int_{v_0}^{v_2} k_s \frac{v_S}{v_0} f_s \mathrm{d}v_s}{\int_{v_0}^{v_2} f_s \mathrm{d}v_s}$$

$$= \frac{\int_{v_0}^{v_2} k_s f_s \mathrm{d}v_s}{\int_{v_0}^{v_2} f_s \mathrm{d}v_s} \tag{10.11}$$

since $v_s/v_o = 1$ for $v_s >= v_o$.

This has been argued (Ingersoll et al. 1956) that v_0 being very much near to v_2 in real situation, the effect of settling velocity distribution among the solids on 'overflow residual efficiency' (ORE) may be neglected.

Let us concentrate on the numerator of Eq. 10.11. For the sake of argument, say, even if it is accepted, though not true, that for the same range of values from v_0 to v_2 the k_s values corresponding to different f_s values for the influent solid concentration remains the same $k_s f_s$ *is* $= 0$ *from* v_{max} *to* v_2. This is obvious from the drawing of frequency distribution diagram for settling velocities. As such ORE for the same overflow velocity will change with frequency distribution. Thus it appears that ORE is affected by the influent concentration of solids. k_s values reflect the impairment of settling from the factors arising out of the overflow velocity and tank geometry.

10.4.4 Exponential Efficiency

A raw water containing settleable solids passes through a settling tank at overflow velocity v_0. The settling velocity distribution frequency diagram is shown in Fig. 10.5 as marked No. 1.

The frequency distribution of settling velocities of the solids in the effluent is shown by curve marked No. 2. The curve marked No. 3 shows the frequency distribution diagram of the effluent solids under ideal performance. From the curve No. 2, the particle having maximum settling velocity in the effluent is v_{max}.

The bases of the two curves (No. 2 and No. 3) differ because of deterioration in performance from the ideal behaviour for the impairment in settling due to short circuiting resulting from the overflow velocity and tank geometry. The more is the difference $(v_{max} - v_0)$, the more marked is the impairment in performance.

Let the measure of performance efficiency be E when the difference is $(v_{max} - v_0)$. Further extension of base by $\mathrm{d}(v_{max} - v_0)$ indicates the impairment of efficiency by $-\mathrm{dE}$.

Let us put a unit free relationship that fractional decrease in performance efficiency $\left(-\frac{\mathrm{dE}}{E}\right)$ is related to the fractional extension of the curve No. (3) $\mathrm{d}\left(\frac{v_{max}-v_0}{v_0}\right)$ as

$(-)\frac{\mathrm{dE}}{E} = k\mathrm{d}\left(\frac{v_{max}-v_0}{v_0}\right)$, where k is a constant.

Integrating between the limits from

$E = 1$ when $v_{max} = v_0$ to $E = E$ when $v_{max} = v_{max}$,

$$\int_{E=1}^{E=E} (-)\frac{\mathrm{d}E}{E} = \int_{v_{max}=v_0}^{v_{max}=v_{max}} k\mathrm{d}\left(\frac{v_{max} - v_0}{v_0}\right)$$

$$\text{i.e. } \log E = (-)k\left(\frac{v_{max} - v_0}{v_0}\right) \tag{10.12}$$

If we set the scale such that $E = e^{-1}$(because E should not be greater than 1) when $v_{max} = 2v_0$.

From Eq. 10.12, $k = +1$ and Eq. 10.12 can be written as

$$\log E = (-)\left(\frac{v_{max} - v_0}{v_0}\right)$$

$$\text{i.e.}\, E = \text{Exp}(-)\left(\frac{v_{max} - v_0}{v_0}\right) \tag{10.13}$$

From Eq. 10.13, the maximum value of $E = 1$ for $v_{max} = v_0$. With the increasing value of $(v_{max} - v_0)$, the value of E decreases. This shows that higher values of E indicate greater removal of solids. The values of E may diminish indefinitely, but its value will never be equal to zero. It is free from the influence of the distribution of settling velocities among the solids in the influent. The value of E can be known from separate and independent observation and can be shown from 10.5.4 that its simple determination can be used to characterise the settling performance of a settling tank.

10.5 Determination of Parameters

In the following, the experimental determination of the parameters discussed is presented.

10.5.1 Ideal Efficiency

Camp (Camp 1946) advocated settling column analysis to find the cumulative frequency distribution diagram for the settling velocities of particles in a suspension. Such determination with the influent suspension can find out curve presented in Fig. 10.1, and removal of solids under ideal performance at an overflow rate v_0 can be calculated from the graph.

10.5.2 Operational Efficiency

Samples are collected from the influent and effluent of the settling tank. Weights of solids present in both of the samples are found out. Dividing the weights by the volumes of samples in which they were present, the concentration of the solids C_i and C_e in the influent and effluent is determined, and the operational efficiency value is calculated.

10.5.3 Overflow Residual Efficiency Determination

10.5.3.1 Overflow Residual Efficiency (ORE) Determination Suggested by Ingersoll et al. (1956): Method 1

Raw water of flow rate Q containing settleable solid concentration C_0 passes through a settling tank (marked 1 in Fig. 10.6) producing effluent solid concentration C_e.

For the determination of 'overflow residual efficiency' portions of flow both from influent and effluent, Q_1 and Q_2, respectively, are bypassed and subjected to upflow clarification through the clarifiers marked 2 and 3. The settling tank and the upflow clarifiers are so operated that

$$\frac{Q}{A} = \frac{Q_1}{A_1} = \frac{Q_2}{A_2} = v_0$$

The flow-through clarifiers continues over the intervals of times t_1 and t_2, say, through the clarifiers marked 1 and 2, respectively. After the flows are discontinued, they are to settle their solids contained in them till settling is complete. It has been argued (Ingersoll et al. 1956) that if the solid accumulations in No. 2 and No. 3 are W_1 and W_2, respectively, they will all consist of particles having settling velocity more than or equal to v_0.

Then the concentration of solids in the influent having $v_S \geq v_0$ is $\frac{W_1}{Q_1 t_1}$ and that in the effluent is $\frac{W_2}{Q_2 t_2}$.

Overflow Residual Efficiency (ORE)

$$= \frac{\left(\frac{W_1}{Q_1 t_1} - \frac{W_2}{Q_2 t_2}\right)}{\frac{W_1}{Q_1 t_1}}$$

One may be sceptic about the dependability of this determination. For most fluid elements will not pass out through the upflow clarifiers with preset velocity $\frac{Q_1}{A_1} = \frac{Q_2}{A_2} = \frac{Q}{A}$, the velocity distribution over the cross sections of the upflow clarifiers not being uniform.

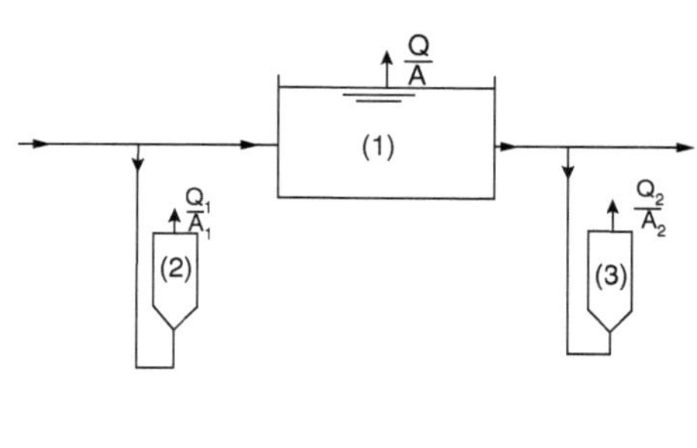

Fig. 10.6 ORE determination

The determined value of ORE will be affected by the efficiencies of the upflow clarification. ORE is likely to be different for the same settling tank at the same overflow velocity with different influent concentration.

10.5.3.2 Graphical Method for the Determination of ORE

Graphical method of determination of ORE is based on the drawing of 'frequency distribution diagram' of the settling velocities of particles in a suspension.

Settling column test is performed with the suspension. From the observations, cumulative frequency distribution diagram of the settling velocities of particles in the suspension may be drawn as presented in Fig. 10.7a.

The corresponding frequency distribution diagram for settling velocities of the particles in the suspension may be prepared as presented in Fig. 10.7b

With reference to Fig. 10.7a, b, let F_s and f_s be the ordinates in CFD diagram and FD diagram, respectively, corresponding to the settling velocity v_s. Then by definition,

$$F_s = \int_{v_0}^{v_2} f_s \mathrm{d}v_s$$

$$\text{i.e.} \frac{\mathrm{d}F_s}{\mathrm{d}v_s} = f_s = \tan\theta$$

i.e. the measure of the tangents at different points of the CFD diagram corresponding to different settling velocities will give the ordinates of the FD diagram at those settling velocities. The FD diagram (Fig. 10.7b), thus, may be drawn from CFD diagram (Fig. 10.7a).

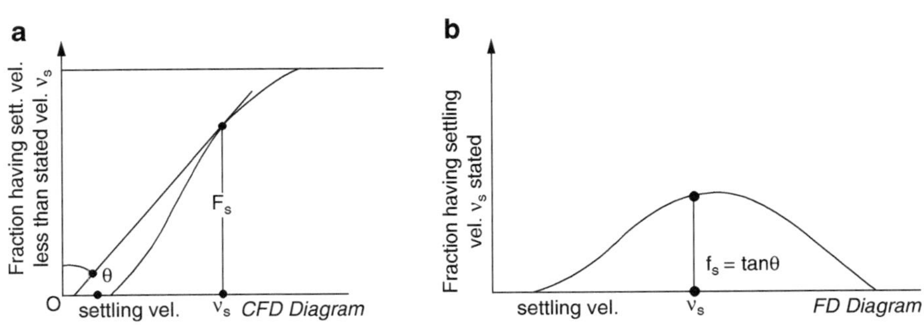

Fig. 10.7 Frequency distribution diagram from cumulative frequency distribution diagram

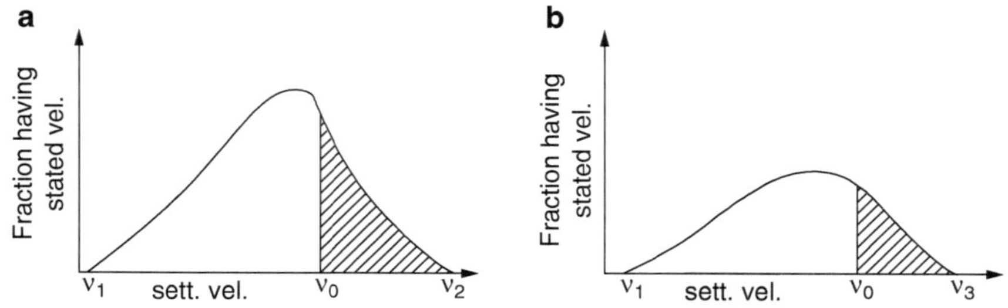

Fig. 10.8 (**a**) FD diagram for influent suspension. (**b**) FD diagram for effluent suspension

10.5.3.2.1 Determination of ORE from FD Diagram

Water containing settleable solid concentration C_0 passes through the settling tank producing effluent with settleable solid concentration C_e at overflow velocity v_0.

The suspensions are collected both from the influent and effluent. They are subjected to settling column analysis, and FD diagrams are drawn as shown in Fig. 10.8a, b.

Solids in the influent having $v_S \geq v_0 = \int_{v_0}^{v_2} f_s \mathrm{d}v_s C_0$ (Fig. 10.8a).

Solids in the effluent having $v_S \geq v_0 = \int_{v_0}^{v_3} f_s \mathrm{d}v_s C_e$ (Fig. 10.8b).

Overflow residual efficiency (ORE)$= 1 - \frac{C_e}{C_0} \cdot \frac{\int_{v_0}^{v_3} f_s \mathrm{d}v_s}{\int_{v_0}^{v_2} f_s \mathrm{d}v_s}$.

Overflow residual efficiency (ORE) can be determined from the graphical integration of the shaded areas of Fig.10.8a, b.

10.5.4 Determination of 'Exponential Efficiency' and Characterisation of Settling Through the Tank

The following steps are to be followed:

1. Exponential efficiency or 'E' values can be determined independent of what the settling tank is clarifying.
2. A large quantity of discrete particles identifiable by colours or otherwise is to be taken. The settling velocity distribution among the particles should be widely varying about the overflow velocity maintained in the tank.
3. These particles are to be dumped into the water flowing into the tank.

4. These particles are to be collected from the effluent as they are found to be coming out with it. The temperature of the effluent water is to be noted.
5. The particles are separated from the suspension collected from effluent. They are dried.
6. A long transparent cylinder is taken with a fairly large length of the column conspicuously marked between two horizontal marks.
7. The column is filled up with water. The temperature of water is noted.
8. The particles are gently sprinkled over the top surface of water in the cylinder in very small batches taking care that the particles sprinkled at a time should simultaneously sit on the water surface.
9. The time taken by the fastest streak of particles to cross past the marked distance of length of the column is noted.
10. The value of $v_{\max}$ of the fastest settling particle coming out through the tank can be calculated by dividing the measured length between the marks by the noted times. If required, temperature correction is employed to get the value of $v_{\max}$ corresponding to the temperature of water that was passing through the tank.
11. 'E' value is calculated from $E = \text{Exp}(-)\left(\frac{v_{\max}-v_0}{v_0}\right)$.
12. By varying the overflow velocity (v_0), different E-values may be determined corresponding to the varying values of v_0 over its selected range of values. The characterisation graph may be prepared from these values to select a particular overflow velocity v_0 depending upon situation.

From the characterisation (v_0 versus E curve) curve, an operator can find out the E-value corresponding to a selected overflow velocity v_0. From these two values, the $v_{\max}$ appearing in the effluent can be found out as

$$v_{\max} = v_0 \log\frac{e}{E} \text{ from Eq. 10.13.}$$

With this value of $v_{\max}$, the removal through the settling tank may be anticipated from the analysis of its influent suspension, and control measures may be taken.

Notations

C_0, C_i	Concentration of influent solids
C_e	Concentration of effluent solids
C_{io}	Concentration of solids in the influent having settling velocity $v_S \geq v_0$
C_{eo}	Concentration of solids in the effluent having settling velocity $v_S \geq v_0$
$v_{\max}$	Maximum settling velocity of particle
v_0	Overflow velocity
v_s	Settling velocity of particle
f_s	Fraction of particles having settling velocity v_S
F_s	Fraction of particles having settling velocities $\leq v_S$

k_s	Fraction to which ideal settling ratio v_S/v_0 will be reduced for the particles of settling velocity v_S through an actual tank
C_{ei}	Concentration of solids in the effluent in ideal operation
C_{ea}	Concentration of solids in the effluent in actual operation
E	Exponential efficiency value
Q	Flow rate
Q_1	Flow bypassed from influent through upflow clarifier
Q_2	Flow bypassed from effluent through upflow clarifier
A	Surface area of the settling tank
A_1	Surface area of upflow clarifier on inlet side
A_2	Surface area of upflow clarifier on effluent side

References

Camp TR (1946) Sedimentation and design of settling tanks. Trans ASCE 111:895

De A (1976) Conceptual studies on discrete and flocculent settling. PhD (Engg) Thesis, Jadavpur University, Calcutta, West Bengal, India

De A (1983) Parameter for settling tank performance comparison. J IPHE India4:21

Ingersoll AC et al (1956) Fundamental concepts and design of settling tanks. Trans ASCE 111:895

Chapter 11
Design of Settling System: An Introduction

Abstract This chapter introduces the concept application of compatibility between the system and the testing on which the system operates.

Keywords Compatible settling system • Settling systems • Compatible design • Development of design • Design procedure

11.1 Settling System and Compatible Design

Sedimentation process provides excellent operation for solid-liquid separation in the clarification of liquid and thickening of sludge.

Water and waste water engineering uses settling tanks and sludge thickeners. These are also employed in other industries too such as chemical, metallurgical, mining, etc.

There are suggestions, recommendations and observed data used as regards their designs. The drawbacks and limitations of those are well known.

Rational design procedure demands not only the proper design but also the compatible operation of the settling system to follow the testing that the system depends on and vice versa. For explanatory mention,

the settling system depends on jar testing in water treatment, settling column test in waste water treatment and sludge column settling in thickener.

11.2 Settling System

In water treatment, raw water contains non-settleables and poorly settleable solids. Coagulation renders them settleables. 'Flash mixing' promotes effective part of neutralisation of negative charges on the solids by the adsorption of positive ions reducing the interparticle forces of repulsion. In slow mixing, a major number of contacts between the particles take place to effect the predominating force of van der Waals' attraction for the coalescence of the particles into settleable flocs.

A. De, *Sedimentation Process and Design of Settling Systems*,
Springer Transactions in Civil and Environmental Engineering,
DOI 10.1007/978-81-322-3634-4_11

It has been deduced (Sect. 6.2.2) that between the two types of particles of diameters d_1 and d_2 of number of particles per unit volume n_1 and n_2, respectively, the number of contacts taking place per unit time per unit volume is

$$N = \frac{1}{6} n_1 n_2 (d_1 + d_2)^3 \frac{\overline{du}}{dy}, \text{ where } \frac{\overline{du}}{dy} \text{ is the mean temporal velocity gradient.}$$

At any instant of time, there may be particle concentrations of n_1,n_2, n_3, …, etc. of diameters d_1, d_2, d_3, …, respectively. It is also true that there may be multiple contacts, not necessarily only those between two types of particles. Then the number of contacts at that instant of time may be put down as

$$N_t = \varnothing_t(n_1, n_2, \ldots\ldots.d_1, d_2 \ldots..) \frac{\overline{du}}{dy} \text{ per unit volume per unit time.}$$

Over a very small interval of time (dt), the total number of contacts/unit volume is

$$= \varnothing_t(n_1, n_2, \ldots\ldots.d_1, d_2 \ldots..) \frac{\overline{du}}{dy}\, dt.$$

The function $\varnothing_t$ also changes at every instant of time due to change in n_1, n_2, … and also d_1, d_2, ….

If $\varnothing_t$ is replaced by a mean temporal value of the function over an interval of time t, that is, 'K' and the mean temporal value of velocity gradient $\frac{\overline{du}}{dy}$ over the interval of time t, by 'G' the total number of contacts over the interval of time t is

$$\int_0^t N_t\, dt = \int_0^t KG\, dt = KGt.$$

'Gt', a dimensionless number, is, thus, indicative of the total number or contacts over the time t in the system.

11.3 Compatible Design

Compatible design demands 'Gt' value in testing should reveal the 'Gt' values in the real settling system. Detention times during testing should reveal also the appropriate detention values in the real settling system.

If the raw water does not contain non-settleables or poorly settleables, 'flash mixing' and 'slow mixing' in the testing are to be left out, and settling time in the jar should relate the detention time in the plain sedimentation tank that is only to be provided for the removal of solids through settling.

In waste water treatment, the solids in the influent are flocculant. Settling column analysis will relate the depth requirement in the primary, intermediate and secondary clarifiers. The test will also relate the detention times in them.

Thickener design should follow the existing theories based on batch settling test with solid sludge.

11.4 Development and Presentation of Design Procedure

Compatible design procedure for settling system will be presented in the following sequence:

1. A real existing settling system should be studied. The settling performance of the system is to be taken up for the compatible design of 'jar testing' procedure for the same.
2. This study should be analysed to evolve design criteria for the compatible performance of the real settling system with regard to the designed jar testing results.
3. Design procedure employing settling column test results.
4. Design of shallow depth settling.
5. Design of sludge thickener.

Chapter 12
Simulation of Real System Settling in Jar Testing

Abstract This chapter demonstrates the design of 'jar testing procedure' in compatibility with real system operation. This has also been shown how to set the real system in operation with a designed jar testing.

Keywords Compatible jar testing • Compatible operation • Compatible flash mixing • Compatible slow mixing • Compatible settling

12.1 Introduction

Coagulation renders poorly settleable and non-settleable solids in suspension settleable. In this process a coagulant is added. It is made to distribute itself uniformly throughout the volume of suspension. This is done at rapid mixing to promote selective adsorption of positive ions, a product of ionisation of the coagulant, to the colloids. Agitation by stirring is provided to the suspension volume to provide temporal velocity gradient for making necessary contacts between the particles and the settleable flocs develop. The process is almost indispensable in the turbidity removal from water.

There are a number of coagulants. Specific coagulant is efficiently effective over specific pH range. Dosage of a particular coagulant depends upon the turbidity, its nature and also the pH and temperature of water. Coagulant dosage to turbid water is known from the 'jar test' procedure.

12.2 Review on Jar Testing Procedure, Its Critical Appraisal and the Objective of the Study

12.2.1 Review

Review on 'coagulant dosage' determination (1, 2, 3, 4, 5, 6, 7, 8, 9 and 10) reveals the following observations:

A. De, *Sedimentation Process and Design of Settling Systems*,
Springer Transactions in Civil and Environmental Engineering,
DOI 10.1007/978-81-322-3634-4_12

1. Standard methods (IS 10500), CPHEEO (1976) and also other cases (Schroeder, Fair et al.) are silent over the procedure for jar testing.
2. Trial determination of coagulant is universally advocated.
3. Trial determination of coagulant in 'jar test' identifies flash mixing, slow mixing and settling.
4. Flash mixing time and the speed of rotation of the stirrer, slow mixing time and the speed of rotation of the stirrer vary widely in different cases.
5. Some (Peavy and Rowe, Manual NEERI) recommend pH control in dosage determination.
6. Practicable parameters on dosage determination may be decided for the consideration of the following statement (Fair et al.).

"Because coagulation depends on so many variables that are themselves interdependent, as many testing parameters as possible should be kept constant. The importance of pH (Sincero and Sincero 1999) in governing the nature of the coagulant or flocculant through the extent of hydrolysis and ionisation (Clesceri et al. 1998) in determining the charge of colloids impurities suggest that the pH be kept constant too. In this connection it is well to remember that pH and alkalinity are changed implicitly when a coagulant is added."

12.2.2 Critical Appraisal of the Review

From the foregoing review, it may be appraised that:

1. Jar testing procedure for the determination of the coagulant dose has not yet been standardised.
2. The time of mixing and the speed of rotation of the stirrer during flash mixing and slow mixing are chosen arbitrarily the rationale behind the choices not being apparent.
3. In spite of pH control in coagulant being an important factor, the practicability of its implementation in real plant operation is not beyond question.
4. Flash mixing distributes the coagulant uniformly throughout the mass for the favoured adsorption of positive ions to the colloids.

Slow mixing produces settleable flocs. Settling time in 'jar test' takes care of the removal of readily settleable agglomeration of solids. Jar test simulates the conditions that are to happen in real settling performance. Strangely enough, nowhere in the above (1, 2, 3, 4, 5, 6, 7, 8, 9 and 10) compatible operation of the clariflocculator has been directed. Needless to say that the coagulant dosage, determined from jar test, without the mention of the compatible paddle speed in mixing chamber and compatible speed of the flocculating paddles in a real system for a particular flow rate appears to ignore the compatibility between dosage determination in 'jar test' and the performance of the settling system and hence makes the test arbitrary and undesirable.

12.2.3 Objective of Present Study

The present study has been undertaken for the compatible design of jar testing procedure for a real system by:

(i) Setting the speed of paddles and the mixing time for flash mixing
(ii) Setting the speed of paddles and the mixing time for slow mixing
(iii) Setting the settling time for turbidity removal, during coagulant dosage determination and suggesting
(iv) Compatible speed of paddles in mixing chamber
(v) Compatible speed of flocculating paddles in real system

Mixing and settling time being fixed for a particular flow rate of water through the system.

This is likely to remove the arbitrariness in the jar testing procedure and make it more meaningful.

12.3 A Real Settling System for Compatible Operation with Jar Test Results

In order to exemplify the design of jar testing procedure for compatible operation of a real settling system, the settling system of Serampore Water Treatment Plant has been taken into consideration. The plant is located in the western bank of river Hugli at Serampore (Exhibit 12.1).

The layout of the plant is shown In Fig. 12.1

12.3.1 Real Settling System

The water metre preceding Parshall Flume (Fig. 12.1) measures the rate of water flowing in per hour. The quantity fixes the amount of coagulants to be added to the water corresponding to the optimum dosage determined from jar test. This is added in the form of solution to the standing wave. This water is flash mixed. It is divided equally through the Parshall Flume following the mixer. The divided streams are led into the clariflocculators. The real settling system operation that should be compatible to jar testing is shown in Fig. 12.2.

Exhibit 12.1 Serampore Water Treatment Plant showing Parshall Flume, flash mixer and clariflocculator

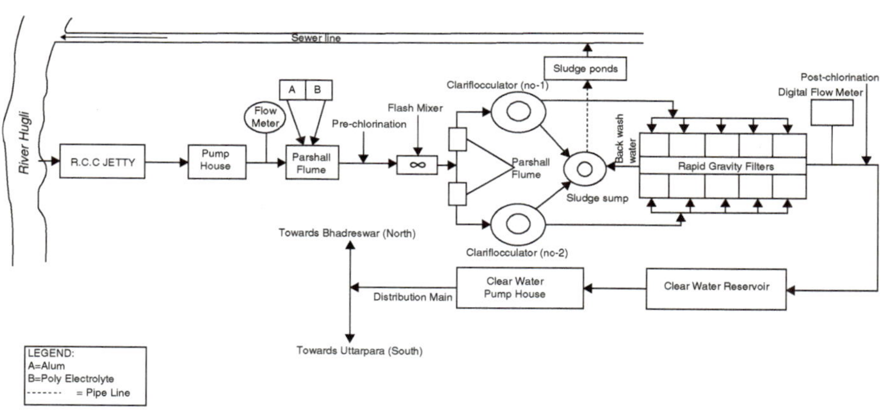

Fig. 12.1 Layout of water treatment plant at Serampore

12.3.2 Components of a Real Settling System

The components of a real settling system may be identified from Fig. 12.2:

1. Flash mixer
2. Parshall Flume
3. Flocculator (Exhibit 12.2)
4. Clarifier

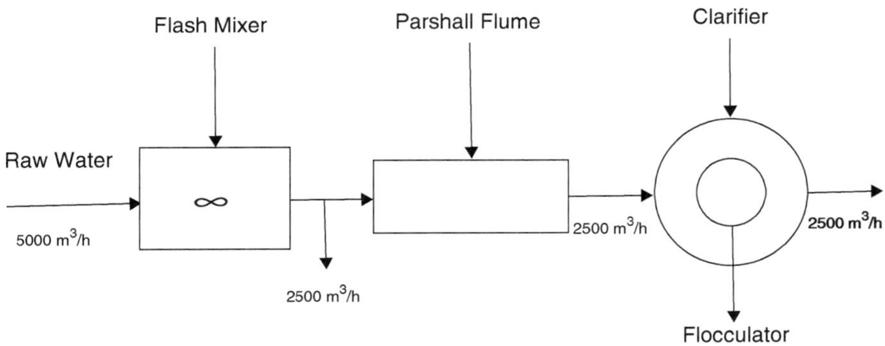

Fig. 12.2 Real settling system

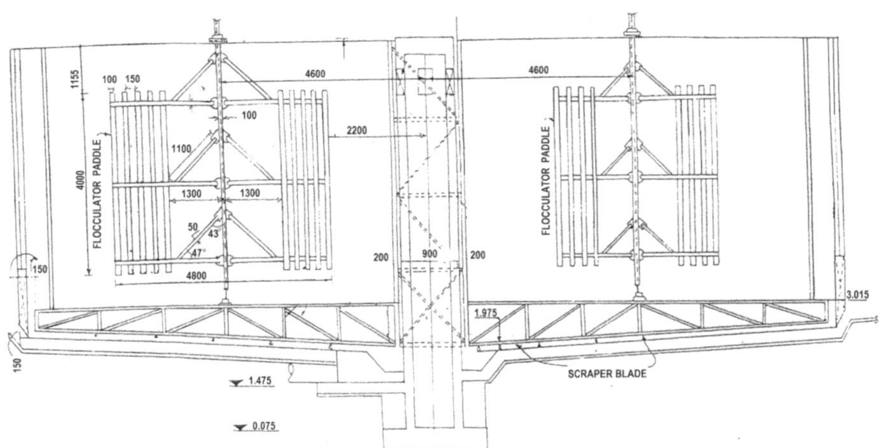

Exhibit 12.2 Flocculator

12.3.3 *Compatible Operation of the Components*

The above components simulate the function that happened during jar test. Compatible operation of the components demands that:

1. Flash mixer should accomplish what jar test achieved during flash mixing, i.e. adsorption of positive ions, resulted from ionisation of the coagulants, and incidental agglomeration of solids.
2. Parshall Flume accomplishes incidental flash mixing whether or not this is to be taken into account may be decided after analysis.
3. Flocculator does slow mixing to the formation of settleable flocs as were formed during slow mixing in jar test.
4. Clarifier removes the settleables in a manner similar to the settling during jar test.

12.4 Materials and Methods

12.4.1 Materials Used for Study

Jar testing apparatus (Fig. 12.3), Tullu Pump Company, Varanasi, India; Digital Nephelo Turbidity Meter, Model No. 132 (Exhibit 12.3), Systronics, India; Digital pH Meter range 0–14, electrically operated, Environmental & Scientific Co., India; Stop watch; Thermometer; Kemmerer Sampler (used to collect sample from all major depth zone of water masses); Ferric alum used as coagulant, pH range varying from 3.31 to 3.42, Alumina content varying from 14 to 14.6 %, soluble iron content range from 0.007 to 0.084 %, insoluble impurities varied from 0.20 to 0.61 %

Alum Solution 1 gm alum dissolved in 100 ml distilled water.

Fig. 12.3 Jar testing apparatus

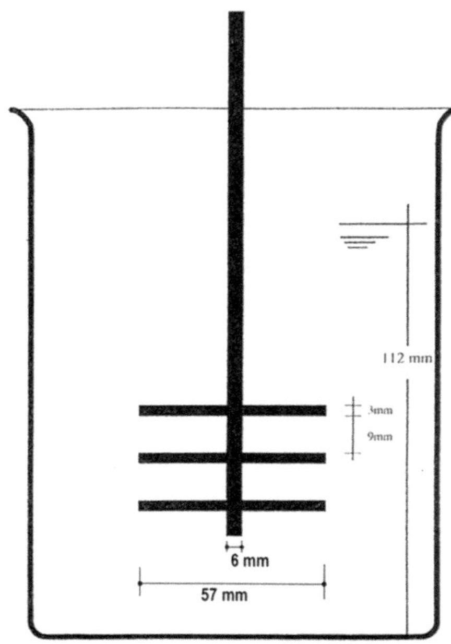

Depth of Water in Beaker	=	112 mm
Dimension of Blades		
No of Blades	=	3
Width of Blades	=	3 mm
Distance Between Two Blades	=	9 mm

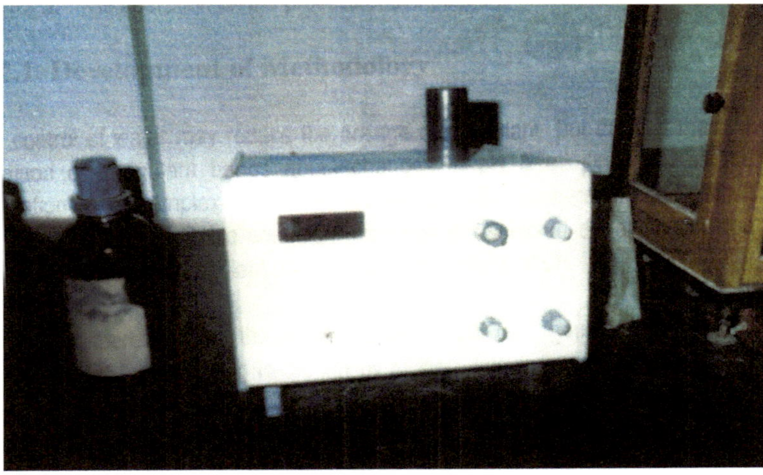

Exhibit 12.3 Turbidity meter

12.4.2 Methods

The dosage of a coagulant depends on (1) the coagulant and (2) the water, i.e. its turbidity, nature of turbidity, colour, pH and alkalinity and the operation on the process of coagulation.

12.4.2.1 Development of Methodology

pH control of water may reduce the dosage of coagulant. But the fact that the addition of coagulant brings about changes in pH and alkalinity makes the operation of the complex process on the bulk of water cumbersome and difficult if not impossible. Suitable coagulant aids may be selected depending on the effectiveness of the coagulant over its application range of pH. It appears, therefore, that the standardisation of jar test should concentrate on the control of operation in the jar by picking definite speed of rotation and duration of mixing both during flash mixing and slow mixing and also the settling time in the jar.

Compatible operation of settling system should ensure similar number of total contacts between the particles both during flash mixing and slow mixing in real settling system as such contacts happened in jar test operation.

For a volume of suspension, the total number of contacts over time interval ‘t’ per unit volume may be put down as KGT where ‘G’ is the mean temporal velocity gradient and ‘K’ is the mean temporal constant over the time ‘t’. Obviously ‘K’ is a function of Gt and particle distribution and changes over total number of contacts which again depends on the initial and final distribution of particles. For compatible operation in a real system to ensure similar contacts, initial and final distribution of particles during flash mixing and slow mixing will be the same as for that in a jar. As such Gt values both during flash mixing and slow mixing in real system operation should be same as the corresponding values of Gt during flash mixing and slow mixing in the jar.

The following steps may be followed:

Step 1: Collect raw water sample.
Step 2: Find its pH, turbidity and temperature.
Step 3: Find the dosage accordingly with flash mixing at 50 RPM for 2 min and slow mixing at 5 RPM for 8 min, settling time 30 min (Exhibits 12.4 and 12.5). These all are by arbitrary choice to start with.
I.S 10500–1983 and CPHEEO Manual recommend maximum turbidity in potable water as 10 NTU. The dose corresponding to residual turbidity just less than 10 NTU is selected.
Step 4: With the selected dose, perform jar test with:

Flash mixing at N RPM for 2 min
Slow mixing at 5 RPM for 8 min
Settling time of 30 min

Observations are to be taken for $N = 50$, 75, 100, 125, 150, 175, 200, 225, 250 ...RPM. The flash mixing speed (FMS) is selected as N_s RPM based on the minimum residual turbidity just below 10 NTU (Exhibit 12.6).
Step 5: Perform the jar test with selected dose:

Flash mixing at N_s RPM for t seconds
Slow mixing at 5 RPM for 8 min
Settling time of 30 min

Exhibit 12.4 Jar test to determine coagulant dosage

Exhibit 12.5 Jar test with flash mixing followed by slow mixing

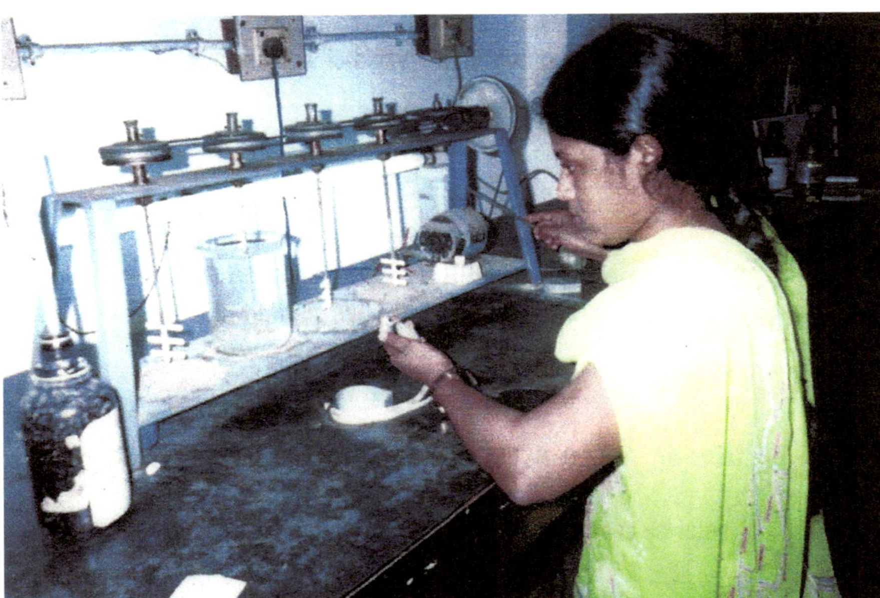

Exhibit 12.6 Setting flash mixing speed with several RPM

Observations are to be taken for $t = 0$, 30, 60, 90, 120, 150, 180, 210, 240, 300. . .seconds. Based on the minimum residual turbidity, flash mixing time (FMT) t_s secs is selected.

Step 6: Perform the jar test with selected dose – flash mixing speed N_s RPM for flash mixing time t_s, slow mixing at 5 RPM for T min, settling time of 30 min. Observations are to be taken for T – 1, 2, 3, 4, 5, 8, 7, 8, 9,. . ..min. Based on the minimum residual turbidity, slow mixing time T_s is selected.

Settling time in jar test has to be selected based on the performance of the clarifier as follows:

Step 7: Observe the residual turbidity in the escaping water from the clarifier flowing through the channel.

Step 8: Find the turbidities at various depths from the surface just in the vicinity of the channel inside the clarifier (Exhibits 12.7 and 12.8).

Exhibit 12.7 Depth-wise sampling using Kemmerer sampler

Exhibit 12.8 Collection of depth-wise samples

Step 9: Find the depth at which depth-wise average turbidity equals the turbidity of the escaping water.

Step 10: Select the settling time in jar test according to

$$\text{Settling time} = \frac{\text{Detention time in the clarifier} \times \text{depth of water in Jar}}{\text{The depth of water found out in step No.9}}$$

Compatible operation: For compatible operation of the real settling system equate

$$(G_j t_j)_{FM} = (G_R t_R)_{FM} \text{ and } (G_j t_j)_{SM} = (G_R t_R)_{SM}$$

where

G_j – Mean temporal velocity gradient in the jar
t_j – Mixing time in the jar
G_R – Mean temporal velocity gradient in the real settling system
t_R – Mixing time in real settling system
FM – flash mixing
SM – slow mixing

These will find out compatible operational speed for flash mixing and slow mixing in real system, mixing times in real system being determined by the volume of its components and the flow rate of water.

12.5 Methodology Applied

12.5.1 Jar Testing

12.5.1.1 Raw Waters Under Study

Three waters of the following description presented in Table 12.1 are taken for study.

12.5.1.2 Selection of dose

The study is conducted according to step 3 outlined in Methodology. The data obtained in accordance with step 3 is presented in Table 12.2.

Table 12.1 Raw water samples

No. 1			No. 2			No. 3		
pH	Temp °C	Turbidity	pH	Temp °C	Turbidity	pH	Temp °C	Turbidity
8.0	30	34 NTU	7.5	29	49 NTU	7.5	30	97 NTU

Table 12.2 Selection of dose (mg/l)
Jar testing parameters: flash mixing time, 2 min at 50 RPM;
Slow mixing time, 8 min at 5 RPM; settling time, 30 min

Water sample no. 1		Water sample no. 2		Water sample no. 3	
Alum dose (mg/l)	Residual turbidity (NTU)	Alum dose (mg/l)	Residual turbidity (NTU)	Alum dose (mg/l)	Residual turbidity (NTU)
2.5	11.5	3.6	11.0	8.2	12.8
2.6	10.8	3.8	10.0	8.4	12.4
2.7	10.0	4.0*	9.1	8.6	12.0
2.8*	9.9	4.2	8.5	8.8	11.6
2.9	8.5	4.5	10.4	9.0*	9.6
3.0	8.0	5.0	11.0	9.2	9.0

Table 12.3 Selection of RPM for flash mixing
Jar testing parameters: Flash mixing, 2 min at N RPM;
Slow mixing, 8 min at 5 RPM; settling time, 30 min

Raw water sample no.	Selected dose of alum (mg/l)	Residual turbidity (NTU) corresponding to flash mixing RPM – N								
		N = 50	75	100	125	150	175	200	225	250
1.	2.8	9.9	9.0	8.7	8.4	7.2	6.8	6.0	5.5	5.0
2.	4.0	9.1	8.6	8.2	7.8	7.4	6.8	6.0	5.8	5.5
3.	9.0	9.6	8.4	7.9	6.7	5.8	5.3	4.9	4.5	4.0

The selected doses of coagulants for water sample No. 1, water sample No. 2 and water sample No. 3 are shown in the table as 2.8 mg/l, 4.0 mg/l and 9.0 mg/l, respectively, and are marked by asterisks. They are selected as being just less than 10 NTU corresponding to residual turbidity.

12.5.1.3 Selection of Flash Mixing Speed

Following step 4 of Methodology, the effect of increasing the flash mixing speed on the residual turbidity for the three waters with their selected doses is presented in Table 12.3.

It may be observed that for all the waters' residual turbidity is increasingly reduced with increase in the flash mixing speed. It is most reduced at 250 RPM.

Mixing just after addition of coagulant induces charge neutralisation of the colloids by the positive ions produced from ionisation of coagulants resulting in subsequent flocculation. At any mixing speed, there is making and breaking of flocs. With the increase in the speed of mixing, the rate of making and breaking the flocs increases. But the rate of making flocs increases at far greater rate than the increase in the rate of breaking the flocs. This is reflected in the decrease in the residual turbidity which attains minimum at 250 RPM. Speed more than this could not be attained by the machine.

Might be that at speed slightly greater than 250 RPM residual turbidity could reach minimum. This shows that at this speed flocculation attains maximum. Further increase in speed would increase the rate of breaking the flocs, and residual turbidity would rise as a result. 250 RPM, thus, may be selected as flash mixing speed.

12.5.1.4 Selection of Flash Mixing Time

Table 12.4 depicts the effect of increasing the time for flash mixing in accordance with step 5.

Residual turbidity attains minimum at mixing time of 180 s in all the three cases. This shows that the total number of contacts for the most flocculation during flash

Table 12.4 Selection of flash mixing time
Jar testing parameters: Flash mixing, *t* secs at 250 RPM;
Slow mixing, 8 min at 5 RPM; settling time, 30 min

Raw water sample no.	Selected dose of alum (mg/l)	Residual turbidity (NTU) corresponding to flash mixing time t secs									
		$t=0$	30	60	90	120	150	180*	210	240	300
1.	2.8	10.5	9	8.2	7.5	5.0	4.8	4.6	4.9	5.1	5.2
2.	4.0	18	10	9.5	6.0	5.5	4.2	3.0	3.8	4.5	9.0
3.	9.0	19.3	10	8.8	5.9	4.0	3.2	2.5	3.5	3.9	4.8

Table 12.5 Selection of slow mixing time
Jar testing parameters: Flash mixing time, 3 min at 250 RPM;
Slow mixing time, *t* minutes at 5 RPM; settling time, 30 min

Raw water sample no.	Selected dose of alum (mg/l)	Residual turbidity (NTU) corresponding to slow mixing time t mins									
		t=0	1	2	3	4	5	6	7	8*	9
1.	2.8	7.8	7.0	6.8	6.6	6.2	5.8	5.3	4.9	4.6	5.0
2.	4.0	6.2	5.0	4.2	3.8	3.6	3.3	3.1	2.8	2.6	3.1
3.	9.0	6.5	5.4	4.8	4.5	3.9	3.5	3.0	2.7	2.5	3.0

mixing in the set-up is complete at 180 s, and subsequent mixing breaks the flocs. The residual turbidity increases as a result.

12.5.1.5 Selection of Slow Mixing Time

For slow mixing, the minimum operational speed (RPM) by the machine is selected at 5 RPM. It could also be some other low speed of rotation. Increasing the time of slow mixing increases the number of contacts between the particles. Total number of contacts leading to most effective flocculation occurs at 8 min as can be seen in Table 12.5 that follows. After that breaking of flocs begins to play the leading role. At some other low speed of rotation, more than this time of slow mixing could probably be reduced.

Needless to say that the operational speed of slow mixing in the jar has to be increased if the paddles of the flocculator in real system are to run at higher low rotational speed. The other parameters will change accordingly.

12.5.1.6 Selection of Settling Time

Table 12.6 presents the effect of increasing the settling time on the residual turbidity which decreases due to obvious reason.

The observations of three studies made according to steps 7, 8 and 9 of Methodology on three different dates are presented in Table 12.7.

Table 12.6 Residual turbidity (NTU) at different settling times
Jar testing parameters: Flash mixing time, 3 min at 250 RPM;
Slow mixing time, 8 min at 5 RPM; settling time, *t* min

		Residual turbidity (NTU) at settling times *t* mins				
Raw water sample no.	Selected dose of alum (mg/l)	$t=0$	30	35	40	45
1.	2.8	4.8	4.6	4.5	4.3	4.2
2.	4.0	3.0	2.6	2.4	2.3	2.2
3.	9.0	2.9	2.5	2.4	2.2	2.1

It appears that the collecting channel of the settling tank carries water turbidity that is depth-wise average turbidity over a depth of 1.5 m. The water on escaping through the flocculator passes through the clarifier volume of 7627 m^3 (clarifier including the flocculator of 17.5 m dia. × 5.81 m SWD is 47.5 m dia. × 4.375 m (including 7.0 cm free board) SWD) at the rate of 2500 m^3 per hour and thus allows settling of solids for $= \frac{7627}{2500} \times 60$ min, i.e. 183 min. This is equivalent to 11.2 cm (water depth in jar) depth-wise turbidity over a settling time of (183/150) × 11.2 = 13.66 min. Thus if compatible operation of the real settling system is ensured, 14 min settling after flash mixing ($t = 0$ in Table 12.6 including SMT) in the jar should give the residual turbidity in the collecting channel of the settling tank. This selects 14 min settling. Thus jar testing parameters for the geometry of the paddle and the jar may be of selected design as:

Flash mixing at 250 RPM for 3 min
Slow mixing at 5 RPM for 8 min
Settling time of 6 min

For the process to follow in jar testing compatible with that in ‘real system’, the testing procedure should allow 14 min of settling. Slow mixing time provides 8 min of settling. Remaining 6 min of settling is to be provided after slow mixing.

Comments: Jar testing procedure involves five variables. These are flash mixing speed (FMS) and flash mixing time (FMT), slow mixing speed (SMS) and slow mixing time (SMT) and settling time (ST).

Innumerable sets of the variables are possible from them. Each set points to a specific pattern and total number of contacts among the flocculating particles. Each set utilises a specific dose of coagulant to the formation of settleable flocs. The dose of coagulant that the particular set can utilise most for the formation of flocs is *the optimum dose of coagulant to the specific set*.

When coming to a specific water, the dose of coagulant that a particular set of variables can utilise for the formation of maximum settleable flocs from the flocculating particles in that water is *the optimum coagulant dosage*. The procedure outlined herein aims at the design of the set of the sets of variables corresponding to the optimum coagulant dosage.

Since the set points to a specific pattern and total number of contacts among the flocculating particles, there may be many other sets from the variables that will meet the similar ends. The designed set from the ‘jar test’ is, therefore, not unique.

Table 12.7 Clarified water turbidity and its depth-wise variation in the settling tank just before escape

Depth (m)	Study no. 1			Study no. 2			Study no. 3		
	Turbidity (NTU)	Depth-wise average turbidity (NTU)	Turbidity of clarified water, NTU	Turbidity (NTU)	Depth-wise average turbidity (NTU)	Turbidity of clarified water, NTU	Turbidity (NTU)	Depth-wise average turbidity (NTU)	Turbidity of clarified water, NTU
0.00	8.5	–	9.20*	9.50	–	10.50*	10.40	–	10.80*
0.25	8.8	8.65		10.00	9.75		10.50	10.45	
0.50	9.1	8.80		10.20	9.90		10.60	10.50	
0.75	9.2	8.90		10.60	10.10		10.70	10.55	
1.00	9.4	9.00		10.80	10.20		10.90	10.62	
1.25	9.5	9.10		11.10	10.36		11.10	10.70	
1.50	9.7	9.20*		11.40	10.50*		11.30	10.78*	
2.00	9.7			11.60			11.50		
2.50	11.6			11.70			11.70		
3.00	12.4			12.90			13.00		
3.50	12.9			13.00			13.20		

12.5.2 Compatible Jar Testing and Operation of Real Settling System

In real system, 'agglomeration of solids' is accomplished through 'flash mixer' and 'flocculator', and 'settling of solids' is affected in 'flocculator' and 'settling tank'.

In 'jar testing', 'agglomeration of solids' takes place through 'flash mixing' and 'slow mixing' in the jar. Agglomerated solids settle in the jar during 'slow mixing' and additional settling time allowed.

Compatible 'jar testing' and 'operation of real system' should aim at similar 'agglomeration' and 'settling' in both the design jar testing and operation of real system.

For compatible agglomeration of solids in both jar testing and real system, operation may be provided

$$\text{either by arranging (1)}\ \left(G_j t_j\right)_{FM} + \left(G_j t_j\right)_{SM} = (G_R t_R)_{FM} + (G_R t_R)_{SM};$$
$$\text{or by arranging (2)}\ \left(G_j t_j\right)_{FM} = (G_R t_R)_{FM}\ \text{and}\ \left(G_j t_j\right)_{SM} = (G_R t_R)_{SM},$$

and for compatible settling it should be

$$\left(\frac{d}{t}\right)_J = \left(\frac{D}{T}\right)_R$$

i.e. the ratio of 'falling through distance' to the 'falling through time' in the jar (J) and real system operation (R) should be similar.

12.5.2.1 Compatible Settling in Settling Tank

Depth of water in the jar = 11.2 cm.
Designed settling time = 14 min.
The particle with maximum settling velocity in suspension = $\frac{11.2\,\text{cm}}{14\,\text{min}}$, i.e. 0.8 cm/min.

12.5.2.1.1 The Theoretical Detention Time

As shown in Fig. 12.4a, flow enters into the flocculator and passes into the settling tank. The settling time that the particle with maximum settling velocity will travel through from the surface of water

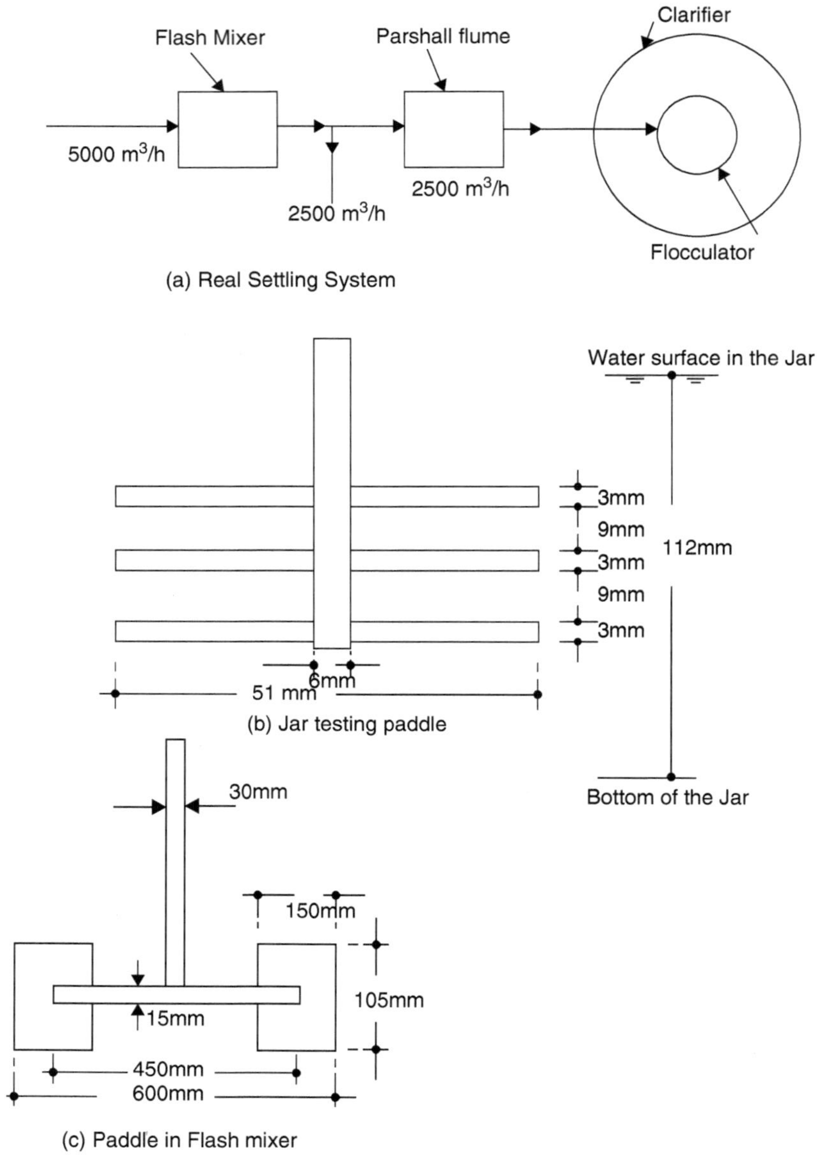

Fig. 12.4 Real settling system and paddles in Jar testing and Flash mixer

$$= \frac{\pi \times 47.5^2 \times 4.305}{4 \times 2500} (70\,\text{cm being the free board})$$
$$= 183\,\text{min}$$

The depth to which this particle will travel = 0.8 × 183, i.e.146 cm or 1.5 m.

It has been observed that the effluent channel of settling tank of the WTP carries the average turbidity over a depth of 1.5 m from the water surface. This turbidity is the same as that in the suspension in the jar after 14 min settling. This is revealed, therefore, that:

(i) The designed 'jar testing' simulates similar settling to that takes place in real system having provided 11.2 cm depth of settling in 14 min, i.e. at the rate provided in real system to fall through 1.5 m in 183 min.
(ii) This being so the depth-wise average turbidity over the depth of 1.5 m in real system is the same as that over the depth of 11.2 cm in jar with the designed testing. This is the turbidity of clarified water (Table 12.7) collected from the effluent channel of the real system.
(iii) This amounts to the statement that the agglomeration provided by the flash mixing speed, flash mixing time, slow mixing speed and slow mixing time in real system is the same as that provided by the designed FMS, FMT, SMS and SMT in the jar.

The designed jar testing provides similar agglomeration and settling to those in real system in compliance with condition (1) for compatible testing an operation (Sect. 12.5.2).

12.5.2.2 Compatible Operation of Flash Mixing and Slow Mixing in 'Real system' in Compliance with Condition (Clesceri et al. 1998) in 12.5.2

Mixing components in the jar and in the real system are shown in Fig. 12.4.
Gt values in the jar are calculated with respect to the Fig. 12.4b.
Gt value in flash mixer is computed with respect to Fig. 12.4c.
Gt value in the flocculator is computed with respect to the Exhibit 12.2.

1. Computation of Gt values of the jar testing:

Assuming – Density of water $\rho = 1\,\text{gm/cm}^3$
Coefficient of viscosity $\mu = 0.8 \times 10^2 \frac{\text{gm}}{\text{cm} \times \text{s}}$
Volume of water – $1000\,\text{cm}^3$

Flash mixing time –180 s; flash mixing speed – 250 RPM

Slow mixing time – 480 s; slow mixing speed – 5 RPM

G (mean temporal velocity gradient imparted by the paddle (Fig. 12.4b) in one litre water) (Fig. 12.4b)

$$= \sqrt{\frac{(2)(3)(1.8)}{(1000)(2)}\left(\frac{\rho}{\mu}\right)\int_{0.3}^{2.55}(0.75)^3\left(\frac{2\pi xN}{60}\right)^3(0.3)\mathrm{d}x} \quad \text{where, drag coeff.}$$

$= 1.8$, relative vel. factor $\dot{-}0.75$

$= 32.2 \times 10^{-3}\sqrt{N^3}$ per sec, Putting the values of ρ and μ

Gt at flash mixing $= 32.2 \times 10^{-3}\sqrt{250^3} \times 180$, i.e. 22911;

Gt at slow mixing $= 32.2 \times 10^{-3}\sqrt{5^3} \times 480$, i.e. 173;

2. Computation of Gt value in flash mixer:

Volume of flash mixer (4.2 m dia. × 6.16 m SWD) = 85.3432 × 10^6 cm
Detention for the rate of flow 5000 m^3/h = 61.4 s
Gt value in flash mixer (Fig. 12.4c) at N RPM

$$= \sqrt{\frac{(2)(1.8)}{(2)(85.3432 \times 10^6)}\left(\frac{1.5\pi}{60}\right)^3\left(\frac{\rho N^3}{\mu}\right)\left[\int_0^{15} x^3(1.5)\mathrm{d}x + \int_{15}^{30} x^3(10.5)\mathrm{d}x\right] \times 61.4}$$

$= 3.11287\sqrt{N^3}$ Putting the values of ρ and μ

$= 22911$, i.e. $N = 378\,RPM$

If this turns out to be very high speed of rotation, the speed of rotation can be reduced by the arbitrary choice of lower flash mixing time and higher speed of rotation for slow mixing in step 3 for the dosage determination in which case it is likely that higher dosage will result.

3. Computation of Gt value in 'Parshall Flume':

Flow rate – 2500 m^3/h = 0.6944 m^3/sec; throat length – 0.6 m

The empirical formula that may be employed (CPHEEO Manual) – $Q = 2.42$ (throat length in m) (upstream gauged depth in m)$^{2.58}$; upstream gauged depth can be calculated 0.75 m.

If 1 % loss of head may be assumed,

G (mean temporal velocity gradient) in the flume

$$= \sqrt{\frac{\rho g h_1}{\mu}}$$

$$= \sqrt{\frac{(1)(981)(0.01)(0.75)(100)}{(0.8 \times 10^{-2})}}$$

$$= 303 \text{ per sec}$$

The flow-through time is a fraction of a second.

Gt value, thus, may be neglected. Otherwise, it may have to be taken into account.

4. Computation of Gt value in flocculator:

   Volume of the flocculator (dia. 17.5 m × 5.81 m) (v) = 1397 × 10^6 cm^3
   For the flow rate of 2500 m^3 per hour, detention time – 2012 s
   G^2 value in the flocculator (Exhibit 12.2)

$$= \left(\frac{\rho}{\mu V}\right)\left(\frac{1.8}{2}\right) \times \left(\frac{1.5\pi N}{60}\right)^3 \left[370 \times 10\,\{(235^3 + 210^3 + 185^3 + 160^3 + 135^3)2\right.$$

$$+ (210^3 + 185^3 + 160^3 + 135^3)2\}$$

$$\left. + \left(10\int_0^{240} x^3 \mathrm{d}x.6 + 10\int_0^{215} x^3 \mathrm{d}x.6 + \int_0^{110\cos 47^\circ} x^3 \frac{5}{\sin 43^\circ} \mathrm{d}x.12\right)\right]$$

$$= 21.77330 N^3$$

i.e. G = 4.66619 $N^{3/2}$
i.e. Gt = 4.66619 $N^{3/2}$ × 2012
=9388 $N^{3/2}$ (in real system)
=173 (in the jar)
i.e. N = 0.07 RPM
i.e. 1 rotation in 14 min

12.5.3 Comparison of Doses

The recommendation in report (NEERI Manual) and the practices on jar testing in Serampore Water Treatment Plant are presented in Table 12.8.

Three waters A, B and C collected on three different dates presented in Table 12.9 are taken up for the comparison of doses that are obtained with practice no. 1, practice no. 2 and with the flash mixing at 250 RPM for 3 min and slow mixing at 5 RPM for 8 min set out in the present study. Settling time has been set 30 min to make the observations comparable with observations with practices.

Table 12.8 Recommendation practice on jar testing parameters

	From verbal enquiry for jar test performance at WTP	
Recommendation in report (NEERI Manual)	Practice no. 1	Practice no. 2
2 min Flash mixing	2 min at 100 RPM	1 min at 250 RPM
Slow mixing	15 min at 15 RPM	2 min at 50 RPM
30 min Settling	Settling time 15 min	5 min at 10 RPM
No mention of the RPM		Settling time 30 min

Table 12.9 Water samples A, B, C

Sample A			Sample B			Sample C		
pH	Temp°C	Turbidity NTU	pH	Temp°C	Turbidity NTU	pH	Temp°C	Turbidity NTU
8.0	30.5	30.5	7.5	30.5	71.0	8.0	30.0	85.0

Table 12.10 Comparison of selected doses according to three procedures

Sample	Jar test results						
A	Alum dose (mg/l)	1.3	1.4	1.5	1.6	1.8	2.0
	Residual turbidity NTU based on present study	11.0	10.3	9.2*	8.8	7.9	4.3
	Residual turbidity NTU based on practice no. 1	12.6	11.8	10.9	9.2*	8.6	5.4
	Residual turbidity NTU based on practice no. 2	12.0	11.2	10.2	9.0*	8.1	5.0
B	Alum dose (mg/l)	1.5	1.6	1.7	1.8	1.9	2.0
	Residual turbidity NTU based on present study	11.8	9.8*	9.2	8.2	7.9	7.5
	Residual turbidity NTU based on practice no. 1	12.5	11.0	10.2	9.4*	8.9	8.2
	Residual turbidity NTU based on practice no. 2	12.2	10.6	10.0	9.2*	8.4	8.0
C	Alum dose (mg/l)	1.5	1.6	1.8	1.9	2.0	2.5
	Residual turbidity NTU based on present study	11.0	10.4	10.0	9.2*	8.8	7.0
	Residual turbidity NTU based on practice no. 1	13.9	13.2	12.8	10.6	9.9*	9.0*
	Residual turbidity NTU based on practice no. 2	12.9	12.5	11.0	10.4	9.8*	8.6

Jar test results according to the present study and practice 1 and practice 2 at the Serampore Water Treatment Plant have been presented in Table 12.10.

The selected doses that correspond to the residual turbidities just below 10 NTU are shown by asterisks. It may be observed that although the selected doses according to practices 1 and 2 are the same, that based on the present study are less, this is true for all three studies.

12.5.4 *Lessons Learnt from the Study*

The lessons learnt from this study provide us with the following understandings. The design procedure for jar testing can set the variable values (FMS, FMT, SMS, SMT and ST) to find optimal coagulant dosage for a particular water. This dose is not unique.

The same dose for the same water may admit many other sets of values of the variables, and still the dose remains optimum.

It is also possible that there are other different coagulant dosages for different other sets of values of the variables, and still all the different coagulant doses are optimum for the same water.

The optimum dose for a particular water with set values of the variables assures maximum contacts for the conversion of the particles into Settleable "flocs". The compatible jar testing and real settling system operation can be revealed in the following:

1. The designed settling time in jar testing has compatible settling in the settling tank.
2. The Gt value in the jar during flash mixing can set the compatible rotational speed in the flash mixer of the WTP.
3. The Gt value in the jar during slow mixing can set the compatible speed of rotation of flocculator paddles.
4. The Gt values during jar testing can help to design new settling system.

Notations

N	Speed of rotation
N_s	Selected speed of rotation
t,T	Time duration
t_s,T_s	Selected time duration
FM	Flash mixing
SM	Slow mixing
G_j	Mean temporal velocity gradient in jar
t_j	Time duration in jar
G	Mean temporal velocity gradient over the interval of time t, T in real system
K	Mean temporal constant over time interval t,T

References

CPHEEO[6] (1976) Manual on water supply and treatment, 2nd edn. CPHEEO, Ministry of Works and Housing

De A (2005) Design of jar Testing procedure and compatible operation of a Real settling system. JIPHE, India (vol-3)

Fair GM, Geyer JC, Okun D[4]. Water and wastewater engineering, vol 2. Wiley, Newyork/London/Sydney

Hammer MJ[9], Hammer Jr. Water and wastewater technology, 3rd edn. Prentice Hall, New Delhi, India

Manual[7] on water and wastewater analysts. NEERI, Nagpur

IS 10500[10]. 1983 Indian Standard Specifications for drinking water Standard Methods[11] for the examination of water and waste water. 29th edn. AWWA, APHA Water Pollution Control federation

Mausumi Das (2005) Compatible Design of Jar Testing procedure with Settling Performance in Real Settling System. MCE thesis. Jadavpur University, Kolkata, India

Clesceri LS, Greenberg AE, Baton AD (eds)[2] (1998) Standard methods for the examination of water and wastewater, 20th edn. American Public Health Association, Washington, DC

Peavy HS, Rowe DR[5]. Environmental engineering. McGraw Hill Book Company

Perfomance[8] and water quality assurance of Serampore Water Treatment Plant at NEERI (2004)

Schroeder ED[3]. Water and wastewater treatment, International Students edition. McGraw Hill/ Kagokosa Ltd

Sincero AF, Sinero GP[1] (1999): Environmental Engineering a Design Approach. Prentice Hall India pvt limited, New Delhi

Chapter 13
Compatible Design of a Real Settling System

Abstract In the following, settling system has been designed following a procedure to actualise in the system the process of appropriate testing on which the system operation depends.

Keywords Settling system design • Flash mixer design • Flocculator design • Settling tank design • Distribution pipe design

13.1 Introduction

For the rational design of a real settling system, the settling system has to simulate the processes that happen in 'jar testing' in the laboratory. Jar testing is accomplished through three phases. These are 'flash mixing', 'slow mixing' and 'settling'.

During flash mixing in the jar, the paddles are rotated at high speed for quick adsorption of ions for charge neutralisation on the non-settleable and poorly settleable solids and keeping the flocs in suspension.

With increasing speed of rotation the rate of adsorption increases. The time duration of flash mixing points to the progress of adsorption which reaches completion at a particular point of mixing time. Beyond this point of completion, breaking of flocs occurs as reflected in the change of decreasing trend of residual turbidity to its increasing trend through a minimum (Table 12.4). For each speed of rotation, there is one such minimum.

During slow mixing, the paddles are rotated at very low speed, the purpose being to make contacts between solids for the formation of flocs taking care about their protection from their breakage. The time duration of slow mixing is indicative of the total number of contacts. With the increase in mixing time, a point is arrived at which the suspension will give minimum turbidity. This is the optimum time for the speed of rotation (Table 12.5).

On discontinuation of slow mixing, some settling time is allowed for the settling of flocs.

A. De, *Sedimentation Process and Design of Settling Systems*,
Springer Transactions in Civil and Environmental Engineering,
DOI 10.1007/978-81-322-3634-4_13

13.2 Design of Settling System

13.2.1 Design of Jar Testing

The design of settling system is initiated with the design of jar testing.

Selection of Coagulant
Take a litre of water in the jar and find its pH, temperature and turbidity. Coagulant is selected.

Selection of Coagulant Dose
Try for varying coagulant doses with arbitrarily chosen set of jar testing parameters such as

FMS (flash mixing speed) = 50 RPM, *t* (Time duration) = 2 min.
SMS (slow mixing speed) = 5 RPM, *T* (Time duration) = 6 min.
Settling time = 10 min.

Minimum dose is selected.

Selection of Flash Mixing Speed
Use this chosen dose and conduct jar testing with *N* (*FMS*) = 50, 75, 100, 125, 150, 175, 200, 225, 250... RPM

with *t* = 2 min, *SMS* = 5 RPM, *T* = 6 min and settling time *ST* = 10 min.

Choose high speed of rotation N for minimum residual turbidity or any other convenient low value not necessarily the minimum one.

Let it be *N* = 200 RPM.

Selection of Flash Mixing Duration t secs
Use the chosen dose and conduct jar testing with

FMS = 200 RPM, *t* = 0, 50, 60, 90, 120, 150, 180, 210, 240, 500... secs
SMS = 5 RPM, *T* = 6 min, Settling time *ST* = 10 min

Select the time *t* secs based on the minimum residual turbidity.

Let this *t* be = 120 s.

Selection of SMS
Unless the need arises *SMS* = 5 RPM may be selected.

Selection of Slow Mixing Time Duration T min
Use chosen dose and conduct jar testing with *FMS* = 200 RPM, *t* = 120 s.

SMS = 5 RPM, *T* = 0, 1, 2, 5, 4, 5, 6, 7, 8, 9, 10 min, settling time = 10 min.

Select the time *T* min based on minimum residual turbidity.

Let this *T* be = 8 min.

Should this *T* min has to be reduced (for the sake of reducing the flocculator volume), increased slow mixing speeds (>5 RPM) are to be tried for observation and a suitable value selected.

Settling Time ST

Varying Settling times should be tried to get at a suitable residual turbidity

Let this be $ST = 0$.

The Designed Jar Testing Procedure

FMS = 200 RPM, t = 120 s, SMS = 5 RPM, T = 8 min, settling time $ST = 0$.

Varying doses should be tried with this designed procedure and minimum dose is to be selected.

13.2.2 Basis of Design

The design of the settling system should be based on the following understandings:

1. The volumes of the flash mixer and/or flocculator has got nothing to do with settling performance except in helping the settling performance by imparting the necessary GT values into the water through them for effective coagulation-flocculation.
2. In Chap. 12 (Ref. Table 12.6), three water with initial turbidities, 34 NTU, 49 NTU and 97 NTU, were subjected to jar testing for $FMS = 250$ RPM, $t = 180$ s, $SMS = 5$ RPM, $T = 8$ min.
   At settling time duration $ST = 0$, the residual turbidities of the above waters were 4.8 NTU, 3.0 NTU and 2.9 NTU, respectively.
   Flash mixing speed being 250 RPM for 3 min, it appears reasonable to neglect settling of particles, if any, during flash mixing. It is further reasonable to conclude, therefore, that almost the entire removal of turbidities took place during slow mixing time of 8 min. If it would require some more time x min, say, of settling after the slow mixing to reduce the residual turbidities to desired level, then the settling time will be counted as $(8 + x)$ min.
3. In Table 12.7 of Chap. 12, it may be seen that the depth-wise average turbidities over 1.5 m of depth adjacent to the effluent channel were found to be equal to that in the effluent channel for all the three waters, weir loading being same of 16.75 m^3/m/h. in all cases. For weir loading less than this residual turbidity of water in effluent channel will be less. (Such data of different 'weir loadings' drawing effluent from the corresponding depth of the tank, thus taking care both weir flow velocity and overflow velocity, are to be generated through extensive research.)
4. None of weir loading or overflow velocity can define any settling performance only by itself. The design of settling system should take care of both weir loading and overflow velocity (Acharya 1990).
5. The system should be configured taking care of minimising short circuiting by controlling the widthwise variation of velocity (Chap. 9).

13.2.3 Procedure for the Compatible Design of Settling System

Settling system comprises of flash mixer, flocculator and settling tank. Compatible design aims at such a design of the system that will simulate the processes taking place in the jar during testing.

After the settling system comes into existence, if the parameters such as flow rate, water quality, etc., suffer some changes, the coagulant dose is to be determined from the compatible design of jar testing procedure, and the speed of rotations of the paddles in flash mixer and flocculator are to be adjusted accordingly.

Design of Flash Mixer
Depending upon the rate of flow, a retention time is to be chosen such that the volume of mixer is minimum, and a suitable paddle may be accommodated into it to impart to the water the necessary Gt value (as it was during the flash mixing in the jar) with proper speed of rotation.

Design of Flocculator
The procedure of its design is same as that of flash mixer.

Design of Settling Tank
From the settling data during jar testing the settling time is to be ascertained as

$ST =$ Time of slow mixing + additional time of subsequent settling.

If 'd' be the depth of water in the jar, the suspension in the jar will contain largest particle of settling velocity $= d/ST$. . . . . (Eq. 13.1) after settling in the jar.

The effluent channel of the settling tank will carry water of residual turbidity that is depth-wise average turbidity over 1.5 m or less of adjacent depth according to whether weir loading is limited to 16.75 m^3/m/h or less, respectively.

If 'T' be the retention time in the tank then 1.5 m depth adjacent to the effluent channel contains suspension with largest particle of settling velocity $\frac{150(\text{cm})}{\text{T(s)}}$ cm/s.

If this suspension over the depth of 1.5 m has to be similar to that in the jar, we have

$$\frac{150(\text{cm})}{T(\text{s})} = \frac{d(\text{cm})}{ST(\text{s})} \tag{13.2}$$

Equation 13.2 takes care both of weir loading and overflow velocity. 'T' can be found out from the Eq. 13.2. The flow rate being Q m^3/h. The volume of the tank is QT (T in hrs) m^3.

Limiting the weir loading to 16.75 m^3/m/h, the diameter of the settling tank

$$= \frac{Q\,\text{m}^3/\text{h}}{16.75\,\frac{\text{m}^3.\pi}{\text{m.h}}}\ \text{i.e.}\frac{Q}{16.75\pi}\,\text{m, Depth of the tank}\,\frac{4QT}{\pi D^2};$$

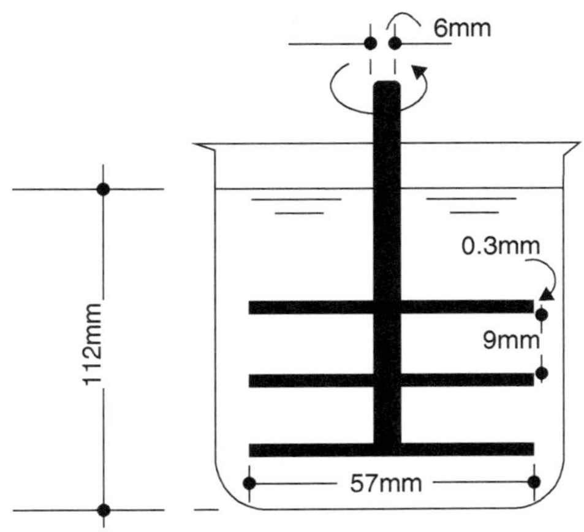

Fig. 13.1 Jar testing

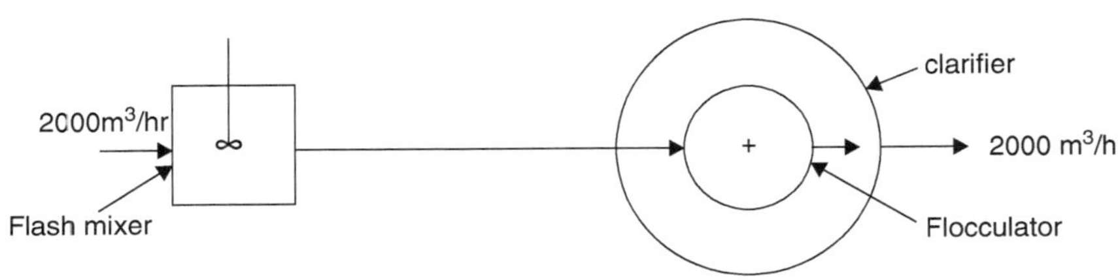

Fig. 13.2 Settling system

Problem 13.1 Design a setting system to process 2000 m^3 of water per hour at 30 °c. Raw water was collected, and the jar testing procedure was designed as follows in a jar as shown in Fig. 13.1 with water volume of 1 litre using alum as coagulant:

$FMS = 200$ RPM, $t = 2$ min; $SMS = 5$ RPM,

$T = 8$ min; settling time = 0 min; design the settling system.

Solution The settling system may be chosen as in Fig. 13.2.

Design of flash mixer

Assume flash mixing time = 1 min.

$$\text{Volume of flash mixer} = 2000\frac{\text{m}^3}{\text{h}} \times \frac{1\,\text{min}}{60\,\text{min/h}}$$

$$= 33.33\,\text{m}^3\ \text{i.e. } 3.0\,\text{m dia, } 4.72\,\text{m SWD}$$

G (Mean Temporal Velocity Gradient) in the Jar

Temp. of water = 30 °C

Coeff. of viscosity = 0.8×10^{-2} g/cm.s

Water density = 1 g/cm^3,
Volume of water in the jar = 1000 cm^3
Surface water depth = 11.2 cm
C_D, the coeff. of drag = 1.8

Relative velocity of the paddles is 75 % of the paddle velocity. The paddle configuration in the jar is shown in Fig. 13.1.

G (mean temporal velocity gradient) at N RPM (Ref. Fig. 13.1 and Sect. 12.5.2.2)

$$= 32.2 \times 10^{-3}\sqrt{N^3}\text{ per sec}$$

Gt value during flash mixing in the jar at $N = 200$ RPM, $t = 2$ min.

$$= 32.2 \times 10^{-3}\sqrt{200^3} \times 120$$
$$= 10929$$

The paddle configuration in flash mixer is shown in Fig. 13.3.

G value in flash mixer

G (mean temporal velocity gradient) in flash mixer at N RPM (Ref. Fig. 13.3)

$$= \sqrt{\frac{(2)(1.8)(1)}{(2)(0.8 \times 10^{-2})(33.33) \times 10^6}\left[\int_0^{15} (0.75)^3\left(\frac{2\pi xN}{60}\right)^3 1.5\,\mathrm{d}x + \int_{15}^{40} (0.75)^3\left(\frac{2\pi xN}{60}\right)^3 (20)\,\mathrm{d}x\right]}$$

$$= \sqrt{\frac{(2)(1.8)(1)}{(2)(0.8 \times 10^{-2})(33.33) \times 10^6}\left(0.75.\frac{2\pi}{60}\right)^3 N^3\left[\frac{1.5 \times 15^4}{4} + \frac{20(40^4 - 15^4)}{4}\right]}$$

$$= 0.00572 \times 10^{-2}\sqrt{N^3}\,(3544.84)$$

$$= 0.2027646\sqrt{N^3}/\text{s}$$

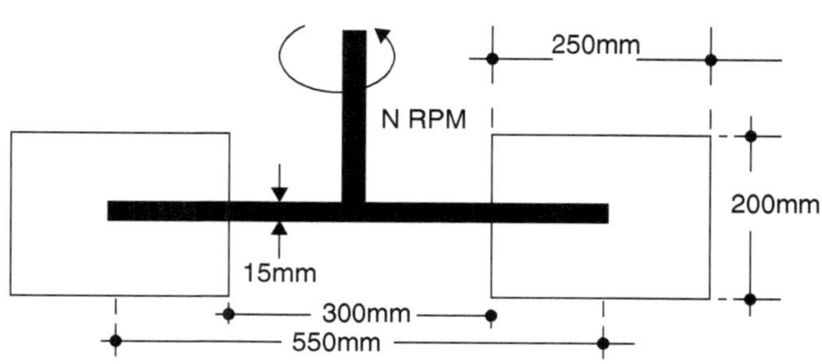

Fig. 13.3 Paddle configuration

Gt value in the flash mixer at *N* RPM

$$= 0.2027646\sqrt{N^3} \times 60 \text{ i.e. } 12.17\sqrt{N^3}$$

Equating the *Gt* value with that in jar testing during flash mixing

$$N = \left(\frac{10929}{12.17}\right)^{1/1.5} \text{ i.e. } 93\,\text{RPM}$$

The speed of rotation of the paddle is 93 RPM.

Design of Settling Tank

The suspension in the jar after 8 min of settling has the fastest particle in the suspension of settling velocity.

$= \frac{11.2\,\text{cm}}{8\times 60\,\text{s}}$, 11.2 cm being the depth of water in the jar.

The effluent channel of the settling tank carries water of residual turbidity that is the depth-wise average turbidity over the 1.5 m depth of adjacent water. If the suspension has to be similar to that in the jar, the fastest particle should fall through 1.5 m during the theoretical detention time *t*.

$$\text{i.e.} \frac{11.2\,\text{cm}}{8 \times 60\,\text{s}} = \frac{150\,\text{cm}}{t \text{ sec}}; \text{ i.e. } t = \frac{150 \times 8 \times 60}{11.2}\,\text{s};$$

$$= 6428.57\,\text{s};$$

$$\text{The volume of the tank} = \frac{2000 \times 150 \times 8 \times 60}{60 \times 60 \times 11.2}$$

$$= 3571.43\,\text{m}^3;$$

Limiting the weir loading to 16.75 m^3/m/h, the diameter of the tank

$$= \frac{\frac{2000\,\text{m}^3}{\text{h}}}{16.75\text{m}^3/\text{m.h}\,(\pi)} \text{ i.e. } 38.0\text{m};$$

Depth of the tank $= \frac{3571.34 \times 4}{\pi(38)^2}$ i.e. 3.15 m;

Use 38.0 m Dia × 3.15 m SWD;

This allows the fastest particle to reach the bottom having settling velocity = $\frac{315 \times 100}{6429}$, i.e. 0.049 cm/s.

The settling tank will contain suspension with fastest particle of settling velocity 0.023 cm/s up to 1.5 m depth and from 1.5 m to 3.15 m the suspension will contain particles having settling velocity ranging from 0.023 cm/s to 0.049 cm/s.

Alum floc of 0.1 cm (Fair) of Sp.Gr 1.002 at 30 °C has settling velocity

$$= \frac{981 \times (0.1)^2 (1)(1.002 - 1)}{18 \times 0.8 \times 10^{-2}}$$
$$= 0.13625\,\text{cm/s} > 0.049\,\text{cm/s}$$

Design of Flocculator

Considering the Gt value during slow mixing in the jar

$$= 32.2 \times 10^{-3}\sqrt{5^3} \times 8 \times 60\,\text{i.e.}\ \ 173;$$

Let us choose the detention time in the flocculator as 4 min.

At the volume rate of flow 2000 m^3/h, the volume of the flocculator

$$= \frac{2000 \times 4 \times 60}{60 \times 60},\ \text{i.e.}\ 133.33\,\text{m}^3;$$

Considering the settling tank 38.0 m dia × 3.15 m SWD

Let us choose the depth of the flocculator.

$$= 3.15\,\text{m} + 1.4\,\text{m}\,(\text{for accommodating it in the clariflocculator})$$
$$= 4.55\,\text{m}$$

The diameter of the flocculator

$$= \sqrt{\frac{133.33 \times 4\,\text{m}}{\pi \times 4.55}},\ \text{i.e.}\ \ 6.1\,\text{m}$$

In the light of small Gt value in slow mixing during jar testing, let us choose the paddle configuration as shown in Figs. 13.4a and 13.4b. Two such will be used symmetrically and simultaneously on either side of the central axis.

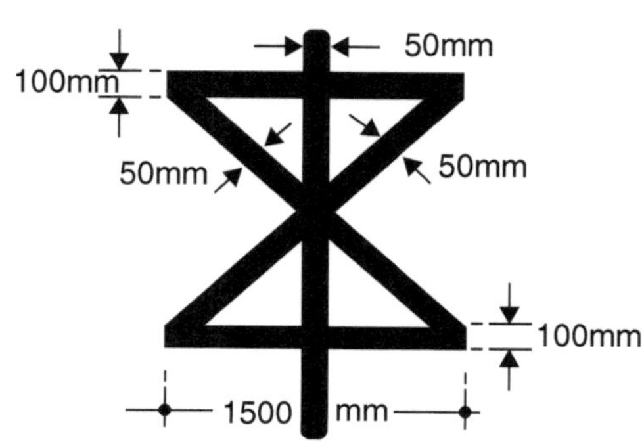

Fig. 13.4a Configuration and paddler details

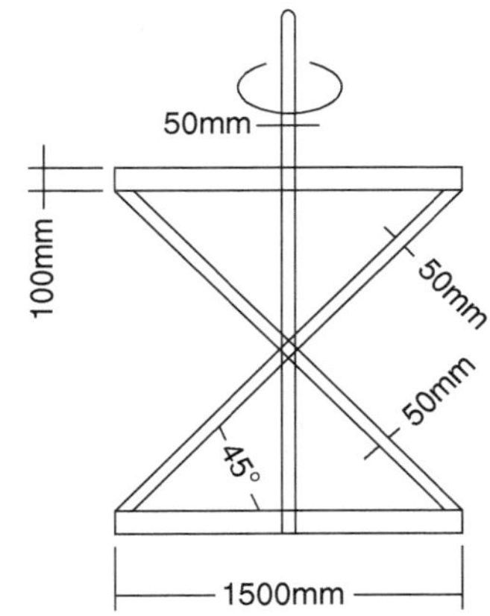

Fig. 13.4b Configuration and paddler details

The G^2 value imparted by the paddles to the water through the flocculator at N RPM

$$= \frac{1.8(1)}{2 \times 0.8 \times 10^{-2} \times 133.33 \times 10^{6}} \left(\frac{(0.75)2\pi N}{60}\right)^{3} \left[\int_{0}^{75} x^{3}.10.\mathrm{d}x.8 + \int_{0}^{75} x^{3}.5\sqrt{2}.\mathrm{d}x.8\right]$$

$$= \frac{1.8 \times 10^{-7}}{1.6 \times 133.33} \left(\frac{1.5\pi}{60}\right)^{3} N^{3} \left(80 \times \frac{75^{4}}{4} + 40\sqrt{2} \times \frac{75^{4}}{4}\right)$$

$$= 0.000409 \times 10^{-7} N^{3} \left(0.108028 \times 10^{10}\right)$$

$$= 0.04418 N^{3} \left(\mathrm{s}^{-1}\right) \text{i.e.}\, G = 0.21\sqrt{N^{3}} \left(\mathrm{s}^{-1}\right);$$

Hence Gt value in the flocculator $= 0.21\sqrt{N^3}.240$, i.e. $50.4\,N^{1.5}$.
Equating the Gt value with that in the jar,

$$N = \left(\frac{173}{50.4}\right)^{1/1.5}, \text{ i.e. } 2.28\,\text{RPM}.$$

Use rotational speed in the flocculator = 3 RPM.
The clariflocculator has been drawn and presented in Fig. 13.5.

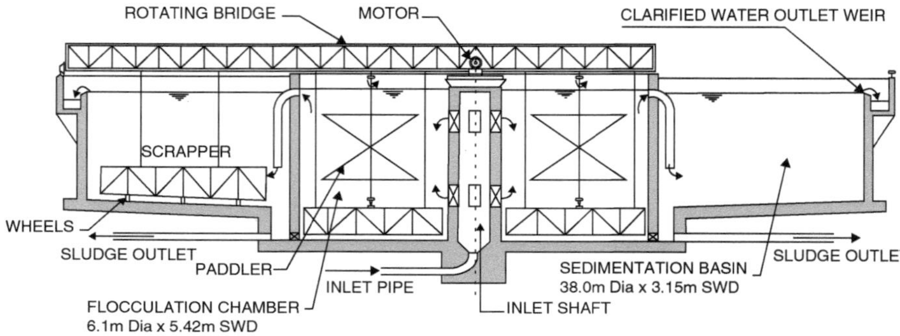

Fig. 13.5 Clari-flocculator

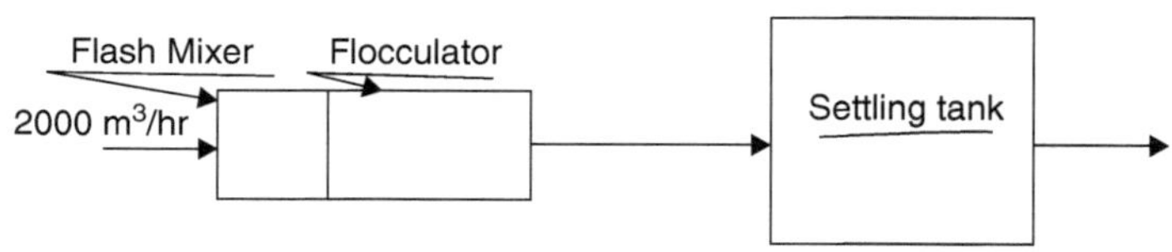

Fig. 13.6 Settling system

13.2.4 Compatible Design of Rectangular Tank

In case of rectangular tank, the length of weir at 16.75 m^3/m/h for the flow rate Q m^3/h is $L = Q/16.75$ m.

The retention time being T the volume of the tank $= QT$ m^3.

Choose a convenient depth D (>1.5 m).

The surface area of the tank $= QT/D$ m^2.

The weir length being L the other dimension is $= \frac{QT}{LD}$ m.

If this dimension is selected as length of small settling tank, to arrange the length L as weir length then, with length-width ratio—n—the width of the tank is QT/LDn metres.

The number of such tanks required $= \frac{L^2Dn}{QT}$.

Problem 13.2 An alternative design to Problem 13.1

Solution The schematic diagram of an alternative solution to Problem 13.1 may be visualised as (Fig. 13.6)

Design of Flash Mixer
Flow rate – 2000 m^3/h,

Assume flash mixing time = 1 min;

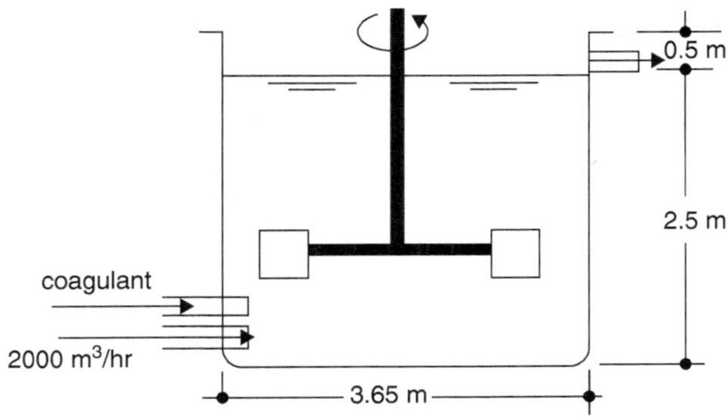

Fig. 13.7 Flash mixer

$$\text{The volume of the flash mixer} = 2000\frac{\text{m}^3}{\text{h}}\cdot\frac{1\,\text{min}}{60\,\text{min/h}}$$
$$= 33.33\,\text{m}^3.$$

Assuming flash mixer of 2.5 m depth of square surface area

$$= \sqrt{\frac{33.33}{2.5}}\,\text{m i.e.}\;\; 3.65\,\text{m} \times 3.65\,\text{m}.$$

Choose the stirrer shown in Fig. 13.3.

The flash mixer may be configured as shown in Fig. 13.7.

The mean temporal velocity gradient G in flash mixer at N RPM

$$= 0.2027646\sqrt{N^3}\left(\text{s}^{-1}\right);$$

Gt value in flash mixer

$$= 0.2027646\sqrt{N^3} \times 60$$
$$= 12.17\sqrt{N^3};$$

Equating this Gt value with the Gt value in the Jar,

$$N = \left(\frac{10929}{12.17}\right)^{1/1.5},\;\text{i.e.}\;\; 93\,\text{RPM}.$$

Fig. 13.8 Paddles

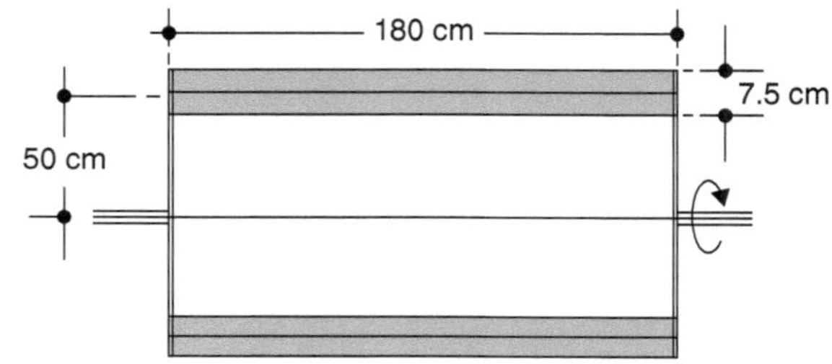

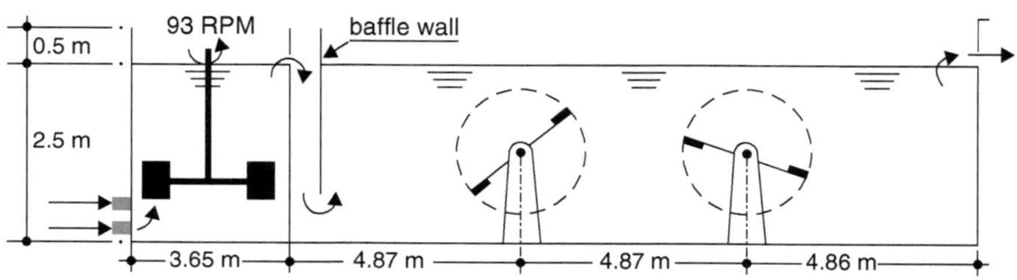

Fig. 13.9 Flash mixer and flocculator

Design of Flocculator

Assuming detention time in the flocculator 4 min, the volume of the flocculator $= 2000\,\frac{\text{m}^3}{\text{h}} \cdot \frac{4\,\text{min}}{60\,\text{min/h}}$, i.e. $133.33\,\text{m}^3$.

Assuming the depth of the flocculator is 2.5 m and its length-width ratio is 4, the flocculator is

$$4 \times \sqrt{\frac{133.33}{2.5 \times 4}}\text{m} \ \ \text{i.e.}\, 14.6\,\text{m} \times 3.65 \times 2.5\,\text{m};$$

Let us choose two paddles of the like as shown in Fig. 13.8.

The mean temporal velocity gradient G per sec. imparted to the water at N RPM can be calculated as (Fig. 13.9)

$$\begin{aligned} G^2 &= \frac{1.8(1) \times 180 \times 7.5 \times 4 \times \left(\frac{0.75 \times 2\pi \times 50N}{60}\right)^3 (\text{s}^{-1})}{2 \times 0.8 \times 10^{-2} \times 133.33 \times 10^6} \\ &= 15.329 \times 10^4 N^3 \left(\text{s}^{-1}\right)^2 \end{aligned}$$

$$\text{i.e.}\, G = 3.915 \times 10^{-2} N^{1.5}$$

$$Gt = 3.915 \times 10^{-2} N^{1.5} \times 240$$

$$= 173\,(Gt\,\text{value in the jar during slow mixing})$$

$$N = \left(\frac{173}{3.915 \times 10^{-2} \times 240}\right)^{1/1.5} \text{i.e.}\,7\,\text{RPM}$$

Design of Settling Tank

In accordance with the designed procedure of jar testing,

the detention time in settling tank is

$$\frac{150 \times 8 \times 60}{11.2}\,\text{s, i.e. } 6429\,\text{s.}$$

Hence, the volume of the settling tank $= \frac{2000}{60 \times 60} \times 6429$, i.e. $3572\,\text{m}^3$.

To limit the weir loading to 16.75 m^3/m/h, the length of the weir required

$$= \frac{2000}{16.75},\ \text{i.e.}\,119.4\,\text{m.}$$

Assuming the depth of the tank is 2.5 m, the surface area of the tank

$$= \frac{3572\,\text{m}^3}{2.5\,\text{m}},\ \text{i.e.}\,1428.8\,\text{m}^2.$$

The surface area has to provide 120 m weir length. Then the other side of the area is =1428.8/120 m, i.e. 11.9 m, and to provide 120 m weir length, this area has to be divided into smaller tanks of lengths 12 metres. Using length-width ratio 4, such tank has 3 m width, i.e. 40 nos of settling tanks of each 12 m $\times$ 3 m $\times$ 2.5 m are required.

Two batteries, each of 20 nos 12 m $\times$ 3 m $\times$ 2.5 m settling tanks in parallel, on either side of centrally running feeder pipe or channel distributing uniformly the water into them are to be used (Fig. 13.10).

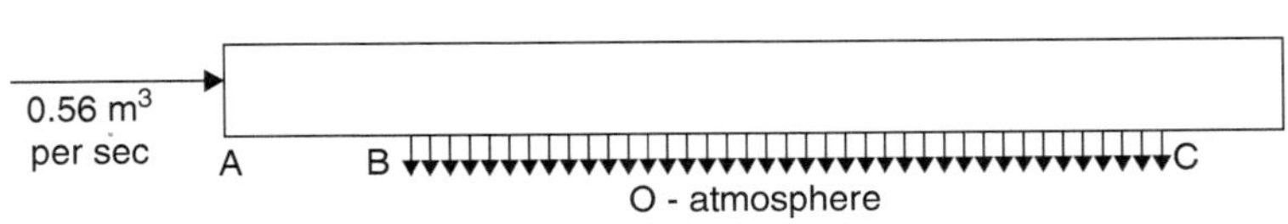

Fig. 13.10 Distributor pipe

Design of Distributor Pipe

$$\begin{aligned}\text{Flow rate} &= 2000\,\text{m}^3/\text{h}\\ &= 0.56\,\text{m}^3/\text{s i.e. } 19.76\,\text{ft}^3/\text{s}\end{aligned}$$

Flow enters into the distribution pipe AC through A and distributes itself through 100 orifices at 60 cm c/c from B to C of length (60 × 99)/100 = 59.40 m into the 40 settling tanks on either sides of distribution pipe.

All orifices are small, circular and identical, the first one being at B and the 100th one at C (Fig. 13.10). Head loss: The frictional head loss through the travel length may be computed with Hazen William's equation.

Hazen William's equation was derived originally for turbulent flow in pipes and open channels, but now it is mostly used for pipe flow.

This is generally written as Eq. 13.4

$$V = 1.318\,\text{C}\,\text{R}^{0.63}\left(\frac{h}{l}\right)^{0.54} \tag{13.3}$$

This may be written as (putting hydraulic radius $R = \frac{D\,(\text{Pipe diameter})}{4}$)

$$h = 3.02\frac{L}{D^{1.17}}\left(\frac{V}{C}\right)^{1.85} \tag{13.4}$$

The exponent 1.85 is often approximated as 2.00.

The loss of head h' across the orifice for the discharge through it is

$$Q = C_d\,\text{a}\sqrt{2gh'}\,\text{i.e.}\,h' = KQ^2 \tag{13.5}$$

where

V = mean velocity in ft/s
C = Hazen William's coefficient
 = 130 for new CI pipe and
 = 120 for concrete surface
h, h' = Head loss in ft
C_d = Coefficient of discharge 0.62–0.65

Considering the travel of water from the entry point to its release into the atmosphere –

Head loss from A to B (h_{AB}) + loss of head across the orifice at B into the atmosphere 'O' (h'_{BO})

= head loss from A to C (h_{AC}) + loss of head across the orifice at C into the atmosphere (h'_{CO})

$$\text{i.e.}\, h_{AB} + h'_{BO} = h_{AC} + h'_{CO} \quad (13.6)$$

$$\text{i.e.}\, h'_{BO} - h'_{CO} = h_{AC} - h_{AB} \text{ i.e. } h_{BC} = \text{Loss of head between points } B \text{ and } C\ (h_{BC}) \quad (13.7)$$

From Eq. 13.5 $h'_{BO} - h'_{CO}$

$= KQ_1^2 - KQ_2^2$, where Q_1 and Q_2 are discharges through 1st orifice at B and last orifice at C respectively.

$= KQ_1^2(1 -- 0.99^2)$, if $Q_2 = 0.99\ Q_1$ for the loss of frictional head from B to C

$= h'_{BO}(1 -- 0.99^2) = h_{BC}$ i.e. $h'_{BO} = 50.3\ h_{BC}$ (13.8)

Flow enters at B at 0.56 m³/s, i.e. 19.76 ft³/s and water released through each orifice at 0.0056 m³/s, i.e. 0.1976 ft³/s.

As such, assuming the draw-off water at $\frac{19.76}{59.4 \times 3.28}$, i.e. 0.1 ft³/ft/s.

The flow rate at a distance $x = (19.76 - 0.1\ x)$ ft³/s.

Diameter of the Distributor

Assuming the mean velocity of flow through, a CI pipe distributor at 0.9 m/s diameter of the pipe may be chosen:

$$= \sqrt{\frac{0.56 \times 4}{0.9\pi}} \text{ i.e. } 0.89\text{m};$$

Use 900 mm dia, i.e. 2.95 ft dia CI pipe.

Head Loss

Cross-sectional area $= \frac{\pi 2.95^2}{4}$, i.e. 6.835 ft²,

$$\text{Mean flow velocity at a distance x ft from the first orifice} = \frac{(19.76 - 0.1x)}{6.835}\ \text{ft/s}$$

$$= (2.89 - 0.015x)\text{ft/s}.$$

The loss of head between B and C

$$= \int_0^{195} (3.02)\left(\frac{1}{2.95^{1.17}}\right)\left(\frac{2.89 - 0.015x}{130}\right)^2 dx$$

$$= 0.027\,\text{ft}$$

Head Available at *B* and *C*

Head available at B:

$h'_{BO} = 50.3 \times 0.027$ ft, i.e. 1.36 ft, (from Eq. 13.6)

Head available at C:

$$h'_{CO} = h'_{BO} - h_{BC}$$
$$= 1.36 - 0.027, \text{ i.e.} 1.33\,\text{ft}$$

Diameter of the Orifice
Discharge through circular orifice of diameter d,

$$Q = C_d \frac{\pi d^2}{4} \sqrt{2gh} \text{i.e.} d = \sqrt{\frac{4Q}{C_d \pi \sqrt{2gh}}}$$

$$= \sqrt{\frac{4 \times 0.0056 \times 10^6}{0.62\, \pi \sqrt{\frac{2 \times 981 \times 1.36 \times 100}{3.28}}}}$$

$$= 6.3\,\text{cm, i.e. } 63\,\text{mm}$$

Alterative design of a concrete feeder channel may be tried in similar way.

Automatic Draining of Sludge
Sludge density being 1.002 (Fair), the sludge will drain under gravity along the 2° sloping floor.

Sedimentation Tank Configured
40 nos of 12 m × 3 m × 2.5 m may form two batteries of tanks in parallel. Each of the batteries consists of twenty such tanks.

If the separating wall between the batteries are eliminated it will reduce two 12 m × 3 m × 2.5 m settling tanks into a single settling tank 24 m × 3 m × 2.5 m with two 3 m weirs along its widths on either sides.

Water from flocculator enters into 90 cm dia, 60 m long feeder pipe running centrally over the trestle between two parallel baffle walls shown in Fig. 13.11. One hundred orifices 63 mm dia at 60 cm c/c over 59.4 m length of the feeder pipe leaving 0.3 m length on either of its ends supply water to the tanks.

13.3 Design of Secondary Clarifier

The design of primary settling tank has been shown in this chapter to be based on the designed jar testing procedure for the compatible operation of the tank to follow the settling that took place in the jar during testing.

Similarly the design of the secondary clarifier will be based on the results of settling column analysis of the waste water to be treated for the operation of the tank to follow the settling that took place in the column during column data collection.

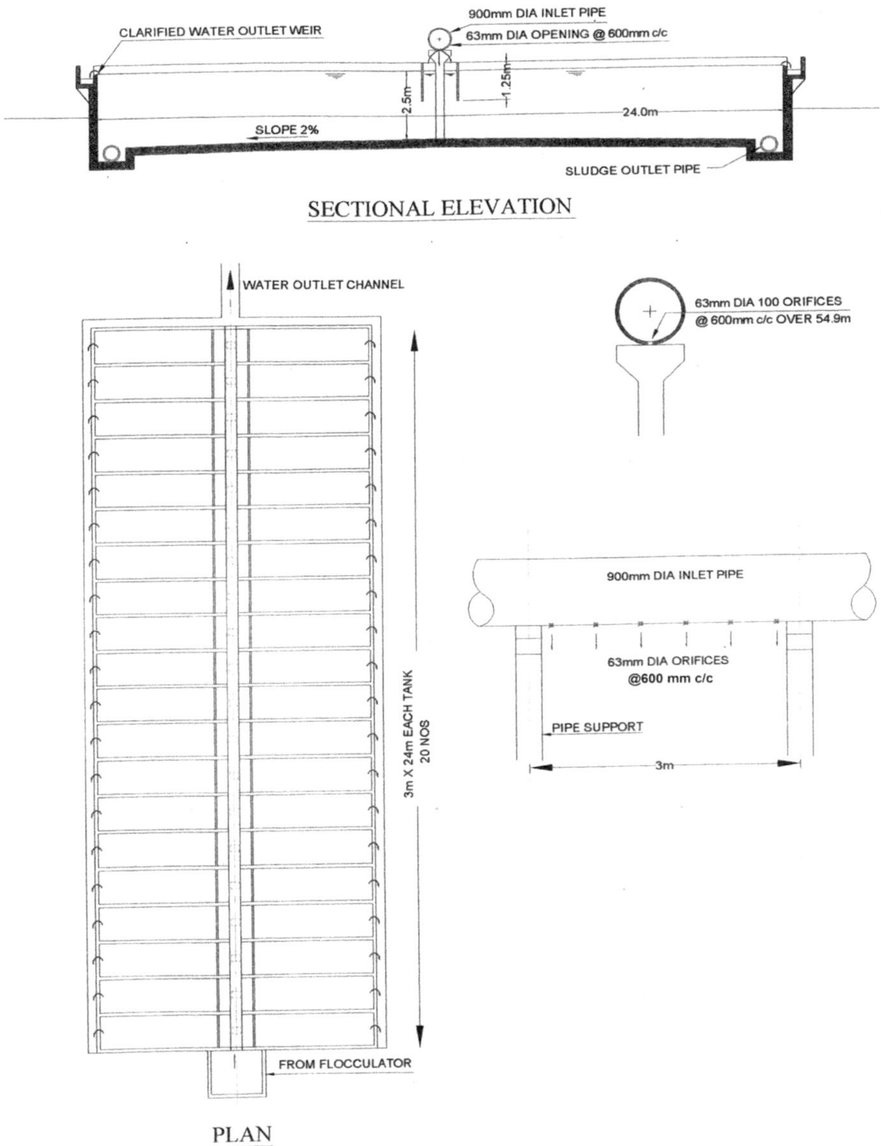

Fig. 13.11 Sedimentation tank

13.3.1 Basis of Design

The following are to be considered for design:

(i) Column settling data should direct the design of the settling tank the objective being the simulation of the column settling in the tank.
(ii) Due to paucity of data, the observation of Table 12.7 should form the basis of design. The solids concentration in the effluent channel of the settling tank will be assumed to be the depth-wise average solids concentration over 1.5 m of the adjacent depth if weir loading is limited to 16.75 m^3/m/h.
(iii) The limitation of weir loading should take care both of weir flow velocity and overflow velocity.

13.3.2 Procedure for the Design of Secondary Clarifier

The design of secondary clarifier may be accomplished following the stepwise procedure as given below:

1. Column settling data should be collected for sufficient number of observations. One of the observations should be selected for the design to follow.
2. From the column settling data, the trajectories of the interface concentrations are to be plotted in depth-time coordinates in accordance with the 'Revised Mode of Analysis of Column Settling Data' (Chap. 8).
3. The length of weir corresponding to the limiting weir loading is found out.
4. From the length of the weir, the diameter D of the circular tank can be calculated.
5. A line at depth 1.5 m and parallel to the time axis is drawn in the diagram for the trajectories of interface concentrations in depth-time coordinates. Verticals are drawn at different times, and residual solid concentrations in the settling column at different times are calculated from the verticals.
6. Retention time T is chosen corresponding to the desired effluent concentration in effluent channel.
7. QT volume of the tank is calculated.
8. From the volume QT and diameter of the circular tank, the depth $d = \frac{4QT}{\pi D^2}$ of the tank is found out. If such computation leads to unsatisfactory and unusual value, a suitable depth d (>1.5 m) may be assumed. Surface area is computed. From the surface area, diameter may be calculated. This reduces weir loading and effluent concentration of solids.
9. In case of rectangular tank, the surface area is calculated from a suitably chosen value of D (>1.5 m) and QT. The surface area is suitably divided into areas to provide the necessary weir length.

Table 13.1 Solids concentrations in mg/l at indicated depths and times
Initial concentration of solids = 540 mg/l
Temperature of the water = 30 °c

Time in min	Depth		
	60 cm	120 cm	180 cm
5	275	362	386
10	189	259	312
20	135	188	232
40	90	119	162
60	92	92	118
120	95	92	93

Table 13.2 Settling column test data with regard to the settleable solids
Initial concentration of settleable solids = 448 mg/l
Temperature of water = 30 °c

Time in min	Depth		
	60 cm	120 cm	180 cm
5	183	270	294
10	97	167	220
20	43	96	140
40	0	27	70
60	0	0	26
120	0	0	0

Problem 13.3 Design a secondary clarifier for an activated sludge system for a design flow of 10,000 m^3/d.

Laboratory settling data of the concentration of suspended solids remaining at indicated depths at varying times were as follows (Table 13.1):

Solution

Step 1: From the observed data, it is apparent that the waste water contains non-settleable solids of ((90 + 92 + 93 + 92 + 92 + 93)/6) = 92 mg/l. The observed data with the settleables may be retabulated.

Step 2: From the data presented in Table 13.2, concentration versus time curves at different depth are prepared and presented in Fig. 13.12.

Step 3: Trajectories of different interface concentrations in depth-time coordinates are drawn from the curves in the Fig. 13.13.

Step 4: At depth of 1.5 m is drawn a line parallel to the time axis in Figs. 13.12 and 13.13. Verticals are drawn up to 1.5 m in Fig. 13.12 at 25 min, 30 min, 35 min and 45 min. Concentrations of solids over the lengths of verticals may be computed as presented in Table 13.3.

Waste water flow = 10,000 m^3/d

$$= 416.7\ \text{m}^3/\text{h}$$

Weir length required at 16.75 m^3/m/h = $\frac{416.7}{16.75}$ i.e. 24.88 m.

Diameter of circular tank = $\frac{24.88}{\pi}$ i.e. 7.9 m, i.e. 8.0 m.

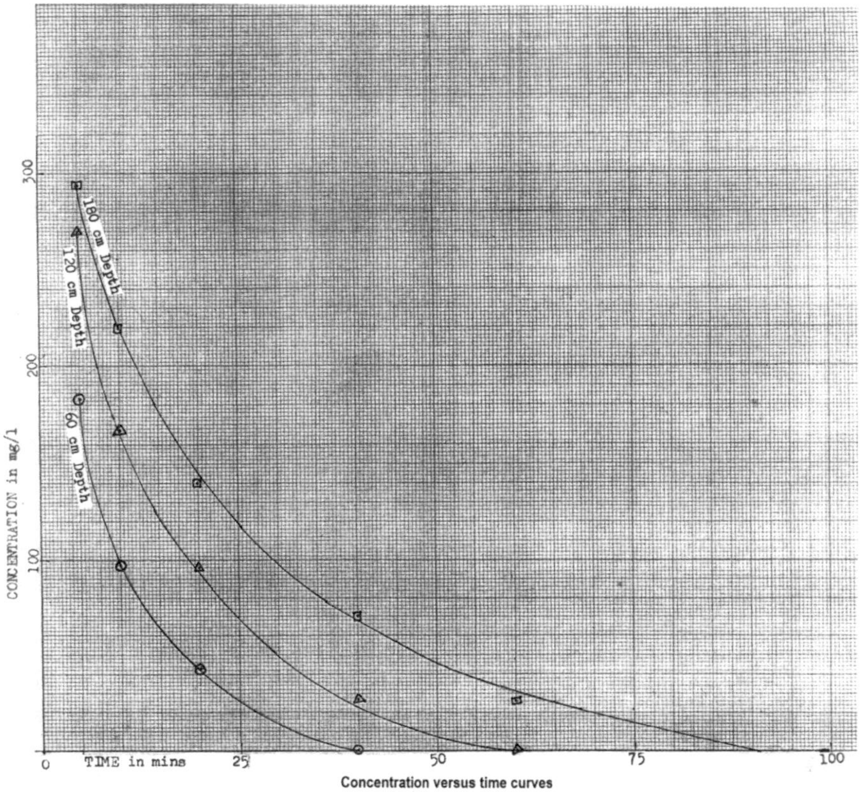

Fig. 13.12 Concentration versus time curves

Volume of the settling tank based on the residual settleable solids concentration of 7.03 mg/l at retention time of 45 min

$$= 416.7 \times \frac{45}{60} \text{m}^3 \text{ i.e.} 312.5 \text{m}^3$$

Depth of the tank $= \frac{312.5 \times 4}{\pi \times 8^2}$, i.e.6.2m.

This depth being on higher side is unsatisfactory.

Assume a reasonable depth greater than 1.5 m = 2.5 m.

Surface area $= \frac{312.5}{2.5}$, i.e. 125.0 m^2.

Diameter of the circular tank $= \sqrt{\frac{125.0 \times 4}{\pi}}$, i.e. 12.6 m.

This implies that the weir loading is reduced. The settleable suspended solids concentration in effluent channel is likely to be less than 7.03 mg/l (i.e. total solids 99.03 mg/l).

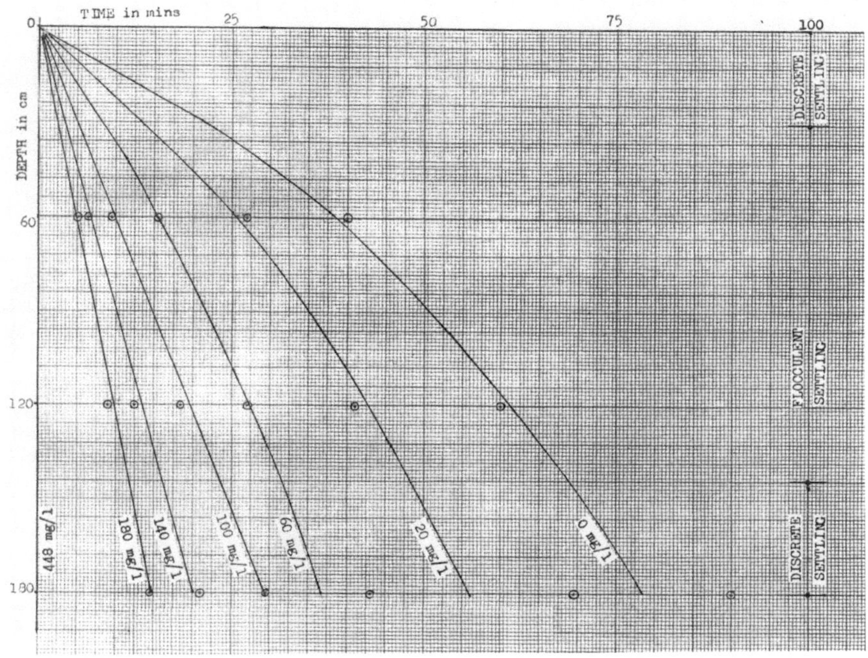

Fig. 13.13 Trajectories of interface concentrations

Table 13.3 Mean residual concentrations of settleable solids over 1.5 m depth at different times

Time in min	Concentration of solids, mg/l
25	37.04
30	18.69
35	17.28
40	11.68
45	7.03

Notations

t	Flash mixing time
FMS	Flash mixing speed
T	Slow mixing time
SMS	Slow mixing speed
d	Depth of water in the Jar
T	Retention time in the tank
Q	Flow rate
G	Mean temporal velocity gradient
D	Depth of the tank
d	Diameter of the tank
L	Length of the weir

References

Acharya TK (1990) Dependence of solids removal through a settling tank on overflow velocity and weir flow velocity. MCE thesis. Jadavpur University, Kolkata, India

Fair GM, Geyer JC, Okun DA. Water and waste water engineering, vol 2. Wiley, New York/London/Sydney

Rich LG. Unit operations of sanitary engineeing. Wiley, New York/London

Chapter 14
Shallow Depth Settling

Abstract With an introductory presentation of the salient literature review on 'shallow-depth sedimentation', this chapter presents complete theory of 'high-rate settling system' without the application of 'Velocity Profile Theorem'.

Keywords High-rate settling • Tube settling theory • Tube settling trajectories • Critical fall velocity • Removal through tube

14.1 Introduction and Literature Review

As early as in 1904, Allen Hazen (1904) spoke in favour of shallow-depth sedimentation principle. Hazen concluded that removal of suspended matters by settling depends upon the floor area of the tank and not upon the tank volume. He considered discrete suspensions and proposed the depth of the tank as little as 1 inch. He realised that the insertion of one horizontal tray would increase the capacity of the basin. He felt that the use of multiple trays spaced at 1 inch interval would be desirable if the problem of sludge removal could be resolved.

One of the first attempts in the application of tray settling principle was patented in 1915 (Barham et al. 1956). Several patents followed it in subsequent years.

Camp (1946) suggested that to resolve the problem of mechanical sludge removal, a minimum of 6 inches vertical spacing interval of horizontal trays would be necessary. He illustrated the design of a settling tank with horizontal trays. Eliassen (Discussion on (Camp 1946)) noted that the tray settling principles had already been used for many years in the chemical and metallurgical industries but only in a few water or sewage treatment systems. Camp (discussion 1946) ascribed this fact to the reluctance of the design engineers to deviate from the conventional basins as regards its shape, size, etc.

The use of trays in the basins of conventional design met with limited success (Hansen et al. 1968) mainly because:

1. Hydraulic conditions were unstable.
2. The minimum tray spacing was limited by the problem of mechanical sludge removal. The theoretic advantage of tray settling principle aroused commercial

A. De, *Sedimentation Process and Design of Settling Systems*,
Springer Transactions in Civil and Environmental Engineering,
DOI 10.1007/978-81-322-3634-4_14

interest which is reflected by the marketing of multistoried tray settling tanks by at least two companies in mid 1940s (Sewage Manual 1946–1947).

In the words of Hansen et al. (1968), 'the status of shallow depth sedimentation in the mid 1940s might be summarised as a process with recognised theoretic advantages, but one whose practical application had been limited by problems associated with distribution of flow to multiple tray units and sludge removal from closely spaced trays'.

In 1941 Frei employed trays in an existing clarifier. The introduction of trays increased the removal efficiency of the basin. Schmitt and Voigt reported the use of a two storied settling basin in a water treatment plant in 1949. Thus from 1940 to 1950, isolated instances of the tray settling principles continued to be reported.

(In 1953 Camp (Camp, T.R. (1953) – Sedimentation Basin Design, Sewage and Industrial Wastes, 25, 1) again spoke of the tremendous advantages of using the tray settling principle with very small vertical clearance for resolving sludge removal problem.

In 1955 Fischerstrom (1955) pointed out that unstable hydraulic condition occurred in applying the tray settling principle because it neglected the importance of maintaining proper hydraulic condition. He felt that Reynolds' number less than 500 (limit of laminar flow at 32 °F) should be maintained in a basin for that purpose.

The introduction of longitudinal baffles horizontal or vertical would increase the wetted perimeter for a given basin, and hence they will reduce the Reynolds' number. Besides reducing the Reynolds' number, the horizontal trays reduce the overflow rate and hence result in the increased removal. Fischerstrom observed that trays spaced adequately (5–6 ft (1.5–1.8 m)) for manual sludge removal gave excellent performance in several installations and he felt that smaller spacing could be used to derive greater benefits. He suggested the use of both horizontal and vertical baffles placed longitudinally to design an efficient solid removal system. Cost analysis revealed that tray settling basins were much less expensive than the conventional ones (Hansen et.al. 1967 (Hansen and Culp 1967)).

In 1967 Hansen (Hansen and Culp 1967) made an excellent literature review on the subject. They showed that the use of small diameter (1–4 inches) tubes 2–4 ft in length could resolve the problem of (Barham et al. 1956) unstable hydraulic condition and (Camp 1946) sludge removal. The following table shows the Reynolds' numbers of tube settlers at different flow rates (Table 14.1).

The following table (Table 14.2) with Table 14.1 shows that the use of small diameter tubes having small lengths can maintain laminar flow condition at a reasonable overflow rate and, thus, can solve the problem of unstable hydraulic conditions. The short detention times can reduce the size of the unit. The cost-saving potentiality of the tube settling becomes apparent.

In the above tables, hydraulic flow rates have been computed per sq.ft of the end area. Surface overflow rates have been calculated at the mid depth of the tube.

Table 14.1 Reynolds' numbers of tube settlers at different flow rates

Hydraulic flow rate gpm/sq.ft	Tube diameter in ins.	Reynolds' number
1	0.5	1
1	1.0	2
1	2.0	5
1	4.0	10
5	0.5	6
5	1.0	12
5	2.0	24
5	4.0	48
10	0.5	12
10	1.0	24
10	2.0	48
10	4.0	96

Table 14.2 Detention times and overflow rates associated with tube settlers

Hydraulic flow rate gpm/sq.ft	Tube length in ft.	Equivalent surface overflow rate gpd/ft^2	Detention time in min
1 in. diameter			
1	2	47	15
1	4	24	13
1	8	12	60
5	2	236	3
5	4	118	6
5	8	59	12
10	2	472	1.5
10	4	236	3
10	8	118	6
2 in. diameter			
1	2	95	15
1	4	48	30
1	8	24	60
5	2	470	3
5	4	235	6
5	8	118	12
10	2	950	1.5
10	4	475	3
10	8	238	6

From the preliminary experiments with single tubes, Hansen et.al (1967) (Hansen and Culp 1967) found that the accumulated sludge could be readily removed by draining the tubes periodically when the tubes were inclined slightly in the direction of flow. An inclination of 5 ° was found suitable for gravity draining of sludge.

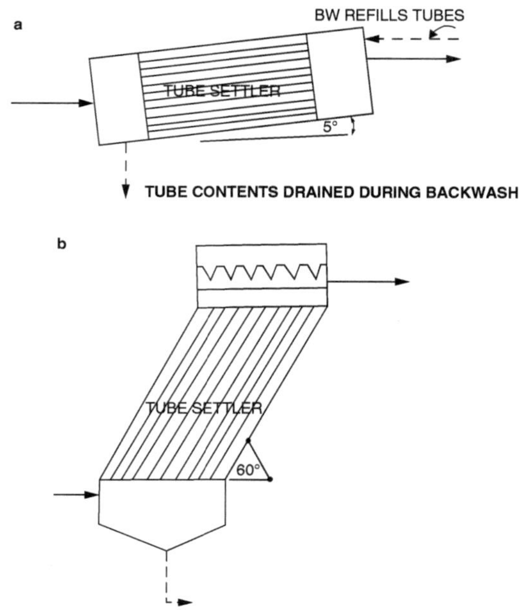

Fig. 14.1 Two basic configuration of tubes (**a**) Essentially horizontal tubes (**b**) Steeply inclined tubes

From the detailed study, they found out that a tube settler and mixed media filter combination could treat the several types of raw water successfully. The effluent water quality through the tube settler was compatible with the filtration capabilities of the mixed media filter at filter rates in excess of 5 gpm/ft^2.

The removal of sludge by gravity drainage eliminated the need of mechanical sludge removal equipment. The tube cleaning cycle could be integrated into the backwash cycle of the filter so that no water was lost.

In their subsequent paper, Hansen et.al (1968) discussed their operating experiences with the tube settling system. Two basic configurations, namely, (a) essentially horizontal (Fig. 14.1a) and (b) steeply inclined (Fig. 14.1b), were considered.

With the essentially horizontal tube, multimedia filter was used in combination. During the filter backwash, the falling water scoured the accumulated sludge in the tubes and they were drained completely. The tubes were inclined with the horizontal at small angle of 5° only to promote the draining of sludges.

Continuous gravity draining of sludges resulted when the tube inclination increased sharply to 45–60 °. The incoming solids settling to the bottom were arrested in the continuous flowing sludge stream sliding downward along the bottom of the tube. Seventeen installations of water treatment plant were listed in which horizontal tubes with multimedia filter were employed. Their capacities ranged from 20 to 2000 gpm with detention times less than 10 min. In a test reported in this paper, a plant produced potable water of 0.1 JU turbidity from the raw water turbidity of 1000 JU using overall detention time of 16 min. This plant provided flocculation, tube settling and mixed media filtration.

The sludge deposits within the tubes resulted in the better distribution of flow than that in the case of tray settling system. This is because if any tube received more flow, the rapid build-up of sludge deposits in that tube caused some flow to divert into the other tube.

Both the laboratory and field tests indicated that 60° inclined tubes provided continuous sludge removal and still performed as efficient settling device. This resulted in the development of modular tube units. These modules could be installed in a clarifier that is existing, increasing its capacity from 1.5 mgd to 3.0 mgd. The coupling of the tube settlers with mixed media filter can increase the capacity of an existing plant and reduces the size and cost of new treatment facilities.

Hansen et al. (1969) presented research data and experiences in plant-scale application of tube settling principles. It has been reported that by converting the secondary clarifier to the aerated biological reactor, installing steeply inclined tubes in the modified clarifier to provide solid separation and subjecting the effluent to filtration, the efficiency of a trickling filter plant could be increased from 85 % to more than 95 % BOD and suspended solid removal.

The steeply inclined tubes could be installed as an integral part of an aeration basin to eliminate separate sludge separation and return system. Additional operating experiences of plant-scale application of the tube settling system have been reported by Hansen et al. (1969) in a different paper.

In 1970, Yao published his paper on 'Theoretical Study of High-Rate Sedimentation'. Yao used the term 'high-rate sedimentation' to refer to the use of shallow-depth gravitational settlers with detention time not more than 15 min.

These settlers achieve comparable or better settling experiences normally obtained in conventional rectangular settling tanks having detention periods of usually not more than 2 h. Yao pointed out that there is no information whether the parameter overflow rate has the same significance in the case of settlers other than those rectangular in shape and also that nothing is known as to how to calculate the overflow rate for inclined tube settlers.

Yao (1970) considered an inclined tube settler shown in Fig. 14.2. The X-axis is parallel to the direction of flow and Y-axis is normal to the direction of flow. θ is the angle of inclination of the tube with the horizontal.

If u is the local fluid velocity at a point where a discrete particle of settling velocity v_s enters, the velocity components of the particle in X and Y directions

$$v_{px} = \frac{dx}{dt} \text{ and } v_{py} = \frac{dy}{dt}$$

can be written as

$$v_{px} = \frac{dx}{dt}$$

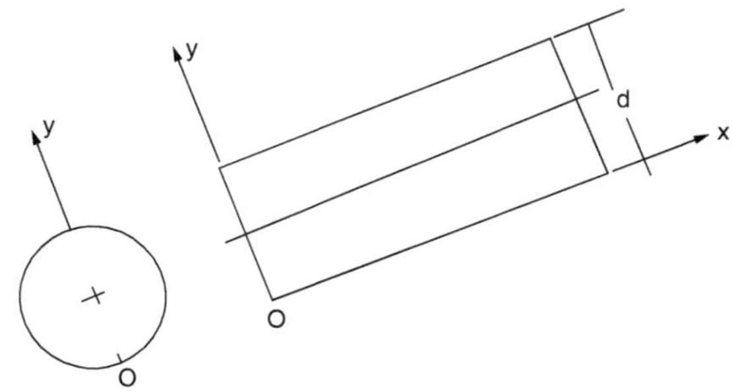

Fig. 14.2 Inclined tube settler

$$= u - v_s \sin\theta \tag{14.1}$$

$$v_{py} = \frac{\mathrm{d}y}{\mathrm{d}t}$$

$$= -v_s \cos\theta \tag{14.2}$$

From the above equations,

$$\frac{\mathrm{d}y}{\mathrm{d}x} = -\frac{-v_s \cos\theta}{\mathrm{u} - v_s \sin\theta} \tag{14.3}$$

$$\text{i.e.}\, u\,\mathrm{d}y - v_s \sin\theta\,\mathrm{d}y + v_s \cos\theta\,\mathrm{d}x = 0 \tag{14.4}$$

Integrating the above equations,

$$\int u\,\mathrm{d}y - v_s y \sin\theta + v_s x \cos\theta = C_0 \tag{14.5}$$

where C_0 is the constant of integration.

Dividing Eq. (14.5) by v_0 the average velocity and d the depth of flow,

$$\int \frac{u}{v_0}\mathrm{d}Y - \frac{v_s}{v_0} Y \sin\theta + \frac{v_s}{v_0} X \cos\theta = C_1 \tag{14.6}$$

where $X = \frac{x}{d}$, $Y = \frac{y}{d}$ and C_1 is the adjusted integration constant.

For circular tube settlers, Yao computed

$$\frac{u}{v_0} = 8\left(Y - Y^2\right) \tag{14.7}$$

Equation (14.6) becomes

$$8\left(\frac{Y^2}{2} - \frac{Y^3}{3}\right) - \frac{v_s}{v_0} Y \sin\theta + \frac{v_s}{v_0} X \cos\theta = C_1 \tag{14.8}$$

Yao claimed that Eq. (14.8) is the general equation for the trajectories of suspended particles in laminar flow through a circular tube.

It is easy to see that Eq. (14.7) is valid only on a diameter parallel to the Y-axis. Equation (14.8) can, therefore, describe the trajectories of particles entering through any point on that diameter. It cannot describe the trajectories of particles entering through any other point not lying on that diameter. Equation (14.8) is, therefore, not a general equation describing the trajectories of particles entering into the tube. Yao's claim is not tenable.

If $X=L$ and $Y=0$, L=relative length of the tube $= \frac{l}{d}$

From Eq. (14.8) one obtains

$$C_1 = \frac{v_s}{v_0} L \cos\theta \tag{14.9}$$

and Eq. (14.8) becomes

$$8\left(\frac{Y^2}{2} - \frac{Y^3}{3}\right) - \frac{v_s}{v_0} Y \sin\theta + \frac{v_s}{v_0}(X - L)\cos\theta = 0 \tag{14.10}$$

Equation (14.10) describes the trajectories of particles entering through any point on the diameter parallel to the Y-axis and reaching the bottom at the end of the tube. For the critical trajectory of a particle entering at the top of the diameter and reaching the bottom at the end of the tube, we put

$X = 0$ and $Y = 1$, and we have from Eq. (14.10)

$$\frac{v_{cr}}{v_0}(\sin\theta + L\cos\theta) = \frac{4}{3} \tag{14.11}$$

where v_{cr} is the critical fall velocity of the particle.

Any particle, $S = \frac{v_s}{v_0}(\sin\theta + L\cos\theta)$ value of which is equal to or greater than the critical value $S_c = \frac{4}{3}$ of the system, will be completely removed. Particles with S-value less than critical S-value of the system will be removed fractionally. Yao considered all particles having the same fall velocity and deduced the fractional removal for systems with horizontal plates and circular tubes in the following way.

The particle trajectory J starts at E_0 at the entrance side and ends at E_2, the bottom point at the exit end (Fig. 14.3). q_1 is the portion of total flow q entering the settler below E_0 and q_2 is the remaining portion entering above E_0. Suspended particles in q_1 will be removed completely in the settler since their trajectories must end up between E_1 and E_2. On the other hand, suspended particles in q_2 will remain in the flow. The fractional removal efficiency is, therefore,

$$= \frac{\int_0^y u\,dy}{v_0 d} \tag{14.12}$$

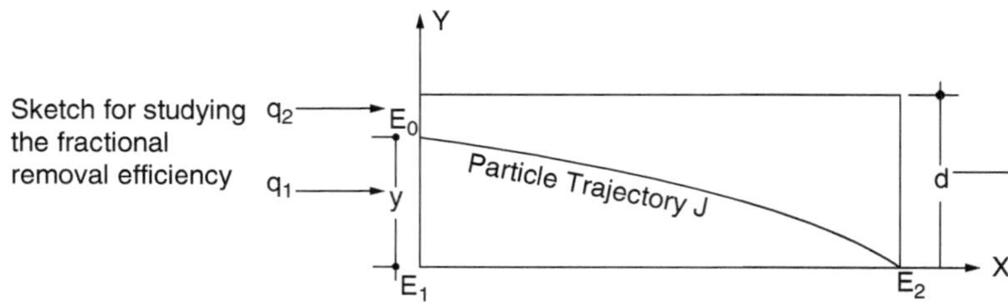

Fig. 14.3 Sketch for studying the fractional removal efficiency

This expression, though true for a parallel plate system, is not true for a circular tube system. This is so because in case of circular tubes, all trajectories of all particles having the same settling velocity v_s which enters into the tube and reaches the bottom of the tube at the tube's end are not identical as assumed in writing Eq. (14.12).

In 1971 Slechta and Conley (1971) described the experiences in plant-scale application of the settling tube concept in primary clarification and secondary clarification of activated sludge and trickling filter solids. They concluded that the tube settler in clarification of activated sludge should be considered as a device for protecting the clarifier against severe loss of solids because of upsets in the biological process for peak flow conditions. Settling tube can improve the settling efficiency of the existing clarifier.

From the review, it appears that tube settling system is a highly efficient solid-liquid separating system. Settling in such system should be theorised in a rationalised way. Such an attempt was made by Yao (1970). There are some drawbacks in his theorization. These have been duly pointed out in the earlier discussion.

De (1976) derived a general equation for the trajectory of a particle that is settling while it is passing through an inclined tube. This generalised equation could show the different parameters affecting settling efficiency in a high-rate settling system. It was also shown how to employ the generalised equation to calculate the percentage removal for a given flow rate through a tube settler if the velocity distribution among the particles in the influent suspension is given.

14.2 Derivation of General Equation and Computation of Removal

14.2.1 Derivation of General Equation

Let us consider a tube of length l and diameter d inclined at an angle θ with the horizontal. We imagine a coordinate system X,Y,α as shown in Fig. 14.4. If a particle with settling velocity v_s enters through a point $(0,y,\alpha)$ into the tube, it

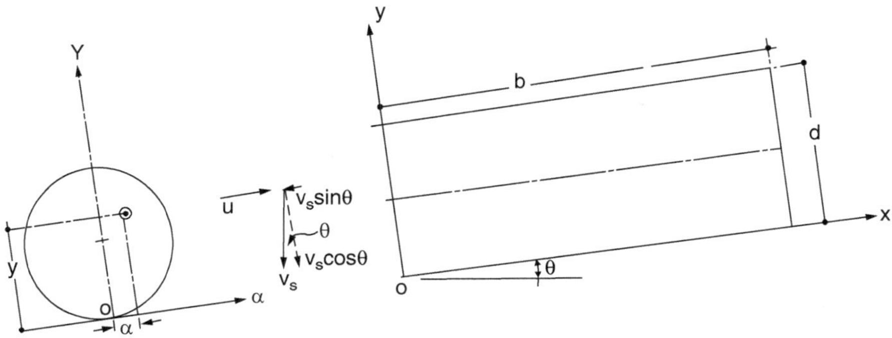

Fig. 14.4 Sketch showing a particle entering through a point $(0, y, \alpha)$ into an inclined tube

will start moving with velocity $(u - v_s \sin\theta)$ in the direction X and $v_s \cos\theta$ in the negative direction of Y.

$$\begin{aligned} v_{px} &= \frac{\mathrm{d}x}{\mathrm{d}t} \\ &= u - v_s \sin\theta \\ v_{py} &= \frac{\mathrm{d}y}{\mathrm{d}t} \\ &= -v_s \cos\theta \end{aligned}$$

where u is the local fluid velocity at the point (0,y,α). u may be written from any standard book on hydraulic s under laminar flow condition:

$$\begin{aligned} u &= \frac{2Q}{\pi R^4}\left(2yR - y^2 - \alpha^2\right) \\ &= \frac{2v_0}{R^2}\left(2yR - y^2 - \alpha^2\right) \end{aligned} \tag{14.13}$$

where Q is the rate of flow through the tube,

R is the radius of the tube $= d/2$,
v_0 is the mean flow through velocity $= \frac{Q}{\pi R^2}$.

If the particle moves through a distance dx in time $\frac{\mathrm{d}y}{v_s \cos\theta}$ during which it falls through from y to y-dy, one can write

$$\mathrm{d}x = (u - v_s \sin\theta)\frac{(-)dy}{v_s \cos\theta}, \quad -ve\,\text{sign stands for decrease in}\, y$$

$$= \left[\frac{2v_0}{R^2}\left(2yR - y^2 - \alpha^2\right) - v_s \sin\theta\right]\frac{(-)dy}{v_s \cos\theta}$$

$$= \frac{2v_0}{R^2 v_s \cos\theta}\left(y^2 - 2yR + \alpha^2\right)dy + \tan\theta\,\mathrm{d}y \tag{14.14}$$

The general equation describing the trajectory of a particle entering at (0,y_1,α) moving to the point (x,y,α) will be obtained by integrating Eq. (14.14) from $x = 0$ to $x = x$ correspondingly from $y = y_1$ to $y = y$, and we have the equation:

$$x = \frac{2v_0}{3R^2 v_s \cos\theta}\left[\left(y^3 - y_1^3\right) - 3R\left(y^2 - y_1^2\right) + 3\alpha^2(\mathrm{y} - y_1)\right] + (\mathrm{y} - y_1)\tan\theta \tag{14.15}$$

The above equation can be rewritten as

$$\frac{2v_0}{3R^2}\left[\left(y_1^3 - y^3\right) - 3R\left(y_1^2 - y^2\right) + 3\alpha^2(y_1 - \mathrm{y})\right] + v_s(y_1 - \mathrm{y})\sin\theta + v_s\mathrm{x}\,\cos\theta = 0 \tag{14.16}$$

14.2.2 Problem Computation of Removal

From the forgoing presentation, the removal may be computed in the following way:

(a) Critical fall velocity:
With critical fall velocity, i.e. $v_s = v_{cr}$, a particle will enter at a point (0,2R,0) and will move to the bottom of the tube at the end of its length, i.e. to the point (l,0,0). The critical fall velocity can be calculated from Eq. (14.16) putting $y_1 = 2R$, $y = 0$ and $x = l$, $v_s = v_{cr,}$ $\alpha = 0$ and can be written as

$$v_{cr} = \frac{8Q}{3\pi R(2R\,\sin\theta + l\,\cos\theta)} \tag{14.17}$$

For flow rate Q through a tube of radius R and length l, all particles having settling velocity equal to and greater than the critical fall velocity given by Eq. (14.17) will be removed completely, and particles having settling velocity less than this will be removed fractionally.

(b) Fractional removal:
If a particle having settling velocity v_s less than the critical fall velocity enters through any point $(0, y_1, 0)$ on the diameter parallel to the y-axis and reaches the

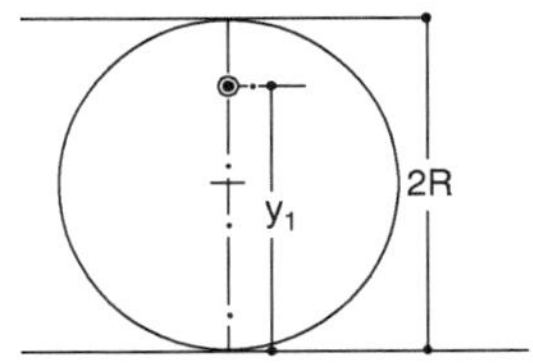

Fig. 14.5 Sketch showing a particle entering through any point (0, y, 0) reaches the bottom at the end of the tube

bottom at the end of the tube, i.e. the point $(l, 0, 0)$, then y_1 will be given by the following equations obtained by putting $x = l$, $y = 0, \alpha = 0$ in Eq. (14.16):

$$y_1^3 - 3Ry_1^2 + \frac{3R^2}{2v_0}.v_s \sin\theta.y_1 + \frac{3R^2}{2v_0}.v_s l \cos\theta = 0 \tag{14.18}$$

All particles having the settling velocity v_s that enter through points lying on the vertical diameter from $y = 0$ to $y = y_1$ shown by Eq. (14.18) will be removed, and those particles entering through points lying on the diameter from y_1 to $y = 2R$ will not be removed. They will be carried with the effluent (Fig. 14.5).

A general equation showing the different y_1 values such that a particle entering through a point $(0, y_1, \alpha)$ will reach the bottom at the end of the tube, i.e. to the point $\left(l, R - \sqrt{R^2 - \alpha^2}, \alpha\right)$, may be derived from Eq. (14.16) by putting $x = l$, $y = \left(R - \sqrt{R^2 - \alpha^2}\right)$ and $\alpha = \alpha$.

$$\frac{2v_0}{3R^2}\left[\left\{y_1^3 - (R - \sqrt{R^2 - \alpha^2})^3\right\} - 3R\left\{y_1^2 - (R - \sqrt{R^2 - \alpha^2})^2\right\}\right.$$
$$\left. +3\alpha^2\left\{y_1 - (R - \sqrt{R^2 - \alpha^2})\right\}\right] + v_s\left\{y_1 - (R - \sqrt{R^2 - \alpha^2})\right\}\sin\theta$$
$$+v_s l \cos\theta = 0 \tag{14.19}$$

If we plot y_1 versus α over the cross section of the tube as shown in Fig. 14.6, an area bounded by this curve and the circumference of the cross section as shown by the ABCOA area will result. Any particle with settling velocity v_s as it is in Equation (14.19) that may happen to enter through the ABCOA area into the tube will be removed. If such particles are uniformly incident on the tube cross section, then fractional removal of such particles may be written as

$$B = \frac{\text{Area ABCOA}}{\text{Area of the tube cross section}}$$

To find the ABCOA area, we find out $y_1^0, y_1^1, y_1^2, \ldots\ldots..y_1^n$ (Fig. 14.6) and compare with the values

Fig. 14.6 Sketch showing the plot of y_1 vs. α over the cross section of the tube

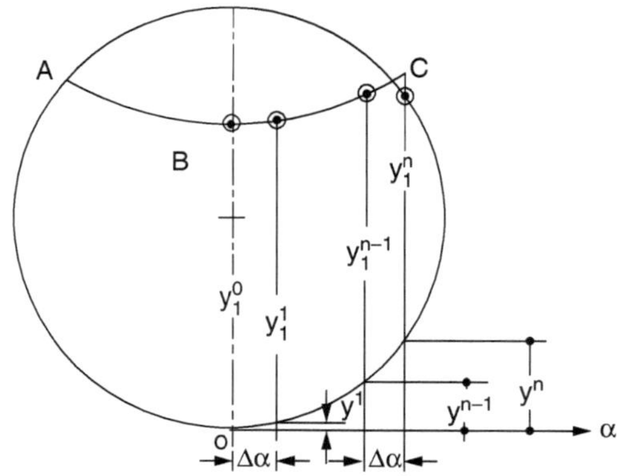

$R+\sqrt{R^2-\alpha_0^2}$, $R+\sqrt{R^2-\alpha_1^2}$, $R+\sqrt{R^2-\alpha_2^2}$, ... $R+\sqrt{R^2-\alpha_n^2}$ at $\alpha_0, \alpha_1, \alpha_2$, ...$\alpha_n$, respectively, till $y_1^n \ngtr R+\sqrt{R^2-\alpha_n^2}$ at α_n, when we put $y_1^n = R+\sqrt{R^2-\alpha_n^2}$, the ordinate at the respective bottoms of the chords being $y^0 = R-\sqrt{R^2-\alpha_0^2}$, $y^1 = R-\sqrt{R^2-\alpha_1^2}$, $y^2 = R-\sqrt{R^2-\alpha_2^2}$ .... $y^n = R-\sqrt{R^2-\alpha_n^2}$.

If α values are chosen equispaced at $\Delta\alpha$, $(n-1)\Delta\alpha = \alpha_n$.

Then the $\frac{1}{2}$ (area ABCOA)

$$= \left[\frac{(y_1^0-y^0)+(y_1^1-y^1)}{2}+\frac{(y_1^1-y^1)+(y_1^2-y^2)}{2}+\dots\dots\dots\right.$$

$$\left.+\frac{(y_1^{n-1}-y^{n-1})+(y_1^n-y^n)}{2}\right]\times\Delta\alpha+\frac{\pi R^2}{180}\tan^{-1}\frac{\sqrt{R^2-\alpha_n^2}}{\alpha_n}$$

$$-\alpha_n\sqrt{R^2-\alpha_n^2}$$

$$= \left[\sum_{n=0}^{n=n}(y_1^n-y^n)-\frac{(y_1^0-y^0)+(y_1^n-y^n)}{2}\right]\Delta\alpha$$

$$+\frac{\pi R^2}{180}\tan^{-1}\frac{\sqrt{R^2-\alpha_n^2}}{\alpha_n}-\alpha_n\sqrt{R^2-\alpha_n^2} \tag{14.20}$$

The ordinates $\left(y_1^n \text{ values}\right)$ may be found out by:

1. Direct solution of Eq. (14.19)
2. Method of differentials, as follows:

1. Direct solution of Eq. (14.19)

   Let Eq. (14.19) be written for convenience as follows:

$$\frac{2v_0}{3R^2}\left[\left\{\left(y_1^n\right)^3-\left(y^n\right)^3\right\}-3R\left\{\left(y_1^n\right)^2-\left(y^n\right)^2\right\}+3\alpha_n^2\left\{y_1^n-y^n\right\}\right]$$
$$+v_s\left\{y_1^n-y^n\right\}\sin\theta+v_s l\cos\theta=0 \quad (14.21)$$

Where a particle enters through $\left(0, y_1^n, \alpha_n\right)$ to reach the bottom of the chord at (l, y^n, α_n), y^n being

$$=R-\sqrt{R^2-\alpha_n^2}$$

Equation (14.21) may be written as

$$\frac{2v_0}{3R^2}\left(y_1^n\right)^3-\frac{2v_0}{3R^2}.3R\left(y_1^n\right)^2+\left(\frac{2v_0}{3R^2}.3\alpha_n^2+v_s\sin\theta\right)y_1^n-\frac{2v_0}{3R^2}\left[\left(y^n\right)^3\right.$$
$$\left.-3R\left(y^n\right)^2+3\alpha_n^2\left(y^n\right)\right]-v_s y^n\sin\theta+v_s l\cos\theta=0 \quad (14.22)$$

i. At chosen equispaced values of α_n separated by small distance $\Delta\alpha$, find out y^n $=R-\sqrt{R^2-\alpha_n^2}$ and set up Eq. (14.22).
ii. Solve the Eqns. to find out y_1^n values and compare the values with the top point of the respective chords $y_t^n=R+\sqrt{R^2-\alpha_n^2}$ till $y_1^n \ngtr R+\sqrt{R^2-\alpha_n^2};.$
iii. Use y_1^n values in Eq. (14.22) to find the $\frac{1}{2}$ (area ABCOA).

Problem 14.1 A 50-cm-long tube of diameter 5 cm, inclined at an angle 30° with the horizontal, is employed for the removal of solids from a flow of 0.06 l/s with concentration of solids of 100 mg/l consisting of particles that are all identical as regards there settling velocities of 0.3 cm/s.

Calculate the solids in the effluent.

Solution:

Choose $\Delta\alpha=0.3$ cm	$\alpha_0=0$cm	$\alpha_1=0.3$cm	$\alpha_2=0.6$cm	$\alpha_3=0.9$cm	$\alpha_4=1.2$cm	$\alpha_5=1.5$cm …………………
$y^n=R-\sqrt{R^2-\alpha_n^2}$ cm	$y^0=0$	$y^1=0.01807$	$y^2=0.07307$	$y^3=0.16762$	$y^4=0.30683$	$y^5=0.5$cm …………………
$y_t^n=2R-y^n$ cm	$y_t^0=5$	$y_t^1=4.98193$	$y_t^2=4.92693$	$y_t^3=4.83238$	$y_t^4=4.69317$	$y_t^5=4.50000$ …………………

$$\frac{2v_0}{3R^2}=\frac{2Q}{3\pi R^4}=\frac{2\times 60}{3\pi 2.5^4}=0.32595,\ v_s l\cos\theta=0.3\times 50\times\cos 30^\circ=12.99038$$

$$y_t^n=\text{Top point of the n th chord}$$

At $\alpha_0 = 0\,\text{cm}$

$$0.32595\left(y_1^0\right)^3 - 0.32595 \times 3 \times 2.5\left(y_1^0\right)^2 + 0.3\sin 30^\circ y_1^0 + 12.99038 = 0$$
$$i.e.\ 0.32595\left(y_1^0\right)^3 - 2.44463\left(y_1^0\right)^2 + 0.15y_1^0 + 12.99038 = 0$$
$$\text{i.e. } y_1^0 = 3.04202\,\text{cm} < 5\text{cm};$$

At $\alpha_1 = 0.3\,\text{cm}$

$$0.32595\left(y_1^1\right)^3 - 2.44463\left(y_1^1\right)^2 + \left(0.32595 \times 3 \times 0.3^2 + 0.15\right)y_1^1$$
$$- 0.32595\left(0.01807^3 - 3 \times 2.5 \times 0.01807^2 + 3 \times 0.3^2 \times 0.01807\right) - 0.3 \times 0.01807\ \sin 30^\circ + 12.99038 = 0$$
$$\text{i.e.}\ 0.32595\left(y_1^1\right)^3 - 2.44463\left(y_1^1\right)^2 + 0.23801y_1^1 + 12.98688 = 0$$

i.e.$y_1^1 = 3.08954 < 4.98193\,\text{cm}$;

At $\alpha_2 = 0.6\,\text{cm}$

$$0.32595\left(y_1^2\right)^3 - 2.44463\left(y_1^2\right)^2 + \left(0.32595 \times 3 \times 0.6^2 + 0.15\right)y_1^2$$
$$- 0.32595\left(0.07307^3 - 3 \times 2.5 \times 0.07307^2 + 3 \times 0.6^2 \times 0.07307\right) - 0.3 \times 0.07307\ \sin 30^\circ + 12.99038 = 0$$
$$\text{i.e.}\ 0.32595\left(y_1^2\right)^3 - 2.44463\left(y_1^2\right)^2 + 0.50203y_1^2 + 12.96662 = 0$$

i.e.$y_1^2 = 3.24329 < 4.92693\,\text{cm}$;

At $\alpha_3 = 0.9\,\text{cm}$

$$0.32595\left(y_1^3\right)^3 - 2.44463\left(y_1^3\right)^2 + \left(0.32595 \times 3 \times 0.9^2 + 0.15\right)y_1^3$$
$$- 0.32595\left(0.16762^3 - 3 \times 2.5 \times 0.16762^2 + 3 \times 0.9^2 \times 0.16762\right) - 0.3 \times 0.16762\ \sin 30^\circ + 12.99038 = 0$$
$$\text{i.e.}\ 0.32595\left(y_1^3\right)^3 - 2.44463\left(y_1^3\right)^2 + 0.94206y_1^3 + 12.89962 = 0$$

i.e.$y_1^3 = 3.55443 < 4.83238\,\text{cm}$;

At $\alpha_4 = 1.2\,\text{cm}$

$$0.32595\left(y_1^4\right)^3 - 2.44463\left(y_1^4\right)^2 + \left(0.32595 \times 3 \times 1.2^2 + 0.15\right)y_1^4$$
$$- 0.32595\left(0.30683^3 - 3 \times 2.5 \times 0.30683^2 + 3 \times 1.2^2 \times 0.30683\right) - 0.3 \times 0.30683\sin 30^\circ + 12.99038 = 0$$
$$i.e.\ 0.32595\left(y_1^4\right)^3 - 2.44463\left(y_1^4\right)^2 + 1.55810y_1^4 + 12.73304 = 0$$

i.e.$y_1^4 = 4.42710 < 4.69317\,\text{cm}$;

At $\alpha_5 = 1.5\,\text{cm}$

$$0.32595\left(y_1^5\right)^3 - 2.44463\left(y_1^5\right)^2 + \left(0.32595 \times 3 \times 1.5^2 + 0.15\right)y_1^5 \\ - 0.32595\left(0.5^3 - 3 \times 2.5 \times 0.5^2 + 3 \times 1.5^2 \times 0.5\right) - 0.3 \times 0.5 \sin 30^{\circ} \\ + 12.99038 = 0$$

$$\text{i.e.}\, 0.32595\left(y_1^5\right)^3 - 2.44463\left(y_1^5\right)^2 + 2.35016 y_1^5 + 12.38571 = 0$$

i.e.$y_1^5 = \mathit{does\,not\,exist.Hence\,put}\, y_1^5 = 4.5\,\text{cm}$;

The y_1^n and y^n values are presented in Table 14.3.

The $\frac{1}{2}$ (area ABCOA) of the diagram

$$= \left[\sum_{n=0}^{n=n}\left(y_1^n - y^n\right) - \frac{\left(y_1^0 - y^0\right) + \left(y_1^n - y^n\right)}{2}\right]\Delta\alpha + \frac{\pi R^2}{180}\tan^{-1}\frac{\sqrt{R^2 - \alpha_n^2}}{\alpha_n} \\ - \alpha_n\sqrt{R^2 - \alpha_n^2}$$

$$= \left[(21.85638 - 1.06559) - \frac{(3.04202 - 0) + (4.5 - 0.5)}{2}\right]0.3 \\ + \frac{\pi 2.5^2}{180}\tan^{-1}\frac{\sqrt{2.5^2 - 1.5^2}}{1.5} - 1.5\sqrt{2.5^2 - 1.5^2} \\ = 5.18093 + 5.79560 - 3.0 \\ = 7.97653\,\text{cm}^2;$$

i.e. Fractional removal $= \frac{7.97653}{0.5 \times \pi 2.5^2}$ i.e. 0.81248

Hence the effluent concentration of solids

$$= 100(1 - 0.81248) \\ = 18.75\,\text{mg/l}$$

Table 14.3 y_1^n and y^n values

n	y_1^n cm	y^n cm
0	3.04202	0
1	3.08954	0.01807
2	3.24329	0.07307
3	3.55443	0.16762
4	4.42710	0.30683
5	4.5	0.5
Σ =	21.85638	1.06559

2. Method of differentials:

To find out the ordinates in the same cross section, let us take differentials of Eq. (14.16) putting $x = l$, and one can write

$$\frac{2v_0}{3R^2}\left[\left(3y_1^2 - 6Ry_1 + 3\alpha^2\right)\mathrm{d}y_1 + 6\alpha y_1 d\alpha\right] + v_s \sin\theta\,\mathrm{d}y_1$$
$$= \frac{2v_0}{3R^2}\left[\left(3y^2 - 6Ry + 3\alpha^2\right)\mathrm{d}y + 6\alpha y\, d\alpha\right] + v_s \sin\theta\,\mathrm{d}y$$

For finite changes, the above equation can be rearranged and written as

$$\Delta y_1 = \frac{\frac{2v_0}{3R^2}\left[(3y^2 - 6Ry + 3\alpha^2) + v_s \sin\theta\right]\Delta y - \frac{2v_0}{3R^2}.6\alpha(y_1 - y)\Delta\alpha}{\frac{2v_0}{3R^2}\left(3y_1^2 - 6Ry_1 + 3\alpha^2\right) + v_s \sin\theta} \tag{14.23}$$

The ordinate of the bottom of the chords is

$$y = R - \sqrt{R^2 - \alpha^2} \tag{14.24}$$

The ordinate of the top of the chords is

$$c = R + \sqrt{R^2 - \alpha^2}\ i.e.\ c = 2R - y \tag{14.25}$$

$$\Delta y = \frac{\alpha\Delta\alpha}{\sqrt{R^2 - \alpha^2}} \tag{14.26}$$

Equation (14.26) cannot be evaluated at α=0 and the accuracy of the determination of Δy increases as $\frac{\Delta\alpha}{\alpha} \to 0$.

Since in the case under consideration $\frac{\Delta\alpha}{\alpha}$ is sufficiently large, Δy may be calculated as

$\Delta y = y^{n+1} - y^n$, y^n being the y value for the n-th chord……..(14.26)

With the help of the equations, the ordinates may be found out according to the following procedure:

1. y_1^0 can be found out from Eq. (14.18) or Eq. (14.19) corresponding to the value of $y^0 = 0$.
2. Choose suitably small value of Δα. Find out $\alpha_0, \alpha_1, \alpha_2, \alpha_3 \ldots\ldots\ldots.\alpha_n$. Correspondingly, find out $y^n = R - \sqrt{R^2 - \alpha_n^2}$ as $y^0, y^1, y^2, y^3, \ldots..y^n$ and also $c^n = 2R - y^n$ as $c^0, c^1, c^2, c^3, \ldots\ldots..c^n$, respectively.
3. Find out y_1^1 from Eq. (14.19) at $y = y^1$.
4. Find out Δy_1^1 from Eq. (14.23) for $y_1^1, y^1, \Delta y^1, \alpha_1$.

5. Find $y_1^2 = y_1^1 + \Delta y_1^1$.
6. Compare y_1^2 with c^2.
7. Find $y_1^n = y_1^{n-1} + \Delta y_1^{n-1}$ values as $y_1^2, y_1^3, \ldots\ldots\ldots y_1^n$ till y_1^n is not greater than c^n, when y_1^n should be put $= c^n$. This may be erroneous but does not involve sufficient inaccuracy of the results particularly when $\Delta\alpha$ is sufficiently small.

Problem 14.2 A 50 cm long tube of diameter 5 cm, inclined at an angle 30° with the horizontal is employed for the removal of solids from a flow of 0.06 l/s with concentration of solids of 100 mg/l consisting of particles that are all identical as regards there settling velocities of 0.3 cm/s.

Calculate the solids in the effluent.

Solution:

$$l = 50\text{cm},\ \theta = 30^{\circ},\ \text{R} = 2.5\text{cm},\ v_s = \frac{0.3\text{cm}}{\text{s}},\ \text{Q} = \frac{60\text{cm}^3}{\text{s}}$$

$$\frac{2v_0}{3R^2} = 0.32595,\ v_s l \cos\theta = 12.99038,$$

$$v_{cr} = \frac{8Q}{3\pi R(2R\sin\theta + l\cos\theta)}$$
$$= 0.44479\,\text{cm/s} > 0.3\,\text{cm/s}$$

$$\text{Choose } \Delta\alpha = 0.3\,\text{cm}-$$

$$\alpha_0 = 0\,\text{cm},\ \alpha_1 = 0.3\,\text{cm},\ \alpha_2 = 0.6\text{cm},\ \alpha_3 = 0.9\,\text{cm},\ \alpha_4 = 1.2\,\text{cm},\ \alpha_5 = 1.5\,\text{cm}\ldots\ldots\ldots$$

The ordinates at the bottom of the chords $y^n = R - \sqrt{R^2 - \alpha_n^2}$ are

$$y^0 = 0,\ y^1 = 0.01807,\ y^2 = 0.07307,\ y^3 = 0.16762,\ y^4 = 0.30683,\ y^5 = 0.5\text{cm}\ldots\ldots\ldots\ldots\ldots$$

$$\Delta y^0 = 0.01807,\ \Delta y^1 = 0.055,\ \Delta y^2 = 0.09455,\ \Delta y^3 = 0.13921,\ \Delta y^4 = 0.19317\ldots\ldots\ldots\ldots$$

The ordinates at the top of the chords $c^n = 2R - y^n$ are

$$c^0 = 5\text{cm},\ c^1 = 4.98193,\ c^2 = 4.92693,\ c^3 = 4.83238,\ c^4 = 4.69317,\ c^5 = 4.5$$

At $\alpha_C = 0\,\text{cm}$

$$i.e. 0.32595\left(y_1^0\right)^3 - 2.44463\left(y_1^0\right)^2 + 0.15y_1^0 + 12.99038 = 0$$
$$\text{i.e. } y_1^0 = 3.04202\,\text{cm} < 5\text{cm};$$

At $\alpha_1 = 0.3\,\text{cm}$

$$i.e. 0.32595\left(y_1^1\right)^3 - 2.44463\left(y_1^1\right)^2 + 0.23801y_1^1 + 12.98688 = 0$$
$$i.e. y_1^1 = 3.08954 < 4.98193\,\text{cm};$$

At $\alpha_1 = 0.3\,\text{cm}$, $\Delta\alpha_1 = 0.3$, $\Delta y^1 = 0.055$, $y_1^1 = 3.08954$, $y^1 = 0.01807$

$\Delta y_1^1 =$

$$\frac{\begin{array}{c}[0.32595(3 \times 0.01807^2 - 6 \times 2.5 \times 0.01807 + 3 \times 0.3^2) + 0.3 \sin 30^\circ] \\ \times 0.055 - 0.32595 \times 6 \times 0.3\,(3.08954 - 0.01807) \times 0.3\end{array}}{0.32595(3 \times 3.08954^2 - 6 \times 2.5 \times 3.08954 + 3 \times 0.3^2) + 0.3 \sin 30^\circ}$$

$$= \frac{(-)0.53237}{(-)5.53370}\, i.e. 0.09621\,\text{cm}$$

$$y_1^2 = y_1^1 + \Delta y_1^1$$
$$= 3.08954 + 0.09621\, i.e. 3.18575 < 4.92693$$

At $\alpha_2 = 0.6\,\text{cm}$, $\Delta\alpha_2 = 0.3$, $\Delta y^2 = 0.09455$, $y_1^2 = 3.18575$, $y^2 = 0.07307$,

$\Delta y_1^2 =$

$$\frac{\begin{array}{c}[0.32595(3 \times 0.07307^2 - 6 \times 2.5 \times 0.07307 + 3 \times 0.6^2) + 0.3 \sin 30^\circ] \\ \times 0.09455 - 0.32595 \times 6 \times 0.6\,(3.18575 - 0.07307) \times 0.3\end{array}}{0.32595(3 \times 3.18575^2 - 6 \times 2.5 \times 3.18575 + 3 \times 0.6^2) + 0.3 \sin 30^\circ}$$

$$= \frac{(-)1.08155}{(-)5.14970}\, i.e. 0.21002\,\text{cm}$$

$$y_1^3 = y_1^2 + \Delta y_1^2$$
$$= 3.18575 + 0.21002\, i.e. 3.39577 < 4.83238$$

At $\alpha_3 = 0.9\,\text{cm}$, $\Delta\alpha_3 = 0.3$, $\Delta y^3 = 0.13921$, $y_1^3 = 3.39577$, $y^3 = 0.16762$,

$\Delta y_1^3 =$

$$\frac{\begin{array}{c}[0.32595(3 \times 0.16762^2 - 6 \times 2.5 \times 0.16762 + 3 \times 0.9^2) + 0.3 \sin 30^\circ] \\ \times 0.13921 - 0.32595 \times 6 \times 0.9\,(3.39577 - 0.16762) \times 0.3\end{array}}{0.32595(3 \times 3.39577^2 - 6 \times 2.5 \times 3.39577 + 3 \times 0.9^2) + 0.3 \sin 30^\circ}$$

$$= \frac{(-)1.68371}{(-)4.38487}\, i.e. 0.38398\,\text{cm}$$

$$y_1^4 = y_1^3 + \Delta y_1^3$$
$$= 3.39577 + 0.38398\, i.e. 3.77975 < 4.69317\,\text{cm};$$

At $\alpha_4 = 1.2\,\text{cm}$, $\Delta\alpha_4 = 0.3$, $\Delta y^4 = 0.19317$, $y_1^4 = 3.77975$, $y^4 = 0.30683$,

$$\Delta y_1^4 = \frac{\begin{array}{c}[0.32595(3 \times 0.30683^2 - 6 \times 2.5 \times 0.30683 + 3 \times 1.2^2) + 0.3 \sin 30^o] \\ \times 0.19317 - 0.32595 \times 6 \times 1.2\ (3.77975 - 0.30683) \times 0.3\end{array}}{0.32595(3 \times 3.77975^2 - 6 \times 2.5 \times 3.77975 + 3 \times 1.2^2) + 0.3 \sin 30^o}$$

$$= \frac{(-)2.41614}{(-)2.95197}\ i.e. 0.81848\,\text{cm}$$

$$y_1^5 = y_1^4 + \Delta y_1^4$$
$$= 3.77975 + 0.81848\, i.e. 4.59823 > 4.5\,\text{cm};$$

Hence put $y_1^5 = 4.5\,\text{cm}$

From above $\sum_0^5 y_1^n = 20.99283\,\text{cm}$, $\sum_0^5 y^n = 1.06559\,\text{cm}$.

The $\frac{1}{2}$ (area ABCOA) of the diagram

$$= \left[(20.99283 - 1.06559) - \frac{(3.04202 - 0) + (4.5 - 0.5)}{2}\right]0.3$$
$$+ \frac{\pi 2.5^2}{180} \tan^{-1} \frac{\sqrt{2.5^2 - 1.5^2}}{1.5} - 1.5\sqrt{2.5^2 - 1.5^2}$$
$$= 4.92187 + 5.79560 - 3.0$$
$$= 7.71747\,\text{cm}^2$$

Fractional removal $= \frac{7.71747}{0.5 \times \pi \times 2.5^2}$, i.e. 0.78610.

Hence the concentration of solids in the effluent

$$= (1 - 0.78610) \times 100 \text{ i.e. } 21.39\,\text{mg/l};$$

14.3 Settling Column Analysis and Tube Settler

Given a suspension for the estimation of removal of solids through a given tube settler for certain rate of flow of the suspension through it, 'settling column analysis' is to be conducted according to the procedure laid down in Chap. 8 of this book.

The interface trajectories of the settling suspension are plotted. Each of the trajectories will have three parts:

(i) Initial discrete settling
(ii) Curvilinear portion of flocculant settling in the middle
(iii) Final discrete settling

Table 14.4 Settling velocity distribution during the initial phase of discrete settling

(1)	(2)	(3)
Concentration in mg/l	Settling velocity	Fractional removal in tube setter
$(c_0 - c_1)$	$(v_1 > v_{cr})$	1
$(c_1 - c_2)$	$(v_2 > v_{cr})$	1
$(c_2 - c_3)$	$(v_3 < v_{cr})$	f_3
$(c_3 - c_4)$	$(v_4 < v_{cr})$	f_4
$(c_4 - c_5)$	$(v_5 < v_{cr})$	f_5
…………	…………	…………

Figure 16.2 of the Chap. 16 exhibits initial discrete settling to a depth of 30 cm for an actual settling data. In fact the diameters of the actual settling tubes are much lesser than the depth to which initial discrete settling of the interface trajectories is exhibited.

Although in the initial phase of settling column the flocculation due to differential settling is negligible, the flocculation due to velocity gradient may come up while the suspension is flowing through the tube settler due to the particles falling from higher momentum region to lower one and vice versa. This factor cannot be taken into account. However, neglecting this flocculation due to velocity gradient and computation of removal with settling column test data will give the conservative estimate to our advantage.

From the settling column analysis the settling velocity distribution among the particles during the initial phase of discrete settling as presented in Table 16.3 of Chap. 16 are found out and may be presented as in Table 14.4.

c_0 is the initial concentration of the suspension. $c_1, c_2, c_3, c_4, c_5 \ldots\ldots\ldots$ are the concentration of the subsequent trajectories in the discrete settling phase.

Concentrations $(c_0 - c_1), (c_1 - c_2), (c_2 - c_3), (c_3 - c_4)(c_4 - c_5) \ldots\ldots$ having settling velocities $v_1, v_2, v_3, v_4, v_5 \ldots\ldots$ are presented Col-(1) and Col-(2) of Table 14.4. Critical settling v_{cr} through the tube settler for the flow rate Q is found out.

If v_1 and $v_2 \geq v_{cr}$ concentrations $(c_0 - c_1), (c_1 - c_2)$ are completely removed. Corresponding to the settling velocities $v_1, v_2, v_3, v_4, v_5 \ldots\ldots$ fractional removal values $f_3, f_4, f_5 \ldots\ldots$ through the settler are computed.

The total removal $= (c_0 - c_1) + (c_1 - c_2) + f_3(c_2 - c_3) + f_4(c_3 - c_4) + f_5(c_4 - c_5) + \ldots.mg/l$

Notations

x,y Two-dimensional coordinates

x,y,α Three-dimensional coordinates

X Coordinate axis in the direction of flow

Y Coordinate axis in direction normal to the direction of flow; also $X = \frac{x}{d}, Y = \frac{y}{d}$

L, l	Length of the tube; also $L = \frac{l}{d}$
R	Radius of the tube cross section
θ	Inclination of the tube with horizontal
Q	Rate of flow
v_0	Average velocity through the tube cross section
v_s	Settling velocity of the particle

References

Allen H (1904) On sedimentation. Trans Am Soc Civil Eng 53:63

Barham WE et al (1956) Clarification, sedimentation and thickening equipment, patent review bulletin No 54, Published by Engineering Experiment Station, Lusiana State University, Baton Rouge La

Camp TR (1946) Sedimentation and the design of settling tanks. Trans Am Soc Civil Eng 111:895

Camp TR (1953) Sedimentation basin design. Sewage Ind waste 25:1

Culp GL, Hsiung KY, Conley WR (1969) Tube clarification process – Operating experiences, ASCE, San. Engg. Div^{n}, SA5,p829

De Alak (1976) Conceptual studies on discrete and flocculent settling Ph.D thesis, Jadavpur University, Kolkata,West Bengal, India

Fischerstrom CNW (1955) Sedimentation in rectangular basin. Proc Soc Civil Engrs, J San Engg Div^{n}

Hansen SP, Culp GL (1967) Applying shallow depth sedimentation theory. J AWWA 59:1134

Hansen SP, Culp GL, Richardson G (1968) High rate sedimentation in water treatment works. J AWWA 60:81

Hansen SP, Culp GL, Stukenberg JR (1969) Practical application of idealised sedimentation theory in waste water treatment. J WPCF 41:1421

Slechta AF, Conley WR (1971) Recent experiments in plant scale application of the settling tube concept. J Water Pollut Control Fed 43:1724–1738

Yao KM (1970) Theoretic study of high rate sedimentation. J Water Pollut Control Fed 72:218

Chapter 15
Verification of Tube Settling Theory

Abstract This is a chapter on the 'experimental verification' of the 'theory of tube settling'.

Keywords Critical length • Experimental set-up • Tube settling and Reynolds' number • Transitional length • Impairment of settling

15.1 Introduction

Yao (1970) published his theoretical study on tube settling in 1970. He deduced the expression for the critical fall velocity through a tube settler under laminar flow condition. It was suggested that the designed length of the tube settler needs to include an initial length for the development of laminarity in the flow. This being the basic to the design of tube settling needs experimental verification. This chapter responds to the need.

Yao (1970) worked out the expression for critical fall velocity of a particle through a tube settler as

$$\text{Critical fall velocity}\,(v_c) = \frac{8Q}{3\pi R(L\cos\theta + 2R\sin\theta)} \tag{15.1}$$

All particles having settling velocity $v_s \geq v_c$ will be removed completely in a tube settler of length L, diameter of cross section of the tube 2R, inclined at an angle θ to the horizontal. Being derived by employing the expression for the local fluid velocity under laminar flow condition, the use of Eq. 15.1, it is believed, imposes the condition of laminarity of flow. An addition of an initial length, transition length or additional length of – $0.58\frac{v_0 d}{\nu}$ (v_0, average velocity through the tube; d, its diameter; ν, kinematic viscosity of the liquid) to the length of the tube settler, in accordance with the provision for the flow to develop into full laminar flow condition, was suggested. Since then, Eq. 15.1 has remained basic to the design of a tube settler. As such it deserved experimental verification. It appears that no such attempt has ever been reported so far.

A. De, *Sedimentation Process and Design of Settling Systems*,
Springer Transactions in Civil and Environmental Engineering,
DOI 10.1007/978-81-322-3634-4_15

Several studies (Mullick 1986; Roy[1] 1986, Roy[2] 1988; Ghosh 1989; Mehera 1989) were conducted to ascertain the following:

If the settling in a tube settler takes place in accordance with the equation

$$(v_c) = \frac{8Q}{3\pi R(L\cos\theta + 2R\sin\theta)}$$

How the deviation of flow from laminarity affects the settling of settleable solids

The importance of adding the additional length of $0.58\frac{v_0 d}{\nu}$ to the length of the tube settler for the development of a fully laminar flow in the tube.

15.2 Approach to the Study

A particle having a certain value of settling velocity may enter through any point on a tube cross section. It will travel through a certain length before it reaches the bottom of the tube. This length will depend upon the tube diameter, inclination of the tube, rate of flow through it and, of course, the settling velocity of the particle itself and also the depth through which it has to fall before it reaches the bottom. For a particle having a particular settling velocity, this length traversed by it through the tube before it reaches the bottom is maximum when it enters through the topmost point of the tube cross section. This length has to be provided for the above particle if it has to be completely absent from the effluent through the tube. This length is, we may choose to call, 'critical length'. This critical length depends upon the diameter ($2R$), angle of inclination (θ) of the tube, rate of flow (Q) through it and the settling velocity (v_s) of the particle. This length can be calculated from Eq. 15.1 as

$$\text{Critical length } (l_c) = \frac{8Q}{3\pi R\ v_s\cos\theta} - 2R\tan\theta. \tag{15.2}$$

If an assortment of particles having varying settling velocities contained in a suspension enter through a tube with a particular rate of flow for a considerable period of time and found to be deposited in the tube, then the extreme end of the deposition will define the critical length for the particles settleable in the tube with the lowest settling velocity. This length is measured and, thus, determined experimentally.

The variation of the critical length and the variation of characteristic parameters of tube settling are studied and compared with that obtained from the theory.

15.3 Materials and Methods

15.3.1 Materials

15.3.1.1 Particles in Suspension

1. Activated carbon: Finely powdered activated carbon particles manufactured by E. Merck (India) Pvt. Ltd. were used.
2. Marble dust: This was locally purchased in the form of finely powdered marble.
3. Fly ash: Finely powdered fly ash particles were collected from CESC Thermal Power Plant at Titagarh. Particles passing through B.S. Sieve No. 100 and retained on B.S. Sieve No. 200 were used.
4. Sand: Medium-sized sand particles passing through B.S. Sieve No. 100 were used.
5. Plaster of Paris: Locally purchased plaster of Paris was used.
6. Pigments: These particles were collected from IEL, Rishra. These are manufactured as colour pigments which are soluble in oil.
7. Kaolin: Locally purchased heavy kaolin particles were used.
8. Glass dust: These particles were prepared by pulverising glass and then passing through B.S. Sieve No. 200.
9. Fine sand: Locally available sand passing through B.S. Sieve No. 200.
10. Sodium hexametaphosphate: This was used to prevent the formation of floc of the particles (used only in the studies of impairment of settling).

15.3.1.2 Accessories

1. Experimental tube: Glass tubes having uniform internal diameters of 10.5 mm, 8 mm, 6 mm and 4 mm were used. The lengths of tubes were about 1000 mm and 1200 mm (Fig. 15.1).
2. Constant head tank: A constant head tank made of galvanised iron sheet was used to maintain a constant head of water during the course of an experimental run. The constant head tank ensured a constant rate of flow through the tube. The details of the tank are shown in Fig. 15.3a.
3. U-tube: A glass U-tube was introduced into the experimental set-up just before the experimental tube in Fig. 15.2. The purpose of this U-tube was to ensure that the particles entered the tube in truly suspended manner. The details of the U-tube are given in Fig. 15.3b.
4. Angle measuring device: A protractor with a plumb bob attached to its centre was used as an angle measuring device. This device directly reads off the inclination of the experimental tubes with the vertical.
5. Measuring cylinder: A graduated glass cylinder was used. The purpose of the cylinder was to facilitate the collection of a measured volume of the effluent over an interval of time.

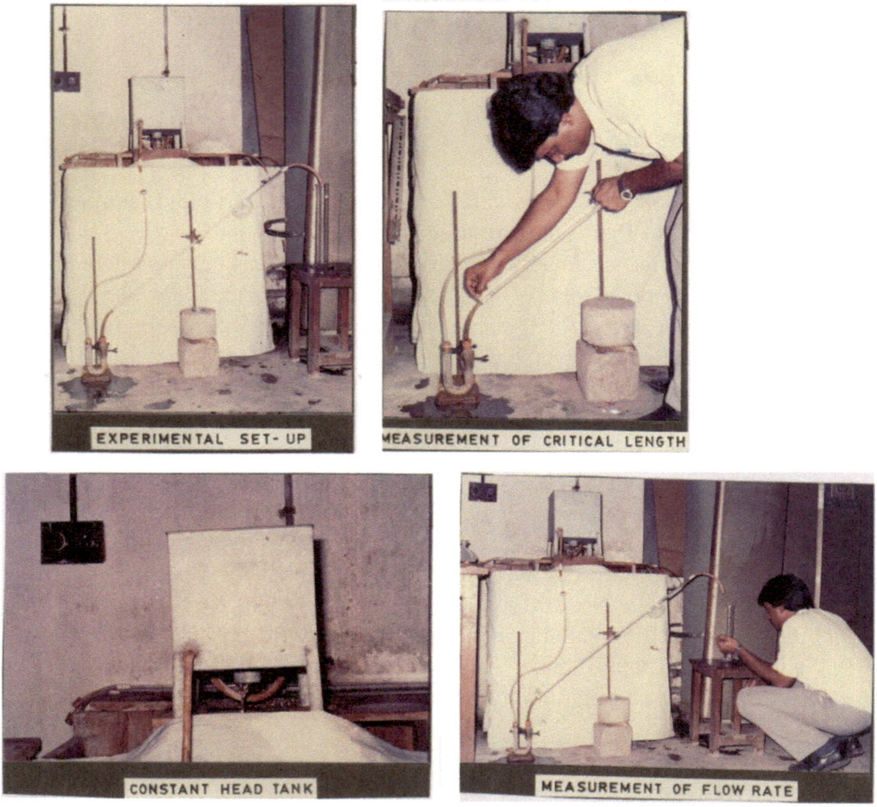

Fig. 15.1 Photographic display of experimental details

6. Retort stands: These were used to hold the U-tube and experimental tubes in desired positions.
7. Connecting tubes and pinch-cock: All connecting tubes were made of rubber. A pinch-cock was inserted in the tube connecting the outlet of the constant head tank and U-tube (Fig. 15.2). This pinch-cock helped in controlling the rate of flow through the experimental tubes.
8. Stop-watch: A stop-watch was used to measure the time over which the volume of effluent was collected and measured.
9. Measuring tape: This was used to measure the critical lengths in the experimental tubes.
10. Thermometer: The temperature of water was determined with a thermometer.

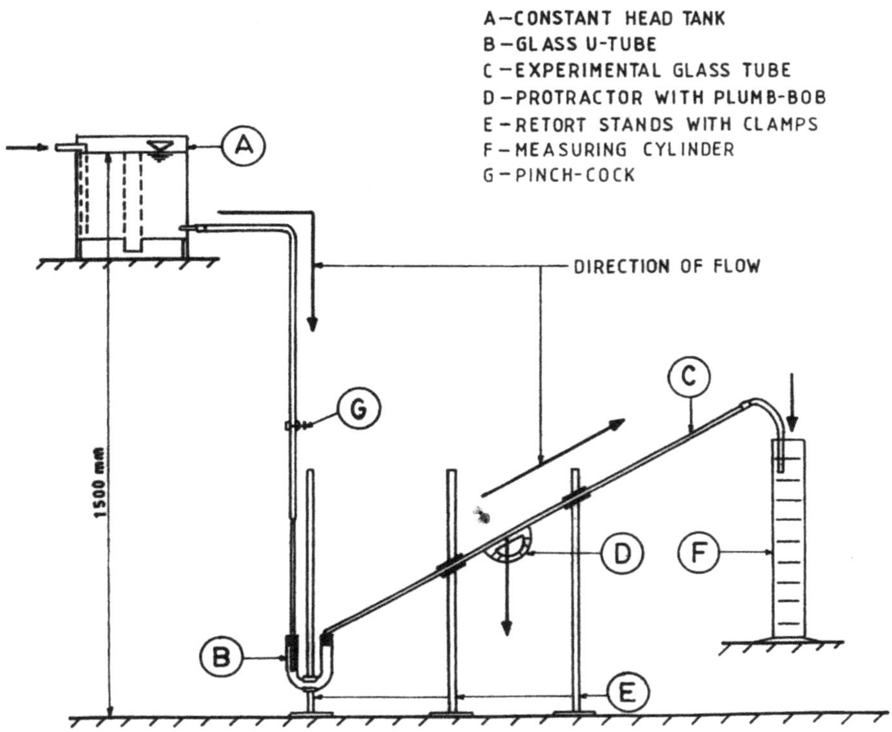

Fig. 15.2 Experimental set-up

15.3.2 Methods: The Experimental Set-Up for Critical Length Determination Is Shown in Fig. 15.2

The tube was used in experimental set-up after the verification of the measure of its diameter from the measurement of the measured volumes of water occupying the different lengths of the same tube.

Figure 15.1 demonstrates the photographic display of the experimental set-up, the determination of the critical length and the measurement of the rate of flow through the tube.

The experimental tube was firmly held in position with the help of retort stands, and the angle of inclination was determined by the angle measuring device.

A quantity of some particles was placed in the U-tube and the end was stoppered firmly. By opening the pinch-cock, varying rates of different flows of water from the constant head tank were made to flow through the U-tube and the experimental tube through a connecting tube between the two.

Water, while passing through the U-tube, threw the particles in the U-tube as suspension, and a true suspension of waterborne particles entered the experimental tube and was carried with the flow.

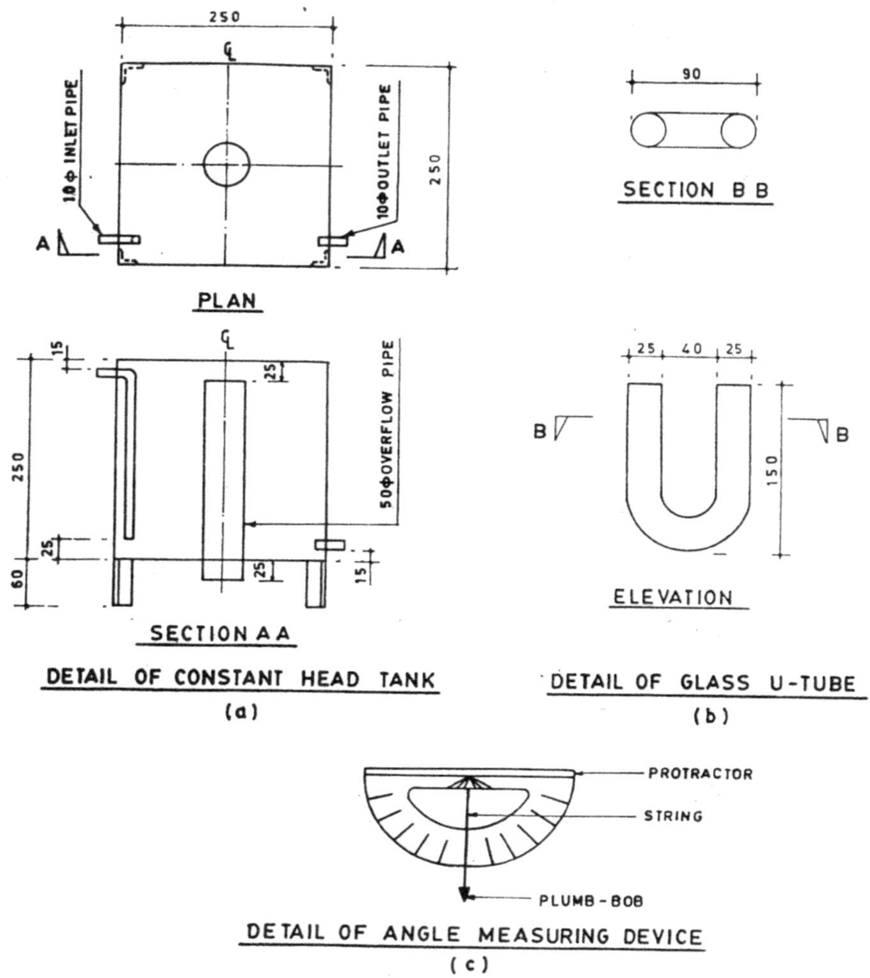

Fig. 15.3 Detail of components (**a**) Detail of constant head tank (**b**) Detail of glass U tube (**c**) Detail of angle measuring device

The particles settleable in the tube then settled to the bottom of the tube after travelling a certain distance depending upon the settling velocity of the particle, flow-through velocity of the suspension, diameter and angle of inclination of tube and the vertical distance it had to fall through in the tube before reaching the bottom.

The particles, thus, formed a thick sludge at the bottom of the tube after some time. The distance between the starting end of the experimental tube and the extreme end of sludge deposition was measured with the help of a measuring tape. This was the critical length for the particle with the lowest settling velocity that settled to the bottom of the tube. The effluent through the experimental tube was collected in the measuring cylinder, and the time interval of collection was measured with the help of the stop-watch. The rate of flow could, thus, be computed.

Before commencing an experimental run, the temperature of the water was determined with the help of a thermometer.

After a particular run the experimental tube, the U-tube and connecting tubes were completely washed, and a separate observation was taken with a higher rate of flow through the suspension.

When the velocity of flow through the suspension caused heavy scour of deposited particles making the determination of critical length difficult, the experimental tube was readjusted at a different angle of inclination, and the whole procedure was repeated.

Different tube inclinations ranging from 0 to 60° were employed for the study.

15.4 Results and Discussions

15.4.1 Results

A typical data sheet is presented in Table 15.1 for activated carbon particles.

It has nine columns. Column 1 gives the temperature of the suspension. Angle of inclination of the experimental tube with horizontal (θ) is given in column 2, and volume of effluent (V) collected in measuring cylinder is represented in column 3 of the table. Values in column 4 are the times (t) in which corresponding volumes V were collected, and those in column 5 are the measured critical length (l_c). Volumetric rates of flow (Q) through the experimental tubes were obtained by dividing the values in column 3 by the corresponding values in column 4 and are presented in column 6 of the table.

Representation of flow-through velocity (v_0) of the suspension, arrived at by dividing values in column 6 by the internal cross-sectional area of the appropriate tubes, is made in column 7.

Table 15.1 Diameter of the tube, 10.5 mm; particles used, activated carbon

T	θ	V	t	l_c	Q	v_0	v_c	N
(°c)	deg.	(cm^3)	(sec)	(cm)	(cm^3/s)	(cm/s)	(cm/s)	Reynolds' number
(1)	(2)	(3)	(4)	(5)	(6)	(7)	(8)	(9)
30.5	60	40	123.2	10.5	0.32	0.37	0.084	48.50
		50	93.8	14.2	0.53	0.61	0.107	79.95
		100	102. 5	30.5	0.98	1.13	0.098	148.11
		100	64.2	46.3	1.56	1.80	0.105	235.93
		100	52.5	55.0	1.90	2.19	0.108	287.04
		100	43.0	60.8	2.33	2.69	0.120	352.58

The critical fall velocity of a particle entering through the topmost point of the tube cross section and settling to the bottommost point travelling a distance l_c may be rewritten from Eq. 15.1 as

$$v_c = \frac{4v_0 d}{3(l_c \cos\theta + d \sin\theta)} \tag{15.3}$$

where

d = diameter of the tube.

Calculated values of critical fall velocity v_c derived from Eq. 15.3 are given in column 8 of the tables.

Reynolds' number of flow N is given by

$$N = \frac{v_0 d}{\nu} \tag{15.4}$$

where

ν = kinematic viscosity of water at the temperature the flow is taking place.

The values of N calculated from Eq. 15.4 are presented in column 9 of the tables.

Data contained in Table 15.1 are the observations that constitute one set of observations with activated carbon at $\theta = 60°$. The complete set of observations with activated carbon consisted of such observations at $\theta = 60°$, 53°, 50°, 45°, 42°, 35° 27°, 20°, 12°, 7° and 0°.

15.4.2 Discussions

Column 8 presents the value of v_c which is the lowest settling velocity of the particles settled in a tube at a particular rate of flow and angle of inclination of tube.

At a higher rate of flow, the value of this lowest settling velocity may increase, the other parameters being kept constant. This value will increase if the lighter particles that settled to the bottom of the tube at lesser flow rate are carried away with the effluent at the subsequent higher rate of flow.

This has been found to be true in most of the observations presented in column 8 with very few, amounting almost no exceptions. But as can be observed from column 8, these variations in the value of v_0 are not appreciable.

Within the limits of experimental errors, therefore, it may be assumed that almost the same particles with the same lowest velocity settled at the extreme end of the depositions through a set of observations.

If this is so, from the theory, the deduced Eq. 15.2 can be written as

$$(l_c) = \frac{8Q}{3\pi R\, v_c \cos\theta} - \mathrm{d}\tan\theta \quad (15.5)$$

$$= K_1 Q - K_2 \quad (15.6)$$

where

K_1 and K_2 are constants for the particular set and $K_1 = \frac{8}{3\pi R\, v_c \cos\theta}$, $K_2 = d \tan\theta$.

Hence the arithmetic plot of l_c versus Q for a particular set will give a straight line. A typical of such plot for the data sheet in Table 15.1 is presented in the plot shown in Fig. 15.4a. For all the sets corresponding to $\theta = 60°$, 53°, 50°, 45°, 42°, 35° 27°, 20°, 12°, 7° and 0°, all such plots were similar and straight lines.

Again Eq. (15.4) may be written as

$$(l_c) = \frac{4\nu N}{3 v_c \cos\theta} - d\tan\theta \quad (15.7)$$

$$= K_3 N - K_2 \quad (15.8)$$

where

$K_3 = \frac{4\nu}{3v_c \cos\theta}$, $K_2 = d\tan\theta$ and K_3 and K_2 are constants.

Hence the arithmetic plot of l_c versus N for a particular set will give a straight line.

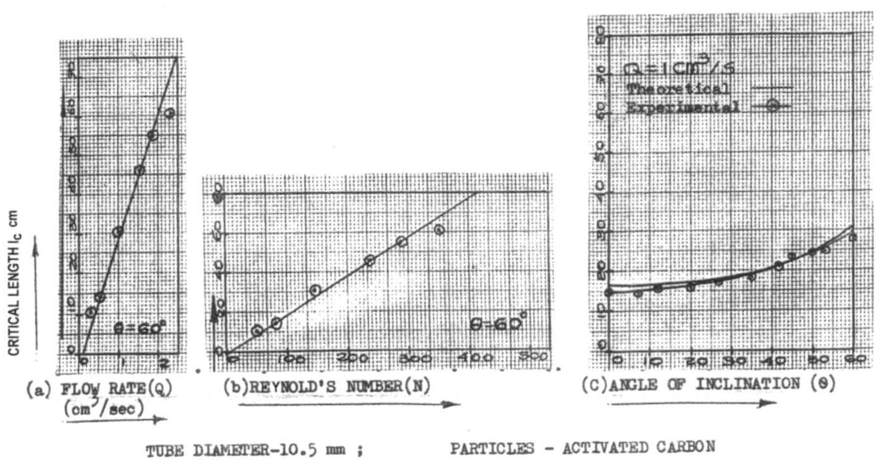

Fig. 15.4 Significance plots

Table 15.2 Diameter of the tube, 10.5 mm; particles used, activated carbon; average v_c, 0.0985 cm/s (From Table 15.1)

Q	θ	l_c cm	l_c cm
(cm^3/s)	(degrees)	(From l_c versus Q plots for activated carbon)	(From Eq. 15.4)
(1)	(2)	(3)	(4)
1	60	28	31
	53	25.0	25.9
	50	24.5	24.3
	45	23.5	22.2
	42	21.0	21.1
	35	19	19.3
	27	17.5	17.9
	20	16.0	17.1
	12	16.0	16.6
	7	14.5	16.4
	0	15.0	16.4

A typical of such plot for the data sheet in Table 15.1 is presented in the plot shown in Fig. 15.4b. For all the sets corresponding to $\theta = 60°$, 53°, 50°, 45°, 42°, 35° 27°, 20°, 12°, 7° and 0°, all such plots were similar and straight lines.

From the 11 sets of l_c versus Q plots l_c, values corresponding to a particular value of Q were scaled and presented in column 3 of Table 15.2

Experimentally determined values of l_c versus θ for that particular value of Q can now be plotted as in Fig. 15.4c. Such plots for various other values Q were also prepared.

The average value of v_c of 11 sets of data for activated carbon as in column 8 of Table 15.1 was determined. The standard deviation indicated that all the values of v_c were extremely close to the average value of v_c.

This average value of v_c of the complete set was employed to calculate l_c values for the particular value of Q (employed to plot experimentally determined l_c versus θ plot) corresponding to different values of θ from Eq. 15.4 and presented in column 4 of Table 15.2. The theoretical plot of l_c versus θ corresponding to that particular value of Q was imposed on Fig. 15.4c for its comparison with the experimentally determined plot.

For a particular tube diameter d, rate of flow Q and particular value of v_c, Eq. 15.4 may be rewritten as

$$(l_c) = \frac{K_4}{\cos\theta} - K_5 \tan\theta \qquad (15.9)$$

where

$K_4 = \frac{8Q}{3\pi R\, v_c}$, and $K_5 = d$ are constants.

The plots conform to Eq. 15.9. Both the theoretical and experimental plots were extremely close and compared well in all cases within the limits of experimental error.

The minor difference between the experimental plots and theoretical curves may be attributed to the fact that individual points in the experimental plot corresponded to the lowest value of v_c in a particular set, while the theoretical curve was done with an average value or v_c of all the values in the complete set.

Four workers (Mullick 1986; Roy[1] 1986, Roy[2] 1988; Mehera 1989) carried out similar studies. Mullick (1986) observed up to Reynolds' number 645. Roy (1986) observed 73 sets up to Reynolds' number 5602.

Roy (1988) observed 70 sets up to Reynolds' number 305. Mehera (1989) observed 62 sets up to Reynolds' number 3797. They all studied with tubes of diameters 10.3 mm, 8 mm, 6 mm and 4 mm and of lengths about 1000 mm and 1200 mm employing different particles.

Paradox In all of the above observations, the lengths l_c determined experimentally were found to be agreeing with their theoretically derived values. High values of Reynolds' number appeared not to affect the lengths l_c to vary. This paradoxical statement demands explanation.

Paradox Explained When the flow is laminar, the directions of the flow vectors of the fluid elements are unidirectional in the direction of flow. Deviation from laminarity makes the distribution of the vectors random. These random vectors may be resolved into their components in the direction of flow and perpendicular to it. The sum of the components in the direction of flow over any cross section will give the same flow rate so long it remains the same. The sum of the perpendicular components amounts to zero as there is no flow in their direction. Under the influence of the perpendicular components, a particle will move as much up as down with its gravitational fall velocity remaining unimpaired throughout its length of travel.

During turbulence, the distribution of component velocity vectors in the flow direction on any particular line on the tube cross section, the line also being on a vertical plane, will be constantly changing. But this depth-wise variation of velocity will not affect settling, and l_c value will remain unaltered (De 1990).

Again in the light of the Velocity Profile Theorem (Alak 2009), the length travelled by the particle in falling through the entire depth depends on the area of the Velocity Profile diagram. So as long as the rate of flow through the line remains the same, the measure of the area of the Velocity Profile diagram through the line remains the same. The shape of the diagram not being of concern l_c value will also remain the same.

The l_c value of the particles is not affected because, as it appears, the particles that ought to have been scoured at very high value of Reynolds' number will continue to remain because of their configuration at the bottom of the sludge, while such other particles are duly scoured away with the flow from the top of the sludge bed.

Ghosh (1989) carried out an experiment with the same experimental set-up using kaolin and marble dust. Seven different angles of inclinations with four different diameters of tubes were employed to study the impairment of settling. A total of 112 runs were observed.

Impairments of settling were compared with the help of

$$\text{E value or Exponential value } (De\ 1983) = \text{Exp}\left[(-)\frac{v_s - v_c}{v_c}\right] \text{ for } v_s > v_c$$

where

v_c = critical fall velocity.
v_s = largest settling velocity of particle appearing in the effluent.
v_c is calculated from l,R,θ and the flow rate Q obtained by dividing a volume of effluent water collected in a measuring cylinder over a time period by the time period itself.

The collected water is filtered. The residue is desiccated. The dried particles are collected with a blade and put into a water column. Dividing the travel length of the foremost, i.e. the fastest streak in the measured time by the time itself, the largest settling velocity v_s of the particle in the effluent is found out.

Flow-through velocity in most (80.4 %) number of cases was greater than the minimum scouring velocity for the particle in the effluent with largest settling velocity. In such cases, therefore, scouring occurred in spite of which the tube gave ideal performance in a large number of cases in the sense that the particles with the largest settling velocity v_s in the effluent were with velocity less than the critical fall velocity.

This is so because, as it appears, those scoured particles in the tube that could get subsequent length of travel to settle within the tube settled. When the particles having settling velocities greater than the critical velocities scoured and could not get the subsequent length to settle within the tube, they were carried into the effluent so that the E-values observed were less than 1.

In some observations, flow-through velocity did not exceed the scouring velocity for the particle with largest settling velocity in the effluent. Even in some of such cases, E-value observed was less than 1. This may be ascribed to the fact that though a deflocculating agent was added, it is possible that the surfaces of the particles got washed so that flocculation might have taken place during determination of the largest settling velocity.

Ideal performance of the tube (E = 1) was observed up to Reynolds' number 1707. In spite of the ideal performance, scouring did occur. Impairment of settling was due to scouring.

15.5 Conclusions

The studies (Mullick 1986; Roy[1] 1986, Roy[2] 1988; Ghosh 1989; Mehera 1989) could conclude the following:

1. Even within a turbulent flow with a high value of Reynolds' number, settling of particles in the tube takes place in accordance with the theory deduced under laminar flow condition.
2. The settling of particles in a tube settler takes place according to the theory without the necessity and provision of an additional or transitional length for the development of laminarity in the flow. This length has been found to be redundant and may be done away with. It appears that no need is there to include the so-called transition length in the designed length of the tube settler.
3. Settling of particles in a tube settler is impaired, while the particles settle according to the theory even at very high value of Reynolds' number. This impairment is due to scouring.
4. In the design of the tube settler, the scouring should be the main consideration and not the Reynolds' number of flow.

Notations

L, l	Length of the tube
$2R$, d	Diameter of the tube
θ	Angle of inclination of the tube
Q	Flow rate through the tube
v_0	Mean velocity or flow-through velocity through the tube
v_c	Critical fall velocity
l_c	Critical length
v_s	Settling velocity, largest settling velocity
ν	Kinematic viscosity of the liquid
K_1, K_2, K_3, K_4	Constants

References

De A (1983) Parameter for settling tank performance comparison. J IPHE 4:21

De A (1990) Effect of short circuiting on the basin efficiency. J IPHE 2:37

De A (2009) Velocity profile theorem – concept for solving settling problem analysis. J IPHE, India

De A et.al (2009) Experimental verification of the theory of 'Tube settling'. J IPHE, India 2009–10(3)

Ghosh A (1989) Impairment of settling in a tube settler. MCE thesis, Jadavpur University, Kolkata, India

Mehera A (1989) An investigation into the extent of adherence of tube settling performance to its theory. MCE thesis, Jadavpur University, Kolkata, India

Mullick S (1986) Critical length determination for a tube settling system using activated carbon and marble dust. MCE thesis, Jadavpur University, Kolkata, India

Roy[1] T (1986) Sedimentation of sand and fly-ash in the light of tube settling theory-MCE Thesis, Jadavpur University, Kolkata, India

Roy[2] K (1988) An experimental study of tube settling parameters and their relationship -MCE Thesis, Jadavpur University, Kolkata, India

Yao KM (1970) Theoretical study of high rate sedimentation. J WPCF 42:218–228

Chapter 16
Residual of the Assorted Solids Through Shallow Depth Settler

Abstract Based on the settling data of raw water suspension, a methodology to compute the residual concentration of solids through a tube settler is developed herein. Laboratory settling data has been employed to illustrate the numerical application of the methodology to work out the effluent concentration of solids through a given tube settler carrying the raw water suspension at a given rate.

Keywords Assorted settling in tube • Residual computation • Flow velocity distribution • Settling velocity distribution • Settleables in effluent

16.1 Introduction

Proper development and design of an efficient settler calls for well-correlated settling theory with design of settler and its performance. Assessment of the likely performance of the settler requires computation of effluent concentration of solids from the settling characteristics of the raw water suspension feeding into it.

Complete theory of tube settling has been presented (De 1976, 2009a). Experimental study of the tube settling has also been accomplished (De 2009c). Design and control of tube settling module has been worked out (De 2009b). Settling characteristics of raw water suspension can be known from the 'revised mode of analysis' of column settling data (De 1998). Analysis on short circuiting (De 2009d) may be instrumental in the development of the computational methodology that this chapter is aiming at.

16.2 Literature Review

Theoretical study on tube settling was initiated by Yao (1970). Yao deduced the equation of trajectory of a particle entering through any point on the vertical diameter of the tube cross section. He could deduce the critical fall velocity through the settler.

A. De, *Sedimentation Process and Design of Settling Systems*,
Springer Transactions in Civil and Environmental Engineering,
DOI 10.1007/978-81-322-3634-4_16

De (1976, 2009a) (Chap. 14) deduced the general expression for particle trajectory through any point on tube cross section. Critical fall velocity through the tube was deduced. He established the complete theory of tube settling by working out the computation of the complete removal of particles having settling velocities equal to/more than the critical fall velocity and also the fractional removal of particles having settling velocities less than the critical fall velocity. The application was illustrated by solving numerical problem. It was shown (De 2009a) that

Total solids removal through the tube cross section

$$= w\Sigma C_s[Lv_s\cos\theta + (y_{1i} - y_{2i})\mathrm{v}_s\sin\theta] \tag{16.1}$$

Where $v_s = v_{ci}$

$$= \frac{Q(\text{Chord length})^3}{3\pi R^4\left(L\cos\theta + (\text{Chord length})\sin\theta\right)}$$

$$y_{1i} = R + \sqrt{R^2 - \alpha_i^2},\ y_{2i} = R - \sqrt{R^2 - \alpha_i^2}$$

for the i-th vertical section when $v_s \geq v_{ci}$; and

$v_s = v_s$, y_{1i} is calculated from

$$ay_{1i}^3 - by_{1i}^2 + cy_{1i} - d = 0 \tag{16.2}$$

where $a = \frac{2Q}{3\pi R^4 v_s\cos\theta}$; $b = 3aR$; $c = \left(3a\alpha_i^2 + \tan\theta\right)$;

$d = \left(ay_{2i}^3 - by_{2i}^2 + cy_{2i} - L\right)$ and $y_{2i} = R - \sqrt{R^2 - \alpha_i^2}$

for the i-th section when $v_s < v_{ci}$.

O'Connor and Eckenfelder (O'Conaor) plotted the concentration of solids obtained at depth '*d*' at time '*t*' in settling column test in depth-time coordinates. So-called isoconcentration curves were run through them. These curves were utilised to compute the total removal of solids corresponding to an overflow velocity employing a conclusion that is strictly valid for discrete suspension only.

The isoconcentration curve so drawn after O'Connor and Eckenfelder does not depict the unique characteristics of the suspension only as they largely depend upon the time of collection of samples from the ports of settling column. These concentrations may not give the interface concentrations crossing the port at the time of collection. The results arrived at, therefore, are liable to be erroneous and misleading.

The inadequacies of the method of the settling column analysis have been pointed out by the (De 1998) (Chap. 8). De plotted concentration versus time curves for each of the port at different depths. From this curve, a particular interface concentration crossing different depths at different times could be scaled out. Plotting these values in depth-time coordinates and subsequently connecting them, the trajectory of movement of the particular interface concentration over the depth-time diagram could be drawn, and these movements for different interface concentration depict the unique characteristics of the particular suspension. At any time, therefore, the different interface concentrations at different depths of setting column being known the total solid in suspension in settling column could be computed directly doing away with any other assumption. Decrease in the percentage of solids present in the column compared with the solid present initially in column, the total percentage removal of solids corresponding to an overflow velocity, computed by dividing the depth of the water column by the time under consideration, could be found out.

De (2009a) (Chap. 3) put forward the concept of 'Velocity Profile Theorem'. This theorem can be of very useful help in solving settling problem analysis. Based on this theorem, 'theory of ideal settling' and complete 'theory of tube settling' were deduced. Numerical problem was also solved in the way of illustration.

Application of the theorem was also demonstrated in the analysis on the phenomenon of short circuiting (De 2009d) (Chap. 9). The analysis shows that flow rate remaining the same the change in variation of flow velocity along the width results in short circuiting that impairs settling but that resulting from the depth-wise variation of flow velocity does not affect settling of the settleables. Short circuiting being the result of both of the above variations, the phenomenon of short circuiting impairs the settling of settleables.

The theory was subjected to experimental study (De 2009c) (Chap. 15). It is observed that:

1. Even within a turbulent flow with high value of Reynolds' number, settling of particles in the tube takes place in accordance with the theory deduced under laminar flow condition.
2. The settling in tube settler takes place according to the theory without the necessity and provision of an additional or transitional length for the development of laminarity in the flow. This length has been found to be redundant and may be done away with. It appears that no need is there to include the so-called transition length in the designed length of the tube settler.
3. Settling of particles in tube settler is impaired, while the particles settle according to the theory even at very high value of Reynolds' number. This impairment of settling is due to scouring.
4. In the design of tube settler, the scouring should be the main consideration and not the Reynolds' number of flow.

All the above experimental observations are well supported by critical analysis of shallow-depth sedimentation process.

16.3 Development of Methodology

16.3.1 Settling Characteristics of Settleables Through Shallow-Depth Settler

The movements of the interface concentrations of any settling suspension can be plotted in depth-time coordinates by performing settling column test (De 1998) (Chap. 8). In all cases, the initial and the trailing portions of the movement trajectories are found to be linear with a curvilinear portion in between. The linearity implies the settling of the interface concentrations at constant rate. The linearity, thus, indicates discrete settling. During the initial phase defined by the initial straight trajectories, no appreciable agglomeration of solids takes place to affect the settling velocity distribution among the particles. The process of agglomeration initiates the accelerated movement when the straight enters into curvature. During the phase defined by the final straight portion, the process of agglomeration ceases to exist because shearing erodes the deposition on the surface of the particles maintaining their shape and size; hence settling velocities are the same.

The depth allowed in shallow-depth settler is well within the depth to which the interface concentrations of any settling suspension exhibit initial linear trajectory of movement in depth-time coordinates.

Shallow-depth settler, therefore, removes settleables of any suspension, while the particles are exhibiting discrete settling in passing through the settler.

16.3.2 Settling Velocity Distribution Among the Particles Exhibiting Discrete Settling in Shallow-Depth Settler

The settling column test data with any raw water suspension may be utilised to plot the movement trajectories of the interface concentrations in depth-time coordinates (De 1998) (Chap. 8). The initial straights of these trajectories may be employed to find the settling velocity distribution among the particles while they are settling in a shallow-depth settler. This is illustrated with the solution of the following Problem 1.

Problem 1 A waste water was subjected to settling column analysis. Laboratory settling data of concentration of suspended solids remaining at indicated depth at indicated times were as follows:

Initial concentration of solids 540 mg/l; temperature of water 30 °C

Solids concentrations in mg/l at indicated depths and times

Time in min	60 cm	120 cm	180 cm
5	275	362	386
10	189	259	312

(continued)

Time in min	60 cm	120 cm	180 cm
20	135	188	232
40	90	119	162
60	92	92	118
120	93	92	93

Find the settling velocity distribution among the particles during initial settling.

Solution

Step 1: From the observed data, it is apparent that the waste water contains non-settleable solids of {(90+ 92+ 93 + 92 + 92 + 93)/6=} 92 mg/l. In the computation of removal by settling, therefore, analysis is to be carried out with the settleables only. The observed data with the settleables may be retabulated as presented in Table 16.1.

Step 2: The data presented in Table 16.1 is utilised to plot the concentration versus time curves at each depth of collecting samples. They are presented in Fig. 16.1. The curve for each depth gives the different interface concentrations crossing the depth at different times.

Step 3: Any particular interface concentration may be chosen. The times at which this interface concentration crosses the different depths can be scaled from curves presented in Fig. 16.1.

In the present case, 180 mg/l, 140 mg/l, 100 mg/l, 60 mg/l, 20 mg/l and 0 mg/l interface concentrations are chosen. The times at which each of the above interface concentrations of settleables crosses the different depths are scaled from Fig. 16.1. For each of the above interface concentrations, the data, so derived, are plotted in depth-time coordinates. Connecting the points for each of the concentrations by smooth curves, the movement trajectories of the above interface concentrations down the depth-time diagram are obtained. They are presented in Fig. 16.2. Each of the curves in 16.2 has initial and final states with a curvilinear portion in between.

Let us consider the curves within the initial phase of discrete settling that extends to a depth of 30 cm. In Fig. 16.2, the movement trajectories of the surfaces of the other interface concentrations C_n between 448 mg/l and 180 mg/l have not been done.

Table 16.1 Settling column test data with regard to the settleable solids
Initial concentration of settleable solids = 448 mg/l; temperature of water = 30 °C
Concentration of settleable solids in mg/l at indicated depth and time

Time in min	60 cm	120 cm	180 cm
5	183	270	294
10	97	167	220
20	43	96	140
40	0	27	70
60	0	0	26
120	0	0	0

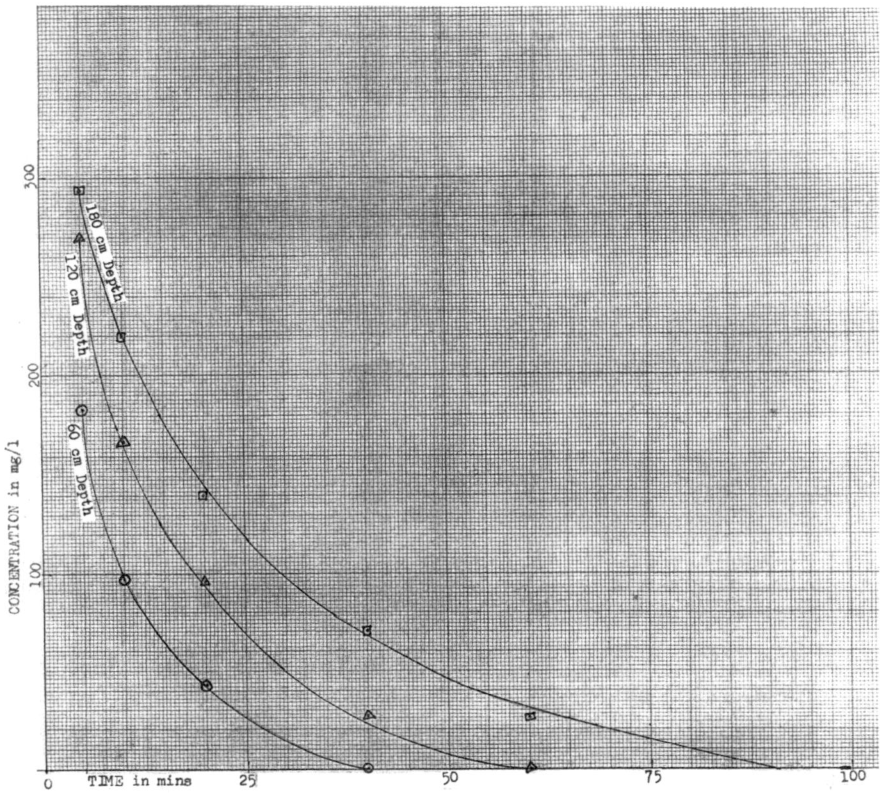

Fig. 16.1 Concentration versus time curves

From the Fig. 16.2 can be scaled the times 3 min, 4 min, 6 min, 10 min, 16 min and 25 min at which the surfaces of the interface concentrations 180 mg/l, 140 mg/l, 100 mg/l, 60 mg/l and 0 mg/l reach the depth 30 cm, respectively.

Let us imagine such other curves for interface concentration $C_1, C_2, C_3, \ldots\ldots\ldots\ldots$ $C_n \ldots\ldots\ldots\ldots\ldots$ 448 mg/l ($C_1 < C_2 < C_3 \ldots\ldots\ldots\ldots$- $< C_n \ldots\ldots\ldots\ldots\ldots < 448$ mg/l) are drawn and their surfaces reach the depth of 30 cm at times t_1, t_2, $t_3, \ldots\ldots\ldots$ $t_n \ldots\ldots(t_1 > t_2 > t_3 \ldots\ldots\ldots\ldots > t_n \ldots)$ and finally 0 min (assuming the surface of concentration 448 mg/l reach the depth almost in no time compared with others).

This implies that the solids composing (C_1–180) mg/l, (C_2–C_1) mg/l, (C_3–C_2) mg/l,......... will move through the depth at 30 cm over time intervals (3–t_1) min, (t_1–t_2) min, (t_2–t_3) min,......etc. at time t_1, t_2 and t_3, respectively.

This shows (448–180) mg/l of solids are composed of different fractions of varying settling velocities that fall through the distance of 30 cm over varying falling through times. It appears reasonable approximation to assume the mean time of fall of 268 mg/l of solids as ½ (0 + 3) min, i.e. 1.5 min. Thus (180–140) mg/l,

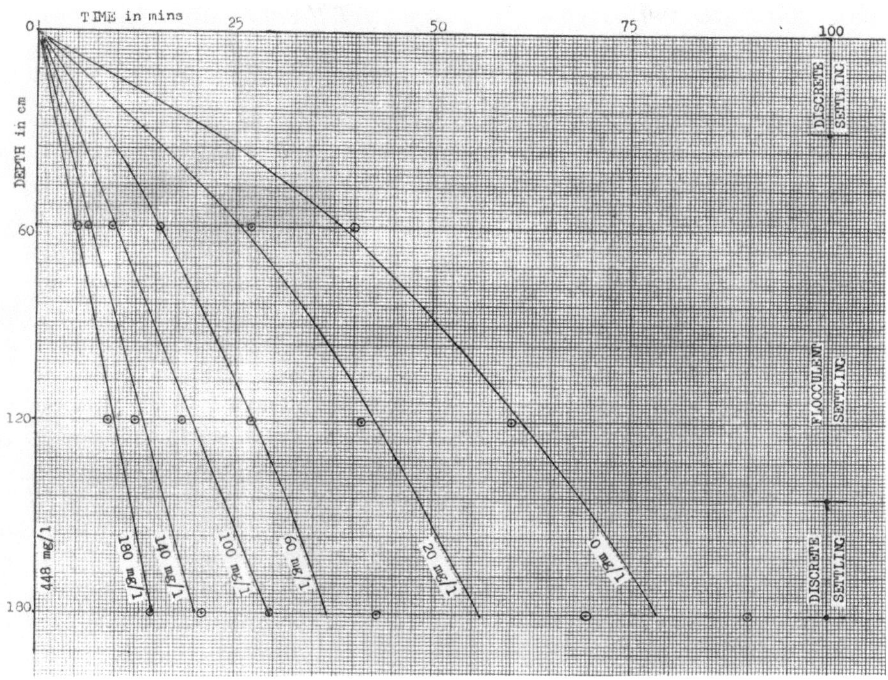

Fig. 16.2 Trajectories of interface concentrations

Table 16.2 Settling velocity distribution during the initial phase of discrete settling

	Concentration in mg/l	Fall through distance, cm	Mean falling through times, min	Settling velocity in cm/s
(a)	268	30	1.5	0.33333
(b)	40	30	3.5	0.14286
(c)	40	30	5	0.10000
(d)	40	30	8	0.06250
(e)	40	30	13	0.03846
(f)	20	30	20.5	0.02439

i.e. 40 mg/l; (140–100) mg/l, i.e. 40 mg/l; (100–60) mg/l, i.e. 40 mg/l; (60–20) mg/l, i.e. 40 mg/l; and 20 mg/l have average falling through times ½ (3 + 4) min, i.e. 3.5 min; ½ (4 + 6) min, i.e. 5 min; ½(6 + 10) min, i.e. 8 min; ½(10 + 16), i.e. 13 min; and ½(16 + 25) min, i.e. 20.5 min, respectively.

Thus the settling velocity distribution among the settleable solids during their initial phase of discrete settling is presented in Table 16.2. Needless to say, the more closely the curves for interface concentrations are spaced, the more accurate details of the distribution will be obtained.

16.3.3 Flow Velocity Distribution Over the Tube Cross Section

In turbulence, the flow vectors are randomly distributed varying in magnitude and direction. The vectors can be resolved into components along the direction of flow and perpendicular to it. If the cross section is divided into vertical strips of infinitesimally small thickness, the sum of the components of flow vectors on any vertical plane through the central line of the strip in the direction of flow through the section will give the flow along it, and the sum of the perpendicular components of the flow vectors vanishes since there is no flow in the perpendicular direction. Any particle under the influence of the perpendicular components of the flow vectors, therefore, will be affected as much upwards as downwards resulting in no net displacement of the particle in the direction perpendicular to the direction of flow. The particle is carried forward by the components of the flow vectors in the direction of flow. The distance to which the particle is carried through depends upon the area of particle velocity diagram and the area of flow velocity diagram of the components of flow vectors along the flow direction between its point of entry and its fall through distance perpendicular to the flow direction. For the particle entering at the top and reaching the bottom, the travel distance along the flow direction depends on the area of the Velocity Profile diagram of the components in the flow direction and not its shape. This is demonstrated (De 2009a, d) (Chaps. 9 and 3) that so long as the area of the flow diagram remains the same, the changing of its shape does not change the distance it is carried through.

It must be borne in mind that the carry through distance of a particle entering through any intermediate point is proportional to the area of Velocity Profile diagram below the point. In this case, even if the flow rate through the vertical containing the particle remains the same, the flow velocity distribution change over it, may change the carry through distance of the particle.

Any change in shape in the diagram will increase the velocity vector at one point or points and decrease the same at one or more other points and vice versa. The sum of the decreases of the vectors is equal to the sum of the increases of the other vectors. This implies that for these changes, the carry through distances of the particles will be distributed about a mean carry through distance within the limits of changes of the shape of the flow diagram.

The decrease of carry through distance increases removal, and the increase of carry through distance decreases the solids removal. On the whole, the removal about the mean is likely to remain the same particularly at low concentration of solids after leaving aside the removal of solids having settling velocities more than equal to the critical velocity through the tube at the given flow rate.

Disregarding the flow condition of laminarity or turbulence, if the shape of the flow velocity diagram is redistributed in accordance with the flow velocity,

$$f_{y,\alpha} = \frac{2Q}{\pi R^4}\left(2yR - y^2 - \alpha^2\right)\text{ i.e.}$$

The Velocity Profile area diagram is redistributed in accordance to the flow Velocity Profile vector:

$$u_{y,\alpha} = \frac{2Q}{\pi R^4}\left(2yR - y^2 - \alpha^2\right) - v_s \sin\theta$$

The distance travelled through by the particle entering at the top remains unaltered. This is in conformity with the experimental observation (De 2009c) (Chap. 15) that even at high values of Reynolds' number of flows, the critical lengths determined experimentally agreed well with the critical length computed with the laminar flow equation.

16.3.4 Computation of Removal of Solids

The tube cross section is divided into even number (for convenience of calculation) of vertical strips of width '*w*'. Application of 'Velocity Profile Theorem' (De 2009a) (Chap. 3) to the discrete settling particles through shallow-depth settler gives

The total removal of solids (through all the strips)

$$= 2w\Sigma C_s[Lv_s\cos\theta + (y_{1i} - y_{2i})v_s\sin\theta] \tag{16.1}$$

C_s – Concentration of solids with settling velocity v_s
When $v_s \geq v_{ci}$
Put $v_s = v_{ci}$

$$= \text{Critical fall velocity through } i - \text{th strip}$$

$$= \frac{Q(\text{Chord length})^3}{3\pi R^4\left(L\cos\theta + (\text{Chord length})\sin\theta\right)}$$

$$y_{1i} = R + \sqrt{R^2 - \alpha_i^2},\; y_{2i} = R - \sqrt{R^2 - \alpha_i^2}$$

and when $v_s < v_{ci}$
put $v_s = v_s$; y_{1i} is calculated from the Eqn.

$$ay_{1i}^3 - by_{1i}^2 + cy_{1i} - d = 0 \tag{16.2}$$

where $a = \frac{2Q}{3\pi R^4 v_s\cos\theta}$; $b = 3aR$; $c = \left(3a\alpha_i^2 + \tan\theta\right)$;

$d = \left(ay_{2i}^3 - by_2^2 + cy_{2i} - L\right)$ and $y_{2i} = R - \sqrt{R^2 - \alpha_i^2}$
for the i-th section.

16.4 Application to Numerical Problem

Problem 2 A 50 cm long tube of diameter 5 cm, inclined at an angle of 30° with the horizontal, is employed for the removal of solids from a flow rate of 0.06 litres/s of waste water; the settling column test data of which is presented in Problem 1. Calculate the effluent concentration of solids.

Solution Divide the cross section into ten vertical strips, each of width 0.5 cm, five strips being on either side of the vertical diameter, marked by the identification numbers of their central chords. The problem is worked out in steps as tabulated in Tables 16.3 and 16.4.

$$\begin{aligned} y_{21} &= R - \sqrt{R^2 - \alpha_i^2} \\ &= 2.5 - \sqrt{2.5^2 - 0.25^2} \\ &= 0.01253; \end{aligned}$$

$$d = ay_{21}^3 - by_{21}^2 + cy_{21} - L$$

$$= 1.12913 \times 0.0125 - 8.46848 \times 0.0125 + 0.78906 \times 0.01253 - 50$$

$$= -49.99144;$$

Equation 16.2 may be written with the above constants:

$$1.12913\, y_{11}^3 - 8.46848\, y_{11}^2 + 0.78906\, y_{11} + 49.99144 = 0$$

The solution of the above Eqn. gives

$$y_{11} = 3.35264;$$

The rest of Table 16.4 is self-explanatory.

Table 16.3 Illustrating the calculation of y_{11} for $v_s = 0.33333\,\text{cm/s}$

$a = \dfrac{2Q}{3\pi R^4 v_s \cos\theta}$	$b = 3aR$	$c = \left(3a\alpha_i^2 + \tan\theta\right)$
$= \dfrac{2\times 60}{3\pi(2.5)^4 \times 0.33333 \cos 30°}$	$= 3 \times 1.12913 \times 2.5$	$= 3 \times 1.12913 \times 0.25^2 + \tan 30°$
$= 1.12913$	$= 8.46848$	$= 0.78906$

Table 16.4 Calculation of settleable solids in the effluent

1. No.	Steps	Chord identification numbers i=				
		(1)	(2)	(3)	(4)	(5)
1.	Distance of the chord from the centre α_i cm	0.25	0.75	1.25	1.75	2.25
2.	Chord length $= 2\sqrt{R^2 - \alpha_i^2}$cm	4.97494	4.76970	4.33013	3.57071	2.17945
3.	The bottom point of the chord $-y_{2i}$ =R-$\sqrt{R^2 - \alpha_i^2}$cm	0.01253	0.11515	0.33494	0.71464	1.41028
4.	Critical velocity v_{ci} through the chordal section in cm/s	0.43824	0.38708	0.29102	0.16456	0.03801
5	$y_{1i} = 2R - y_{2i}$, where $v_s \geq v_{ci}$	4.98747	4.88485	4.66506	4.28536	3.58972
6.	y_{1i} is calculated from Eq. 16.2 where $v_s < v_{ci}$; and $y_{1i} = R + \sqrt{R^2 - \alpha_i^2}$ cm where $v_s \geq v_{ci}$	4.98747	4.88485	4.66506	4.28536	3.58972
	(a) $v_s = 0.33333\,\text{cm/s}(268\text{mg/l})$	3.35264	3.71975	4.66506	4.28536	3.58972
	(b) $v_s = 0.14286\,\text{cm/s}\,(40\text{mg/l})$	1.87277	2.04148	2.43898	3.44538	3.58972
	(c) $v_s = 0.10000\,\text{cm/s}(40\,\text{mg/l})$	1.51894	1.66903	2.01130	2.73270	3.58972
	(d) $v_s = 0.06250\,\text{cm/s}(40\text{mg/l})$	1.16749	1.30283	1.60467	2.18971	3.58972
	(e) $v_s = 0.03846\,\text{cm/s}(40\text{mg/l})$	0.89845	1.02438	1.30163	1.81765	3.58972
	(f) $v_s = 0.02439\,\text{cm/s}(20\text{mg/l})$	0.70723	0.82724	1.08948	1.56700	2.69876
7.	Velocity v_s with which particle falls from y_{1i} to y_{2i} travelling through the tube length L, in cm/s	(a) 0.33333	0.33333	0.29102	0.16456	0.03801
		(b) 0.14286	0.14286	0.14286	0.14286	0.03801
		(c) 0.10000	0.10000	0.10000	0.10000	0.03801
		(d) 0.06250	0.06250	0.06250	0.06250	0.03801
		(e) 0.03846	0.03846	0.03846	0.03846	0.03801
		(f) 0.02439	0.02439	0.02439	0.02439	0.02439
8.	Computation of $\sum (C_s L v_s \cos\theta)_i$	$\sum_i (C_s L v_s \cos\theta)_i$ $= L\cos\theta\left[(C_s \sum_i v_s)\text{for(a)} + (C_s \sum_i v_s)(\text{for (b), (c), (d), (e)}) + (C_s \sum_i v_s)(\text{for (f)})\right]$ $= 50\cos 30\theta°(268 \times 1.16025 + 40 \times 0.60945 + 40 \times 0.43801 + 40 \times 0.28801 + 40 \times 0.19185 + 20 \times 0.12195)$ $= 16215.408\,\text{cm} \times \text{mg}/1000\,\text{cm}^3 \times \text{cm/sec}$ i.e.16.215408 mg/cm.sec				

(continued)

Table 16.4 (continued)

1. No.	Steps	Chord identification numbers i=				
		(1)	(2)	(3)	(4)	(5)
9	Computation of $(y_{1i} - y_{2i})$cm from step 3 and step 6	(a) 3.34011	3.60460	4.33012	3.57072	2.17944
		(b) 1.86024	1.92633	2.10404	2.73074	2.17944
		(c) 1.50641	1.55388	1.67636	2.01806	2.17944
		(d) 1.15496	1.18768	1.26973	1.47507	2.17944
		(e) 0.88592	0.90923	0.96669	1.10301	2.17944
		(f) 0.01694	0.01737	0.01840	0.02079	0.03143
10	Computation of $(y_{1i} - y_{2i})v_s$	(a) 1.11336	1.20152	1.26015	0.58760	0.08284
		(b) 0.26575	0.27520	0.30058	0.39011	0.08284
		(c) 0.15064	0.15539	0.16764	0.20181	0.08284
		(d) 0.07219	0.07423	0.07936	0.09219	0.08284
		(e) 0.03407	0.03497	0.03718	0.04242	0.08284
		(f) 0.00041	0.00042	0.00045	0.00051	0.00077
11	Computation of $\sum_i (C_s(y_{1i} - y_{2i})v_s \sin\theta)_i$	$\sum_i (C_s(y_{1i} - y_{2i})v_s \sin\theta)_i$ $= \sin\theta\left[\left(C_s\sum_i (y_{1i} - y_{2i})v_s \text{for(a)}\right) + \left(C_s\sum_i (y_{1i} - y_{2i})v_s \text{for, (b), (c), (d)(e)}\right) + \left(C_s\sum_i (y_{1i} - y_{2i})v_s \text{for (f)}\right)\right]$ $= \sin 30^\circ\,(268 \times 4.24547 + 40 \times 2.70509 + 20 \times 0.00256)\text{mg}/1000\,\text{cm}^3 \times \text{cm} \times \text{cm/s} = 0.624\,\text{mg/cm.s}$				
12	Solids removed					
	$= 2w\sum_i C_s[(Lv_s\cos\theta) + (y_{1i} - y_{2i})v_s \sin\theta] = 2.0.5\,(\text{cm})(16.215 + 0.624)\text{mg/cm.s} = 16.839\,\text{mg/s}$					
13	Concentration of settleable solids in the effluent					
	$= \frac{(\text{Settleable solids entering per sec.} - \text{Settleable solids removed per sec.})}{\text{Flow rate}} = \frac{\left(\frac{448\,\text{mg}}{\text{l}}\right)\left(\frac{0.061}{\text{s}}\right) - (16.839\,\text{mg/s})}{\left(\frac{0.061}{\text{s}}\right)} = 167.35\,\text{mg/l}$					
14	Total solid concentration in the effluent					
	$= (\text{Settleable solids concentration} + \text{non} - \text{settleable solids concentration}) = (167.35 + 92)\text{mg/l i.e.} 259.35\,\text{mg/l}$					

16.5 Conclusions

It follows, therefore, to conclude that:

1. Relevant literature for the development of methodology for computing the concentration of residual solids through a tube settler has been presented.
2. Computational methodology for the residual settleables through tube settler has been developed.
3. A numerical problem has been worked out, in the way of illustration, based on the methodology developed herein to find the concentration of residual solids through a tube settler for a waste water flowing through it the laboratory settling data being given.

Notations

L	Length of tube settler
$2R$	Tube diameter
θ	Inclination of the tube with horizontal
Q	Flow rate
w	Width of the vertical strip
α_i	Distance of the central line of the i-th strip from the vertical diameter of the tube
v_s	Settling velocity of particles
$f_{y,\alpha}$	Flow velocity at (y,α) of tube cross section
$u_{y,\alpha}$	Profile velocity at (y,α) of tube cross section
C_s	Concentration of solid particles with settling velocity
v_{ci}	Critical velocity for particles flowing through the i -th strip
y_{1i}	'y'-coordinates of the particles entering through the i-th strip and just reaching the bottom travelling through the entire length L of the tube settler
y_{2i}	Particle entering through y_{1i} and settles to the bottom at y_{2i}

References

De Alak Kumar (1976) Conceptual studies on discrete and flocculent settling, Ph.D Thesis (Engg), Jadavpur University, Kolkata

De A (1998) Revised mode of analysis of column settling data. Indian Chem Eng, Section B, 40:392–400

De A (2009a) Velocity profile theorem- concept for solving settling problem analysis. J IPHE India 2009–2010(1)

De A (2009b) Theoretic study on the control of design parameters for tube settling. J IPHE India 2009–2010(2)

De A (2009c) Experimental verification of the theory of 'Tube Settling'. J IPHE India 2009–2010 (3)

De A (2009d) Analysis of the effect of short-circuiting on settling – an application of 'Velocity Profile Theorem', J IPHE India 2009–2010(4)

De A (2010) Computation methodology for residual solids through tube settler. J IPHE India 2010–2011(1)

O'Connor, Eckenfelder WW, Jr. Evaluation of laboratory settling data for process design, biological treatment of sewage and industrial wastes, vol 2. Reinhold Publishing Corporation, New York

Yao KM (1970) Theoretical study of high hate sedimentation. J WPCF 42:218

Chapter 17
Control Application on Design Parameters

Abstract The removal of solids through a tube settler depends on the parameters like tube length (L), tube radius (R), tube inclination (θ) and flow rate through the tube (Q). This chapter studies the limits of the parameters and the influence of their changes on critical velocity. The control application to the design of tube settler has been illustrated with a worked out example.

Keywords Design parameters • Limitation of inclination • Limitation of mean flow • Limitation of length • Limitation of diameter

17.1 Introduction

Settling is an important operation in solid-liquid separation. The cost-saving potentiality of shallow-depth settling system has been known for a long time. But it is only in the mid-1960s of the last century that the shallow-depth sedimentation could be implemented with inclined tubes, trays, etc.

For controlling the parameters of tube settling presented herein is a procedure to control the design parameters of tube settling system to fix their coordinated values for optimised design.

Settling process finds indispensable application in solid-liquid separation in the treatment of water, waste water and also in chemical and mining industries. The application consumes a large proportion of the cost of investment and operation of the total application system. Optimisation of this cost calls for minimising the volume and maximising the efficiency of operation of the application unit.

As early as in 1904, Allen Hazen (Hansen and Culp 1967) advocated shallow-depth sedimentation for its cost-saving potentiality with insertion of horizontal trays in settling tank. This could not be implemented because shallow-depth flow resulted in unstable hydraulic condition and created unsurmountable problem as regards the installation of sludge removal mechanism.

Shallow-depth sedimentation in tubes, trays, etc. resolved the above problems. Maintaining proper Reynolds' number, stable hydraulic condition was obtained. Inclined configuration of the tubes, trays, etc. resulted in automatic draining of

A. De, *Sedimentation Process and Design of Settling Systems*,
Springer Transactions in Civil and Environmental Engineering,
DOI 10.1007/978-81-322-3634-4_17

sludge and thus doing away with the necessity of the installation of the sludge removal mechanism.

The existing procedure for designing tube settling system is based on arbitrary choice of the values of its control parameters and using empirical relations and values. Such procedure runs into the risks of all those associated with the use of empirical formulae and as such provides very poor control on setting the parameters for efficient settling performance.

This chapter investigates into the theoretic study on fixing the values of the design parameters for their control for efficient settling.

17.2 Design Parameters

Through a tube of length L, diameter $2R$, inclined at an angle θ with the horizontal, all particles having settling velocity $v_s \geq v_c$ will be removed completely from the flow rate Q, where

$$\text{Critical fall velocity } (Yao) v_c = \frac{8Q}{3\pi R(L\cos\theta + 2R\sin\theta)} \tag{17.1}$$

$$= \frac{8 v_{\text{mean}} R}{3(L\cos\theta + 2R\sin\theta)} \tag{17.2}$$

To maintain v_c through the tube, the parameters $\theta, v_{\text{mean}}, R, L$ are to be controlled.

17.3 Influence of the Changes in the Parameters on the Critical Fall Velocity (De 2009)

From Eq. 17.2 $v_c = \phi(\theta, v_{\text{mean}}, R, L)$

$$\mathrm{d}\phi = \frac{\partial\phi}{\partial\theta}\mathrm{d}\theta + \frac{\partial\phi}{\partial v_{\text{mean}}}\mathrm{d}v_{\text{mean}} + \frac{\partial\phi}{\partial R}\mathrm{d}R + \frac{\partial\phi}{\partial L}\mathrm{d}L \tag{17.3}$$

and also (from Eq. 17.2)

$$\frac{\partial\phi}{\partial\theta} = \frac{v_c(L\sin\theta - 2R\cos\theta)}{(L\cos\theta + 2R\sin\theta)} \tag{17.4}$$

$$\frac{\partial\phi}{\partial v_{\text{mean}}} = \frac{8R}{3(L\cos\theta + 2R\sin\theta)} \tag{17.5}$$

$$\frac{\partial\phi}{\partial R} = \frac{v_c L\cos\theta}{\mathrm{R}(L\cos\theta + 2R\sin\theta)} \tag{17.6}$$

$$\frac{\partial \phi}{\partial L} = (-)\frac{v_c \cos\theta}{(L\cos\theta + 2R\sin\theta)} \tag{17.7}$$

Putting Eqs. 17.4, 17.5, 17.6 and 17.7 in Eq. 17.3 change in v_c for known small changes in the parameters can be obtained. The same change can be calculated from the Eq. 17.1.

But the real significance of the Eq. 17.3 lies in the fact that the necessary changes in parameters may be adjusted to the desired change in v_c which may require several trials with Eq. 17.1 to ones' inconvenience. Assigning any arbitrary values to the parameters, say,

$$L = 100\,\text{cm},\ 2R = 5\,\text{cm},\ \theta = 10^\circ,\ Q = 60\,\text{cm}^3/\text{s};$$
$$v_c(\text{Calculated from Eq.17.1}) = 0.20505\,\text{cm/s};$$
$$v_{\text{mean}}(\text{Computed}) = 3.05577\text{cm/s};$$
$$\frac{\partial \phi}{\partial \theta} = 0.02568\text{cm/s/degree (positive value)};$$
$$\frac{\partial \phi}{\partial v_{\text{mean}}} = 0.06710\,(\text{positive value});$$
$$\frac{\partial \phi}{\partial R} = 0.0813/\text{s(positive value)};$$
$$\frac{\partial \phi}{\partial L} = (-)0.00203/\text{s (negative value)}$$

It is at once apparent that with increase in θ, v_{mean} and R, the critical velocity v_c increases and clarification deteriorates. Increasing L decreases v_c, and clarification improves as a result.

The expression may provide the means for the necessary adjustments of the parameters in quantitative measure for setting the critical velocity for desired removal of solids when the parameters happen to change within limits.

17.4 Control Limitations of the Design Parameters (De 2009)

17.4.1 Limitation of the Angle of Inclination (θ) of the Tube

The inclination of the tube (θ) with the horizontal aims at promoting automatic draining of sludge. Increase of the angle of inclination (θ) deteriorates clarification. For efficient settling, therefore, the angle (θ) of inclination should be kept down to a minimum subject to the condition that it should maintain automatic gravity draining of sludge.

In limiting case the automatic draining of sludge will initiate if the angle of inclination just exceeds the angle of repose of the same. When in water and the sludge particles are at the point of movement, the resistance to the movement due to interlocking among the particles appears to be reduced to minimum due to lubrication by water. With high degree of precision, angle of repose may be approximated to angle of friction (α).

Leaving aside gravels (60–2 mm) of soil classification, sludge may appear as various combinations of sand (2–0.06 mm), silt (0.06–0.002 mm) and clay (<0.002 mm).

Presented in Table 17.1 are the angles of friction (Vazirani, Chandola) for some classified soils.

It may be observed from the Table. 17.1 that:

1. Coarser grains have greater angle of friction (α).
2. In mixture relative fractions of coarser and finer grains determine the angle (α) of friction.
3. Finer particles produce lubricating effect to the coarser ones to reduce the angle (α) of friction.
4. Water produces lubrication and brings down the angle of internal friction.

From preliminary experiments with single tubes, Hansen et.al. (Hansen and Culp 1967) observed that an inclination of 5° was found suitable for gravity draining of sludge.

From Table 17.1, sludge composition with predominantly clays could result gravity draining of sludge with the angle of inclination of 5°.

It appears imperative to decide the minimum value of the angle (θ) of inclination of the tube with gravity draining of sludge after due experimentation.

In the absence of the same, three values of the angle of inclination of the tube θ $(=\alpha) = 5°$, 10° and 30° may be identified.

Any other intermediate values of θ may not be beneficial. For coarser minorities will be carried riding on the finer majority, their angle of internal friction is being reduced to that of the finer majority grains as the coarser grains are lubricated by the flow of the fines.

The angle of inclination (θ) of the tube should be minimum and should be equal to the angle of internal friction of the major fraction of particles, i.e. if the major fraction of particles are sand, $\theta = \alpha = 30°$; if it is silt, $\theta = \alpha = 10°$; and if it is clay, $\theta = \alpha = 5°$. The finer fraction will readily undergo gravity draining because they have lesser angle of friction.

Table 17.1 Angle of internal friction

Soil classification	Angle of internal friction (α)
Sand and gravel mixture	34°
Sand and clay mixture	30°
Wet sand	34°
Silt	10°
Clay-liquid	0°
Clay-soft	2°–4°
Clay-stiff	4°–8°

17.4.2 Limitation of Mean Flow-Through Velocity (v_{mean})

Figure 17.1 shows the tube cross section with the sludge draining through. If h be the loss of head across the length of the tube L, the stress τ exerted by the flowing water on its contact surface

$$\tau = \frac{(hw)(\pi R^2)}{(L)(2\pi R)} = \text{wri},$$

$$i\,(\text{hydraulic gradient}) = \frac{h}{L};$$

$$r\,(\text{hydraulic radius}) = \frac{R}{2};$$

This stress τ tends to drag the weight component of the unit areas of the layer of single particles of diameter d at porosity p down the slope α, in contact with water. w and w_s are the specific weights of water and solid particles. In limiting case,

$$\text{wri} = (1 \times 1 \times d)(1 - p)(w_s - w)\sin\alpha$$

$$v_{\text{mean}} = C[d(1 - p)(S - 1)\sin\alpha]^{\frac{1}{2}}$$

$$\text{From}\; v_{\text{mean}} = c\sqrt{ir}$$

Putting Manning's evaluation of Chezy's constant

$$C = \frac{1.49}{n} r^{\frac{1}{6}}$$

and also $k = (1 - p)\sin\alpha$, where k is a constant depending on sediment character. k ranges 0.04 (for most clean) to 0.8 (for most gritty sediment)(Fair).

In order that the sediments of diameter d may not be lifted and the gravity draining of sludge is not impaired, v_{mean} is to be limited to

$$v_{\text{mean}} = \frac{1.49}{n} r^{\frac{1}{6}}[k(S - 1)d]^{\frac{1}{2}}$$

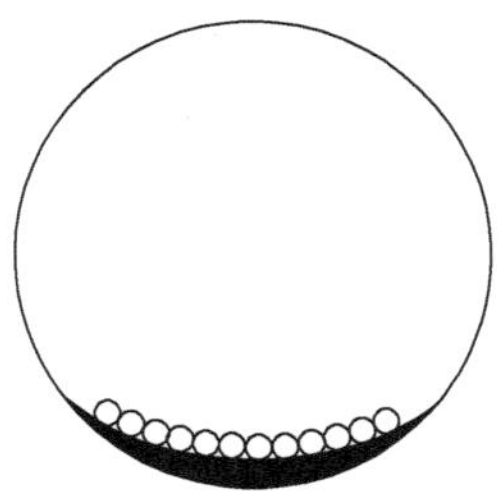

Fig. 17.1 Tube cross section showing single layer particles in contact with water

The diameter d should be the lowest diameter of particles to be removed completely.

17.4.3 *Limitation of Diameter (2R) of the Tube*

Clarification improves with the decrease of the diameter of the tube. But such a decrease in diameter is limited by the condition of laminarity with reasonable rate of flow, if it has to be maintained. For laminarity of flow, Reynolds' number has to be limited within (R_e=) 2300. Let Reynolds' number be limited to (R_e=) 2000.

$$\text{i.e. } \frac{v_{\text{mean}}(2R)}{\nu} = 2000$$

$$\text{i.e. } 2R = \frac{2000\nu}{v_{\text{mean}}}$$

17.4.4 *Limitation of the Length (L) of the Tube*

Increase in the length of the tube results in better clarification. But increased length increases the loss of head across the same. Hence the length of the tube is limited by the calculated length for the desired performance.

Having decided $\theta, v_{\text{mean}}, 2R$, the length of the tube can be calculated from Eq. 17.1 with chosen value of v_{cr} as

$$L = 2R\left(\frac{4v_{\text{mean}}}{3\,v_c\,\cos\theta} - \tan\theta\right)$$

17.5 Control Application to the Design of Tube Settler (De 2009)

Problem Find the controlled dimensions and configuration of tube settler to remove solids from 5 MLD of water. The analysis on sludge suspension revealed that the composition of the sticky sludge was

Gravels (60–2 mm)	10 %
Sands (2–0.06 mm)	70 %
Silts (0.06–0.002 mm)	8 %
Clays (0.002 mm)	12 %

Step 1: The major fraction of solids is sand (70 %). The angle of inclination (θ) of the tube settler is decided to be the angle of friction (α) of sand, i.e. $\theta = \alpha = 30\,^{\circ}$. The silts and clays will readily drain. The coarser gravels will be carried by the finer fractions being lubricated by the fines and water.

Step 2: The lowest diameter to be removed completely is 0.06 mm. Let us use $2R = 6$ cm dia tube:

$$\text{For}\, n = 0.013,\, d = 0.006\,\text{cm},\, k = 0.8,\, S = 2.65$$

$$v_{\text{mean}} = \frac{1.49}{0.013}\left(\frac{6}{4}\right)^{\frac{1}{6}}[0.8(2.65-1)0.006]^{\frac{1}{2}} = 10.91325\,\text{cm/s}$$

Step 3: Limiting the Reynolds' number to 2000 for limiting diameter $2R = 6$ cm for $\nu_{20\,^{\circ}\text{C}} = 1.0105 \times 10^{-2}\text{cm}^2/\text{s}$:

$$v_{\text{mean}} = \frac{2000 \times 1.0105 \times 10^{-2}(\text{cm}^2/\text{s})}{6\,\text{cm}} = 3.36833\,\text{cm/s} < 10.91325\,\text{cm/s}$$

$$\text{Reynolds' number} = \frac{3.36833 \times 6}{1.0105 \times 10^{-2}},\ \text{i.e.}\,2000 < 2300.$$

Step 4: At 20 °C the maximum diameter of such particle that will obey Stoke's law is 0.48 mm. The settling velocity of particle of diameter 0.006 cm by Stoke's law $v_c = 0.320$ cm/s.

$$L = 6\left(\frac{4 \times 3.36833}{3 \times 0.320 \cos 30^{\circ}} - \tan 30^{\circ}\right) = 93.77121\,\text{cm, Adopt}\,100\,\text{cm}$$

The tube settler and its controlled set-up:

The tube settler:

For 5 MLD the number of tubes

$$= \frac{5 \times 10^6 \times 1000\,(\text{cm}^3)}{(24 \times 60 \times 60)\,(\text{secs})\,(\pi 3^2\,\text{cm}^2)(3.36833\,\text{cm/s})} = 607.64,\ \text{Adopt}\,600\,\text{nos;}$$

Use 3 nos of modules, each containing 200 nos of 100 cm × 6 cm dia PVC tubes inclined at angle of 30° with the horizontal, the tubes being arranged and nestled together in 10 nos of layers and placed one above the other, 20 nos of tube forming a layer being placed side by side.

Controlled set-up:

The modules are to be set up on a stable structure.

Incoming water should be made to pass through an inlet through which the mean velocity should be limited within the mean velocity of flow through the tubes.

In case the flow rate exceeds beyond the zone of laminarity through any tube both during the transition of flow and turbulence, on entering into the tubes, the flow vectors are erratically oriented. They, too, will be in continuously changing state.

These vectors may be resolved into components in the direction of flow and perpendicular to it.

At any cross section at any instant of time, the sum of the components in the direction of flow will give the flow through the section.

Since there is no flow in direction perpendicular to the flow direction, the sum of the components perpendicular to the direction of flow vanishes. This fact amounts to the movement of any particle as much upward as downward under the influence of those components. The net effect is that the net movement of any particle downward is with its settling velocity.

Experimentation has shown conclusively (Chap. 15) that so-called transition length or initial length mentioned by Yao (1970) does not exist.

This may leave one with a question as to the necessity of maintaining the Reynolds' number limited. With the increase of Reynolds' number, the local flow velocity of water in contact with solids will increase the velocity of water relative to the draining sludge, and scouring will result in the deterioration of settling performance.

Little divergence from laminarity that may occur during transition is likely not to produce such effect. This divergence is minor because this system has the self-adjusting uniform distribution of flow through the tubes. For if greater flow through any tube occurs, the head loss through the tube will increase, and as a result, through some other tubes, flow will decrease resulting in lesser loss of head through them. More flow in the other tubes will now find flowing through the tubes with lesser flow.

17.6 Conclusion: What Follows from the Foregoing Discussions Are Presented Below (De 2009)

1. A theoretic study has been pursued in search of a procedure for fixing the optimised coordinated values of the design parameters for tube settlers.
2. Quantitative changes in the critical fall velocity for small changes in one or more of the design parameters have been worked out. This provides solution for adjustment for small changes in the values of the parameters.
3. It has been shown that increase in the angle of inclination, the diameter of the tube and also the mean velocity of flow through the tube settler, settling performance deteriorates. With the increase in its length, performance of the tube settler improves.
4. Limitations of the values of the design parameters for optimised design of the tube settler have been worked out.
5. An example has been solved to demonstrate how to control the values of the parameters for the optimised combination of the same.

Notations

L	Length of the tube settler
R	Radius of the tube
θ	Angle of inclination of the tube
v_s	Settling velocity of the particle
v_c,v_{cr}	Critical fall velocity
Q	Flow rate
v_{mean}	Mean velocity of flow
α	Angle of internal friction
h	Head loss across the length
r	Hydraulic radius
d	Diameter of solid particle in contact with flowing water
w	Specific weight of water
w_s	Specific weight of solids
p	Porosity
C	Chezy's constant
S	Sp.Gr. of solids
n	Friction factor
i	Hydraulic gradient
k	Characteristic constant of sediment
ν	Kinematic viscosity
τ	Stress exerted on contact surface

References

De A (2009) Theoretic study on the control of design parameters for tube settling. Jour IPHE, India 2009–10(2)

Fair GM, Geyer JC, Okun DA. Water and waste water engineering, vol 1. Wiley/Toppan Company Limited, New York/Tokyo

Hansen SP, Culp CL (1967) Applying shallow depth sedimentation theory. J AWWA 59:1134

Vazirani VN, Chandola SP (1993) Concise hand book of civil engineering. S.Chand & Company Ltd, New Delhi

Yao KM (1970) Theoretical study of high rate sedimentation. J WPCF 42:221

Chapter 18
Design of High-Rate Settlers

Abstract This chapter presents the design methodology for a high-rate settling system and demonstrates its application with worked-out problems.

Keywords Design recommendations • Observed field data • Basis of design • Design method • Overloading accommodation

18.1 Introduction

Water and waste water settling has found in it a very convenient application of high-rate settlers with regard to its cost-saving potentiality, better control and flexibility in application of the system.

The system is more attractive because of its capability of increasing the capacity of an existing clarifier. The system is unique in its self-adjusting system operation with regard to uniform distribution of flow. If flow through any tube is limited by solids that build up into it, the surplus water is diverted through another tube across which loss of head is less than that through previous one. This process continues to promote uniform distribution of flow through all the tubes.

18.2 Recommendations and Observations

In quest of a suitable procedure for the design of 'high-rate settlers', the recommendations (Culp and Culp 1970) and observations (Hernandez and Wright 1970) noted are put down in Tables 18.1, 18.2, 18.3, 18.4 and 18.5.

A. De, *Sedimentation Process and Design of Settling Systems*,
Springer Transactions in Civil and Environmental Engineering,
DOI 10.1007/978-81-322-3634-4_18

Table 18.1 Loading for horizontal basins, raw water turbidity 0–100 JTU

Overflow rate based on total clarifier area		Overflow rate based on portion covered by tubes		
m/min	gpm/sq.ft	m/min	gpm/sq.ft	Probable effluent turbidity JTU
0.08	2.0	0.10	2.5	1–3
0.08	2.0	0.12	3.0	1–5
0.08	2.0	0.16	4.0	3–7
0.12	3.0	0.14	3.5	1–5
0.12	3.0	0.16	4.0	3–7

Table 18.2 Loading for horizontal basins, raw water turbidity 100–1000 JTU

Overflow rate based on total clarifier area		Overflow rate based on portion covered by tubes		
m/min	gpm/sq.ft	m/min	gpm/sq.ft	Probable effluent turbidity JTU
0.08	2.0	0.10	2.5	1–3
0.08	2.0	0.12	3.0	3–7

Table 18.3 Loading for upflow clarifier

Overflow rate based on total clarifier area		Overflow rate based on portion covered by tubes		
m/min	gpm/sq.ft	m/min	gpm/sq.ft	Probable effluent turbidity JTU
0.08	2.0	0.08	2.0	1–3
0.08	2.0	0.12	3.0	1–5
0.08	2.0	0.16	4.0	3–7
0.10	2.5	0.10	2.5	3–7
0.10	2.5	0.12	3.0	5–10

18.2.1 Recommendations

Analysis and Discussions For tube length l and radius R, inclined at an angle θ with the horizontal and carrying a flow rate q through it, a particle having settling velocity v_c entering at the top of its cross-sectional area that moves through the length to reach just the bottom of its cross-sectional area is given by

$$v_c = \frac{8q}{3\pi R(l_c \cos\theta + 2\text{R} \sin\theta)} \tag{18.1}$$

where v_c is the critical fall velocity and l_c.its critical length.

All particles having settling velocity $v_s \geq v_c$ will be settled within the length 0 $-l$ completely, and particles having settling velocity $v_s < v_c$ will be settled fractionally within the length $0 - l$. The rest will flow out of the tube. The lesser the critical velocity is, the more is the removal accomplished.Equation 18.1 may be written as

Table 18.4 Laboratory data for 5° inclined tube settler with influent turbidity 450 JTU without any polyelectrolyte dose

Flow rate gal/min/ft^2	Effluent turbidity in JTU	Tube length in ft	Tube diameter in ins	$\frac{v^2 R \times 10^{-7}}{L}$
2	18	2	½	1.1
2	22	2	1	2.0
2	40	2	2	4.2
2	76	2	4	8.2
2	18	4	½	0.4
2	18	4	1	1.1
2	12	4	2	2.0
2	36	4	4	4.2
2	18	8	½	0.2
2	13	8	1	0.4
2	18	8	2	1.1
2	18	8	4	2.0
5	200	2	½	6.4
5	250	2	1	12.6
5	288	2	2	25.8
5	350	2	4	52.3
5	140	4	½	3.8
5	196	4	1	6.4
5	207	4	2	12.9
5	250	4	4	25.8
5	76	8	½	1.8
5	108	8	1	3.8
5	158	8	2	6.4
5	162	8	4	12.9
8	265	2	½	16.4
8	333	2	1	32.8
8	337	2	2	65.6
8	369	2	4	133.0
8	250	4	½	8.2
8	300	4	1	16.4
8	316	4	2	32,8
8	337	4	4	65.6
8	144	8	½	4.2
8	216	8	1	8.2
8	247	8	2	16.4
8	264	8	4	32.8

$$\frac{v_c}{v_l} = \frac{4}{3\left(\frac{1}{2R}\cos\theta + \sin\theta\right)} \tag{18.2}$$

where $v_l = \frac{q}{\pi R^2}$ is the mean longitudinal, axial or flow velocity.

Table 18.5 Table 18.4 is rewritten after due conversion

Mean flow vel.cm/s	Length cm	Diameter cm	$\frac{v_c}{v_l}$	v_c cm/s	Removal %
0.136	61	1.27	0.02781	0.00378	96
,,	122	,,	0.01392	0.00189	96
,,	244	,,	0.00696	0.00095	96
,,	61	2.54	0.05553	0.00755	95
,,	122	,,	0.02781	0.00378	96
,,	244	,,	0.01392	0.00189	97
,,	61	5.08	0.11066	0.01505	91
,,	122	,,	0.05553	0.00755	95
,,	244	,,	0.02781	0.00378	96
,,	61	10.16	0.21972	0.02988	83
,,	122	,,	0.11066	0.01505	92
,,	244	,,	0.05553	0.00755	96
0.340	61	1.27	0.02781	0.00948	56
,,	122	,,	0.01392	0.00473	69
,,	244	,,	0.00696	0.00237	83
,,	61	2.54	0.05555	0.01888	44
,,	122	,,	0.02781	0.00946	57
,,	244	,,	0.01391	0.00473	76
,,	61	5.08	0.11066	0.03762	36
,,	122	,,	0.05553	0.01888	54
,,	244	,,	0.02781	0.00946	65
,,	61	10.16	0.21972	0.07471	23
,,	122	,,	0.11066	0.03762	44
,,	244	,,	0.05553	0.01888	65
0.543	61	1.27	0.02781	0.01510	41
,,	122	,,	0.01392	0.00756	44
,,	244	,,	0.00696	0.00378	63
,,	61	2.54	0.05553	0.03015	26
,,	122	,,	0.02781	0.01510	33
,,	244	,,	0.01392	0.00756	52
,,	61	5.08	0.11066	0.06009	25
,,	122	,,	0.05553	0.03015	32
,,	244	,,	0.02781	0.01510	45
,,	61	10.16	0.21972	0.11931	18
,,	122	,,	0.11066	0.06009	25
,,	244	,,	0.05553	0.03015	41

The plan area of the tube $= 2R\,l\cos\theta$

$$= R^2\left(\frac{16}{3}\cdot\frac{v_l}{v_c} - 4\sin\theta\right) \tag{18.3}$$

written from Eq. 18.2.

The overflow based on the portion covered by the tube, i.e. surface loading (per unit area),

$$= \frac{q}{2Rl\cos\theta}$$

$$= \frac{\pi R^2 v_l}{R^2\left(\frac{16}{3}\cdot\frac{v_l}{v_c} - 4\sin\theta\right)}$$

$$= \frac{3\pi}{16}v_c. \tag{18.4}$$

$$\text{Neglect}\,4\sin\theta,\ \text{since}\frac{16}{3}\cdot\frac{v_l}{v_c} \gg 4\sin\theta.$$

This is going to show that the surface loading fixes the critical fall velocity through the system irrespective of the flow rate, length l, diameter $2R$ and inclination θ. Surface loading having been fixed the critical fall velocity determines the removal for a particular water.

The recommendations in Tables 18.1 and 18.2 provide flexibility in design. Let us take one value for surface loading, say, 0.1 m/min. This fixes

$$v_c = 0.1 \times \frac{16}{3\pi}\ \text{i.e.}\ 0.16977\,\text{m/min}.$$

As such, recommendation of 0.08 m/min overflow rate based on total clarifier area is, it appears, likely to act as finisher, its value being 0.5 × 0.16977 m/min = 0.084 m/min. This recommendation appears to be misleading for the assembly of tubes as a whole because of the overlapping of plan areas of the tubes.Putting $\theta = 90°$ in Eq. 18.1,

$$v_c = \frac{8q}{3\pi R(l_c\cos 90° + 2R\sin 90°)} = 1.33\,v_0$$

where v_0 is the overflow velocity.

Nandi (1990) studied upflow clarification through vertical tubes and found the mechanism of removal of settleable solids through them to be distinctly different from that through inclined tube system.

The particle with maximum settling velocity (v_c) in the effluent was found to be

$$v_c = 1.81\,v_0^{0.71} \tag{18.5}$$

where v_0 is the overflow velocity or mean flow-through velocity.

This is so because the particles are not settling to the inside surface of the tube but are settling vertically, and the escaping particles are dragged out by the weir flow action over the periphery of the end area.

This reporting is in contradiction with the recommendation in Table 18.3.

18.2.2 Observations

Hernandez and Wright (1970) studied a large number of laboratory and field data and reported for the purpose of design of a tube settler. Some of the data are presented in the following table.

Analysis and Discussions Data in Table 18.4 are presented under five columns. Removals that are defined by the flow rate, diameter and length of the tube have been attempted for correlation with a function $\frac{v^2 R \times 10^{-7}}{L}$ with an intention to arrive at the desired design of flow rate, length and diameter of the tubes.

For a tube settler, the parameters critical fall velocity (read 'removal'), length, diameter $2R$, flow rate q and inclination of the tube θ are bound by definite relationship. Any function made out of these variables can be correlated with any of the variables. But the question remains as to how this correlation can lead to the design values of the parameters.

Critical velocity v_c controls removal. Mean longitudinal flow velocity v_l controls scour. $\frac{v_c}{v_l}$ ratio is dimensionless. The ratio holds the tube characteristics $\frac{1}{2R}$ and tube configuration characteristic θ.

Table 18.4 is rewritten in Table 18.5 with due conversion of flow rate gal/min/sq.ft into cm/s in column 1, diameter in cm in column 3 and length in cm in column 2 together with $\frac{v_c}{v_l}$ ratio calculated from Eq. 18.2 in column 4. v_c is calculated from column 1 and column 4 and is placed in column 5. Removal percentage is in column 6. Critical velocity is the maximum velocity in the largest fraction of particles present in escaping water.

Observed removals in the laboratory are calculated in percentage from Table 18.4 and placed in column 6 of Table 18.5. The data in Table 18.4 are for tubes inclined at $\theta = 5°$.

The effect of polyelectrolyte dose is left out. Because it has been demonstrated (Chap. 16) that in tube settler, discrete settling takes place. Flocculation cannot contribute towards removal because of small settling time in the tube. Polyelectrolyte dose improves removal through the tube because of the conversion of solids with higher velocity distribution among the settling particles.

From the test data in Table.18.5, the following may be observed:

(i) Critical fall velocity (v_c) increases with the increase in flow-through velocity (v_l).
(ii) Critical fall velocity (v_c) varies almost inversely with the length (l) of the tube.
(iii) Critical fall velocity (v_c)varies almost directly with the diameter($2R$) of the tube.
(iv) v_c/v_l ratio varies almost directly with the diameter ($2R$) and inversely with the length of the tube (l).
(v) With decrease in critical fall velocity (v_c), percentage removal increases.

These observations are very apparent if one looks at the critical fall velocity expression. All these observations are in conformity with the theory of tube settling presented already and verified experimentally in Chap. 15 of this book.

18.3 Design of Tube Settling System

18.3.1 Basis of Design

The design of tube settling system is based on the experience gained through the experimental verification of the tube settling theory and the theoretical study on the subject (Chaps. 15 and 17). They are as follows:

(i) There is no need for the provision of transition length as additional length since the particles settle according to theory even in turbulence.
(ii) Mean flow-through velocity (v_l) is limited by the scouring velocity, and proper Reynolds' number may be maintained to keep down the loss due to erosion of solids from the surface of the deposited sludge to minimum.
(iii) Inclination of settlers should take care of the angle of repose of the deposited sludge.
(iv) In the design of the tube settler, the scouring should be the main consideration and not Reynolds' number.
(v) Design methodology should aim at maintaining the desired critical fall velocity through the system to control the removal through it.

18.3.2 Design Method

From the foregoing discussion, the author proposes the following 'method of projection' for the design of tube settling system.

With the limiting values of $2R$ for laminarity and v_l for scour, the methodology suggests the following steps of design:

Step 1: From the column settling observation interface, settling time to a depth of about 15 cm is noted, and the interface settling velocity is found out. This settling velocity is to be maintained as 'critical fall velocity' (v_c) through the system.
Step 2: Choose θ based on the angle of repose of the deposited sludge.
Step 3: Choose any base values of the length and diameter for the tube settler.
Step 4: Compute flow rate 'q' through the tube settler defined by the base values as

$$q = \frac{3\pi}{8} v_c R(l\cos\theta + 2R\sin\theta).$$

This computation with chosen base values l and $2R$ is done to arrive at the design values of the parameters. Depending upon the area to be covered by the settling system with regard to the design flow rate Q, if the base flow rate q through a single tube is to be increased to mnq, the tube diameter $2R$ is to be increased to m($2R$) and l is to be increased to nl because it is very apparent from the expression for q (Eq. 18.1) that the increase in the product Rl will increase the value of q in the same proportion $2R\sin\theta$ being negligibly small in comparison with $l\cos\theta$.

Step 5: Compute the flow rate through the tube of length nl, diameter m2R and inclined angle θ. This value will be mnq approximately.

Step 6: Check for critical fall velocity with the values in step 5.

Step 7: Find the total number of tubes N by dividing the design flow Q by mnq calculated in step 5.

Step 8: If the tubes N are arranged 'p' nos along the width and 'sp' nos along the length one above the other, $p = \sqrt{\frac{N}{s}}$, $\mathrm{sp} = s\sqrt{\frac{N}{s}}$.

Step 9: If the thickness of the PVC tubes be t, the settling system is

$$\left[\frac{2s(R+t)}{\sin\theta}\sqrt{\frac{N}{s}} + l\cos\theta\right] \times 2(R+t)\sqrt{\frac{N}{s}} \times l\sin\theta;$$

Problem 18.1 A tube settling system has to be designed to process waste water at 2000 m^3 per hour. In a settling cylinder, the solid-liquid surface of separation was found to cross 15 cm mark at 10 mins and 45 s.

The interface concentration is 40 % of the original turbidity at 30 °C. Design the system.

Solution The design of the system is as follows:

Critical fall velocity of the interface $= \frac{15\ \text{cm}}{10\ \text{mins}\ 45\ \text{s}}$, i.e. 0.02326 cm/sec; corresponding diameter of the particle = 0.00144 cm.

Based on the angle of repose of deposited sludge, $\theta = 10°$ is selected (40 % solids being $\leq$ 0.00144 cm diameter).

Choosing base values $2R = 1\,\text{cm}$, $1 = 50\text{cm}$.

Flow through each of the tubes $= \dfrac{3\pi}{8} v_c R(l\cos\theta + 2R\sin\theta)$

$$= \frac{3\pi}{8} \times 0.02326 \times 0.5(50\cos 10° + 1.\ \sin 10°) \quad = 0.67704\ \text{cm}^3/\text{sec}.$$

Let the flow through a single tube be increased to 12 times of 0.67704 cm^3/s.

Choose $2R = 1 \times 3$ cms, $l = 4 \times 50$ cms, i.e. 200 cmsec.

Then

$12q = \frac{3\pi}{8} \times 0.02326 \times 1.5(200\cos 10° + 3\sin 10°)$, i.e. 8.11728 cm^3/sec.

$$\text{Critical fall velocity } v_c = \frac{8 \times 8.11728}{3\pi \times 1.5(200\cos 10° + 3\sin 10°)} = 0.02326\,\text{cm/s}.$$

$$v_l = \frac{8.11728}{\pi(1.5)^2},\ \text{i.e. } 1.14836\ \text{cm/s}.$$

Check Allowable

$v_{\text{mean}} = \frac{1.49}{0.013} \cdot \left(\frac{3}{4}\right)^{\frac{1}{6}} [0.8(2.65 - 1)0.00144]^{\frac{1}{2}}$, i.e. $4.76308\frac{\text{cm}}{\text{s}} > 1.14836\frac{\text{cm}}{\text{s}}$.

And Reynolds' number $= \frac{1.14836 \times 3}{0.8039 \times 10^{-2}}$, i.e. $429 < 2300$.

Total number of tubes $N = \frac{2000 \times 10^6}{60 \times 60 \times 8.11728}$, i.e. 68441 nos.

With 3mm thick PVC tubes, let p nos be arranged along the width and 3p nos be arranged along the length.

The length of the system $= 2 \times \frac{3(1.5+0.3)}{\sin 10°} \sqrt{\frac{68441}{3}} + 200 \cos 10° = 9590.97\,\text{cm}$, i.e. 95.91 m.

The width of the system $= 2 \times (1.5 + 0.3)\sqrt{\frac{68441}{3}} = 543.75\,\text{cms}$, i.e. 5.44 m.

Depth $= 200 \sin 10° = 34.7\,\text{cms}$, i.e. 0.35 m.

That is, the system is 95.91 m $\times$ 5.44 m $\times$ 0.35 m.

The length and the width can be adjusted to the desired dimension. For example, the length 95.91 m can be reduced to $\frac{1}{3} \times 95.91\,\text{m}$, i.e. 31.97 m, by rearranging the tubes and by increasing the width to 3×5.44 m, i.e. 16.32m.

The tentative sketch of the system is shown in Fig. 18.1. The shaded portion of the sketch is filled up with closed tubes to prevent the short circuiting through the space.

Alternatively. Choose $2R = 4 \times 1$, i.e. 4 cm; $l = 3 \times 50$, i.e. 150 cm; and $\theta = 10°$.

Flow through the single tube

$$= \frac{3\pi}{8} v_c R(l \cos\theta + 2R \sin\theta)$$

$$= \frac{3\pi}{8} \times 0.02326 \times 2(150 \cos 10° + 4 \sin 10°), \text{ i.e. } 8.13394 \frac{\text{cm}^3}{\text{s}}.$$

Arranging p nos along the width and 3p nos along the length, the width of the system with 3 mm thick PVC tubes

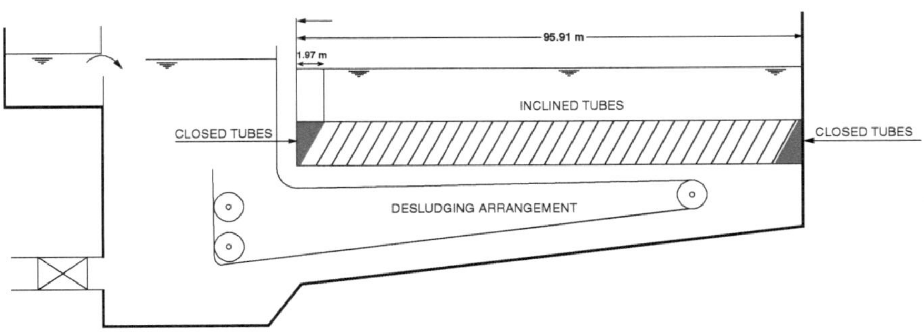

Fig. 18.1 Line diagram of tube settler

$$= 4.6 \times \sqrt{\frac{2000 \times 10^6}{60 \times 60 \times 8.13394}} = 1202.18\,\text{cm, i.e.}\,12.0\,\text{m.}$$

$$\text{Length} = \frac{3}{\sin 10^\circ} \times 1202.18 + 150 \cos 10^\circ = 20916.95\,\text{cm, i.e.}\,209.17\,\text{m.}$$

$$\text{Depth} = 150 \sin 10^\circ\,\text{cm, i.e.}0.26\,\text{m.}$$

That is, the system is 209.17 m × 12.0 m × 0.35 m.

This shows that it is better to increase the base length of the tube by bigger multiplier if the plan area covered by the system has to be smaller.

Problem 18.2 Redesign the existing system in Problem 18.1 to carry 30 % overload.

Solution Using PVC tube of thickness 3 mm and diameter $2R = 4$ cm, and the angle of inclination of the tube $\theta = 10°$.

The plan area of the existing system = 5.44 m ×95.91 m.
Along the width arranging 544/4.6, i.e. 118 tubes.
Along the length arranging $(9591 - 200 \cos 10^\circ) \frac{\sin 10^\circ}{4.6}$, i.e.355 tubes;.
Flow through the system $= \frac{1.3 \times 2000 \times 10^6}{60 \times 60}\,\text{cm}^3/\text{s}$.
Flow through a single tube $= \frac{1.3 \times 2000 \times 10^6}{60 \times 60 \times 118 \times 355}\,\text{cm}^3/\text{s}$, i.e. $17.24092\text{cm}^3/\text{s}$.
Mean flow-through velocity $= \frac{17.24092}{\pi 2^2}$, i.e. $1.37199 \frac{\text{cm}}{\text{s}} < 4.76308\,\text{cm/s}$.
Critical fall velocity $= 0.02326\,\text{cm/s}$.
Flow through a length of l cm and of diameter $2R = 4.0$ cm
$= \frac{3\pi}{8} \times 2 \times 0.02326\,(l \cos 10^\circ + 4 \sin 10^\circ)$, i.e. $17.24092\,\text{cm}^3/\text{s}$.
That is, length of each tube $= \frac{17.24092 - 0.05481 \times 4 \sin 10^\circ}{0.05481 \cos 10^\circ} = 318.70518\,\text{cm}$, using = 319 cm.

The other suitable combinations of l, R and θ may be tried.

Notation

l Length of tube
l_c Critical length of tube
R Radius of tube
θ Inclination of tube with horizontal
q Flow rate through a single tube
v_c Critical fall velocity

v_l Mean flow-through velocity
v_0 Overflow velocity

References

Culp GL, Culp RL (1970) New concepts in water purification. Van Nostrand Reinhold, New York

Hernandez JR, Wright JR (1970) Design parameters for tube settlers. Indus Waste Conference Purdue University, Engg. series No.141, P67

Nandi S (1990) Upflow clarification – its design and mechanism of removal. M.C.E. Thesis, Jadavpur University, Calcutta, India

Chapter 19
Design of System Module for Couette Flow Settler

Abstract Couette flow settling may prove to be an efficient application of shallow-depth sedimentation. The proper design of the settling system calls for the working out of the theory of settling in the system and the study of its control parameters.

The theory of couette flow settling has been developed, and the control parameters have been studied. Design procedure for the system has been presented and illustrated with the help of solution to a design problem for the system.

Keywords Couette flow settling • Control of parameters • Limitation of inclination • Limitation of mean flow • Limitation of spacing

19.1 Introduction

The application of shallow-depth sedimentation has been gaining importance since the mid-sixties of the last century. This is mainly due to the cost-saving potentiality and easy, convenient operatability of the settling system and also due to its potentiality in its application for the enhancement of capacity of an existing clarifier. In most of the cases, tubes of circular cross section have been employed. Cross sections other than the circular one may also prove to be beneficial.

Clarification of settleable solids by passing water containing the settleables through inclined stationary parallel plates may be a loveable solution to the clarification problem. To design such a system for its efficient working, the theory of settling through the system has to be developed, and the influence of changing the parameters on its performance should be studied. Based on these, a design procedure for its system module may be worked out, and the system may be controlled.

19.2 Theory of 'Couette Flow' Settling (De 2009c)

With reference to the coordinate axes *X*, *Y* and α, two parallel plates L × W separated by vertical distance '*h*' and inclined at an angle θ to the horizontal are shown in Fig. 19.1. Water containing settleables at concentration C_s of particles,

A. De, *Sedimentation Process and Design of Settling Systems*,
Springer Transactions in Civil and Environmental Engineering,
DOI 10.1007/978-81-322-3634-4_19

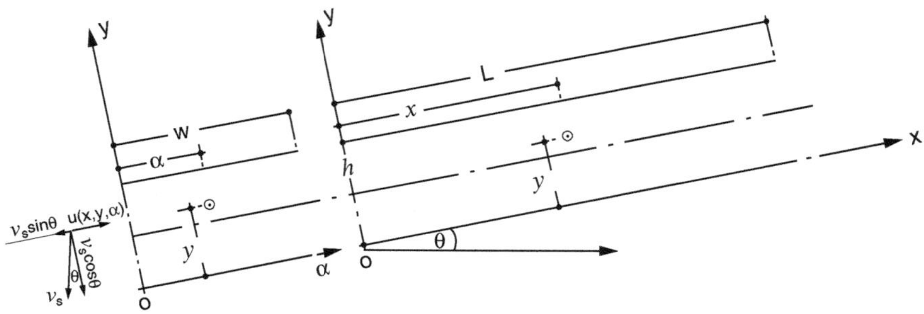

Fig. 19.1 Couette flow settler

each of which has settling velocity v_s, flows in the direction of X-axis at the rate Q per unit width.

The local fluid velocity at any point (x, y, α)

$$u_{x,y} = \frac{6Q}{h^3}\left(hy - y^2\right) \tag{19.1}$$

Maximum velocity $u_{\max}$ occurs at $y = \frac{h}{2}$, and it is

$$= \frac{6Q}{4h}. \tag{19.2}$$

By Velocity Profile Theorem (De 2009a), the 'critical fall velocity' for the couette flow is obtained from

$$L = \frac{\frac{2}{3} \times \frac{6Q}{4h} \times h \; - \; v_c \sin\theta \times h}{v_c \cos\theta}.$$

$$v_c = \frac{Q}{L\cos\theta + h\sin\theta}.$$

$$v_c = \frac{h.v_{\text{mean}}}{L\cos\theta + h\sin\theta} \tag{19.3}$$

Removal of solids: All particles having settling velocity $v_s \geq v_{cr}$ will be completely removed. If C_0 be the concentration of such particles, the total of such solids removed from the flow rate Q per unit width is QC_0.

For particles having settling velocity $v_s < v_{cr}$, one of such particles entering through a point $(0, y_s, \alpha)$ will move to $(L, 0, \alpha)$, and all other such particles entering through the depth from $(0, y_s, \alpha)$ to $(0, 0, \alpha)$ will be removed completely.

By 'Velocity Profile Theorem' (De 2009a), the area of flow velocity diagram between $(0, y_s, \alpha)$ and $(0, 0, \alpha)$

$$= v_s(L \cos\theta + y_s \sin\theta) \tag{19.4}$$

If C_s be the concentration of such particles having settling velocity v_s, the total of such solids that will be removed per unit width

$$= C_s v_s (L \cos\theta + y_s \sin\theta). \tag{19.5}$$

Taking all the different other particles having settling velocities $v_s < v_{cr}$, the total removal of such solids

$$= L \cos\theta \Sigma C_s v_s + \sin\theta \Sigma C_s v_s y_s, \tag{19.6}$$

and the total removal of solids with flow through unit width

$$= QC_0 + L \cos\theta \Sigma C_s v_s + \sin\theta \Sigma C_s v_s y_s, \tag{19.7}$$

i.e. the concentration of settleable solids in the effluent is reduced by

$$= \frac{1}{Q}(QC_0 + L \cos\theta \Sigma C_s v_s + \sin\theta \Sigma C_s v_s y_s). \tag{19.8}$$

Evaluation of y_s in Eq. 19.8:

$$\text{From the definition of } y_s \; -\int_{y_s}^{0} \left[\frac{6Q}{h^3}\left(hy - y^2\right) - v_s \sin\theta\right] \frac{(-)dy}{v_s \cos\theta} = L. \tag{19.9}$$

$$a y_s^3 - b y_s^2 + c y_s + L = 0. \tag{19.10}$$

where $a = \frac{2Q}{h^3 v_s \cos\theta}$, $b = 1.5\,ha$ and $c = \tan\theta$.

19.3 Control of Couette Flow Settling Parameters

From Eq. 19.3 $v_c = \phi(\theta, v_{\text{mean}}, h, L)$,

$$d\phi = \frac{\partial\phi}{\partial\theta} d\theta + \frac{\partial\phi}{\partial v_{\text{mean}}} dv_{\text{mean}} + \frac{\partial\phi}{\partial h} dh + \frac{\partial\phi}{\partial L} dL \tag{19.11}$$

From Eqs. 19.3 and19.11,

$$\frac{\partial\phi}{\partial\theta} = v_c \frac{(L \sin\theta - h \cos\theta)}{(L \cos\theta + h \sin\theta)} \tag{19.12}$$

$$\frac{\partial \phi}{\partial v_{\text{mean}}} = \frac{v_c}{v_{\text{mean}}}. \tag{19.13}$$

$$\frac{\partial \phi}{\partial h} = \frac{v_c L \cos \theta}{h(L \cos \theta + h \sin \theta)}. \tag{19.14}$$

$$\frac{\partial \phi}{\partial L} = (-)\frac{v_c \cos \theta}{(L \cos \theta + h \sin \theta)} \tag{19.15}$$

In reality, the combination of L, h and θ is such that the RHS of Eq. 19.12 is always positive, L being very much larger than h. RHS of Eqs. 19.13 and 19.14 are always positive. An increase in θ, v_{mean} and h, therefore, deteriorates settling,

The RHS of Eq. 19.15 is negative.

An increase of L improves settling. Eq. 19.11 provides control measures for adjusting for the small changes of the system parameters.

19.4 Basis of Design

The design of module reported herein is based on the following:

(i) Inclination θ should take care of the angle of repose of the major fraction of the sludge material to assure the automatic draining of sludge (De 2009c).
(ii) In the design of module, "scouring should 'be the main consideration and not Reynolds' number (De 2009b)" (Chap. 15). The mean velocity of flow 'v_{mean}' should take care not to lift the particle of lowest velocity to be removed through the system.
(iii) The distance of separation between the plates should prevent the impairment of settling (De 2009b) (Chap. 15) with allowable mean velocity of flow at suitable Reynolds' number. Although this may prevent the impairment, it may not prevent the flow through the cross flow area between the plates from becoming two-dimensional in which case the assessment of removal with regard to the settling analysis is likely to be erroneous. This should be taken care of.
(iv) The length of the plates in the direction of flow is determined by the critical length (l_C) for the particle to be removed, computed from the expression, deduced under laminar flow condition. This length is invariant with change of 'Velocity Profile distribution' so long as the rate of flow remains the same.
(v) The so-called initial length need not be added to the designed length (De 2009b) (Chap. 15).

19.5 Problem Design of 'Couette Flow Module' for the Removal of Solids

The design procedure for a couette flow settling module may be followed through the solution of the following problem.

Problem Design a couette flow module for the removal of solids from 5 MLD of waste water. The settling column test data of the waste water are indicated as given below:

Initial concentration of solids = 540 mg/l.
Temperature of the water = 30°C.
Solids concentrations in mg/l at indicated depths and times.

	Depth		
Time in mins	60 cm	120 cm	180 cm
5	275	362	386
10	189	259	312
20	135	188	232
40	90	119	162
60	92	92	118
120	95	92	93

Analysis of sludge solids indicated that the major fraction (70 %) of the solids was silty clay.

Solution

(i) *Limitation of the angle of inclination* θ:
To ensure the automatic draining of sludge, the angle of inclination (θ) should be restricted to the angle of repose of the major fraction of sludge. The resistance due to interlocking of grains may be neglected due to lubrication by water. The inclination angle θ is restricted to the angle of friction 10 °(De 2009c) of the major fraction of sludge which is silty clay.

(ii) *Limitation of the average flow through velocity* v_{mean}:
The average velocity v_{mean} in the couette flow should be limited not to lift the lowest diameter of particles to be removed. For the problem, let the lowest diameter of the particles not to be lifted be the lowest diameter composing the interface concentration of 0 mg/l in their settling column test of the settleables. Determination of this diameter requires the analysis of settling column test data (De 2010b) as follows:

Step 1: From the observed data, it is apparent that the waste water contains non-settleable solids of ((90 + 92 + 93 + 92 + 92 +93)/6=) 92 mg/l. The observed data with the settleables may be retabulated as presented in Table 19.1

Step 2: The data presented in Table 19.1 is utilised to plot the concentration versus time curves at each depth of collecting samples. They are presented in Fig. 19.2. The curve for each depth gives the different interface concentrations crossing the depth at different times.

Step 3: Any particular interface concentration may be chosen. The times at which this interface concentration crosses the different depths can be scaled from curves presented in Fig. 19.2.

In the present case, 180 mg/l, 140 mg/l, 100 mg/l, 60 mg/l, 20 mg/l and 0 mg/l interface concentrations are chosen. The times at which each of the above

Table 19.1 Settling column test data with regard to the settleable solids
Initial concentration of settleable solids = 448 mg/l
Temperature of water = 30 °C

Time in mins	Depth		
	60 cm	120 cm	180 cm
5	183	270	294
10	97	167	220
20	43	96	140
40	0	27	70
60	0	0	26
120	0	0	0

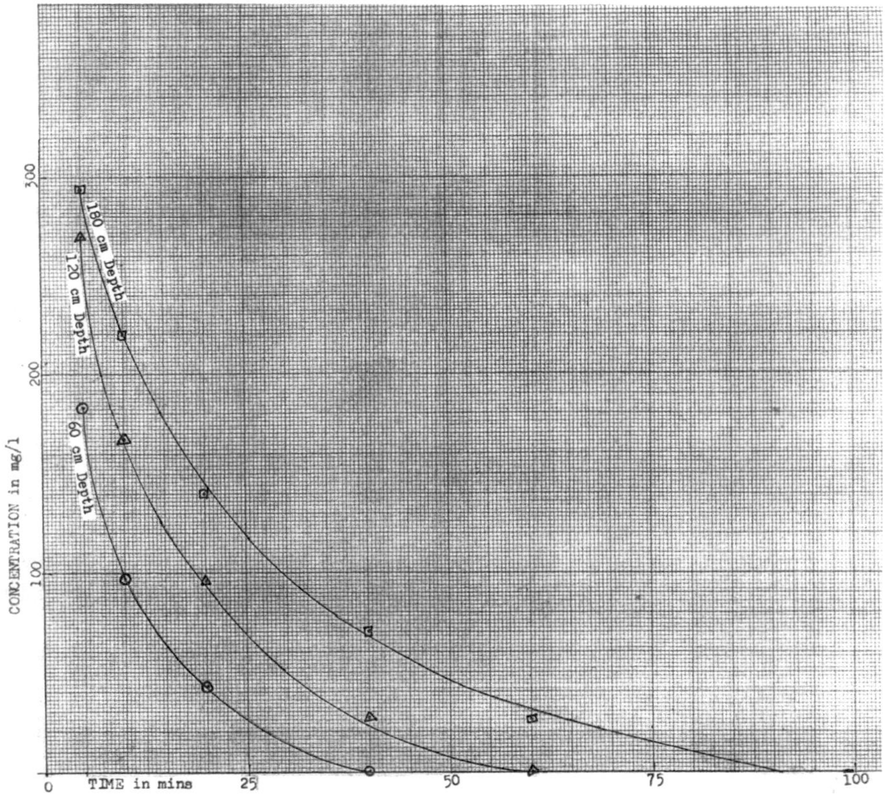

Fig. 19.2 Concentration versus time curves

interface concentrations of settleables crosses the different depths are scaled from Fig. 19.2. For each of the above interface concentrations, so derived, the concentrations are plotted in depth-time coordinates. Connecting the points for each of the concentrations by smooth curves, the movement trajectories of the above interface concentrations down the depth-time diagram are obtained. They are presented in Fig. 19.3.

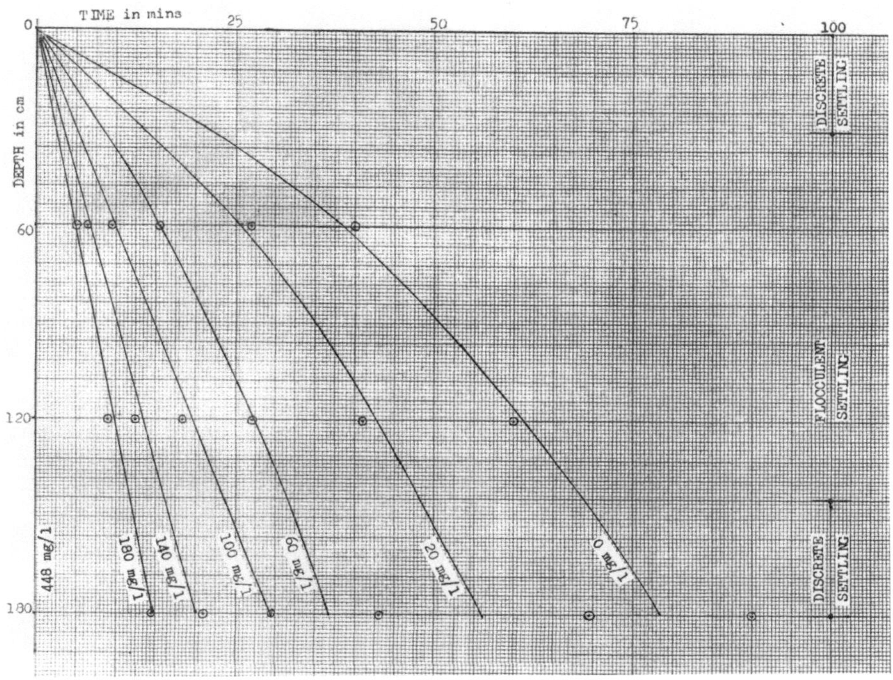

Fig. 19.3 Trajectories of interface concentrations

Let us consider the curves within the initial phase of discrete settling that extends to a depth of 30 cm. In Fig. 19.3, the movement trajectories of the surfaces of the other interface concentrations C_n between 448 mg/l and 180 mg/l have not been done.

From Fig. 19.3, the times 3 mins, 4 mins, 6 mins, 10 mins, 16 mins and 25 mins at which the surfaces of the interface concentrations 180 mg/l, 140 mg/l, 100 mg/l, 60 mg/l and 0 mg/l reach the depth 30 cm, respectively, can be scaled.

Let us imagine that such other curves for interface concentration C_1, C_2, C_3, ..., C_n...448 mg/l ($C_1 < C_2 < C_3 \ldots C_n \ldots$448 mg/l) are drawn and their surfaces reach the depth of 30 cm at times t_1, t_2, t_3, ..., t_n..., $t_1 > t_2 > t_3 \ldots > t_n \ldots$ and finally 0 mins (assuming the surface of concentration 448 mg/l reach the depth almost in no time compared with others).

This implies that the solids composing (C_1-180) mg/l, (C_2-C_1) mg/l, (C_3-C_2) mg/l... will move through the depth at 30 cm over time intervals ($3-t_1$) mins, (t_1-t_2) mins, (t_2-t_3) mins, ..., etc. at time t_1, t_2, t_3, respectively.

This shows that (448 – 180) mg/l of solids are composed of different fractions of varying settling velocities that fall through the distance of 30 cm over varying falling through times. It appears reasonable approximation to assume the mean time of fall of 268 mg/l of solids as ½ (0 + 3)min, i.e. 1.5 min. Thus (180 – 140) mg/l, i.e. 40 mg/l; (140 – 100)mg/l,

Table 19.2 Settling velocity distribution during the initial phase of discrete settling

	Concentration in mg/l	Fall through distance, cm	Mean falling through times, mins	Settling velocity in cm/sec
(a)	268	30	1.5	0.33333
(b)	40	30	3.5	0.14286
(c)	40	30	5	0.10000
(d)	40	30	8	0.06250
(e)	40	30	13	0.03846
(f)	20	30	20.5	0.02439

i.e. 40 mg/l; (100 − 60) mg/l, i.e. 40 mg/l; and (60 − 20)mg/l, i.e. 40 mg/l and 20 mg/l, have average falling through times ½ (3 + 4) min, i.e. 3.5 min; ½ (4 + 6) min, i.e. 5 min; ½(6 + 10) min, i.e. 8 min; ½(10 + 16), i.e. 13 min; and ½(16 + 25) min, i.e. 20.5 min, respectively.

Thus the settling velocity distribution among the settleable solids during their initial phase of discrete settling is presented in Table 19.2. Needless to say, the more closely the curves for interface concentrations are spaced, the more accurate details of the distribution will be obtained.

The lowest diameter of the particles composing the interface concentration of 0 mg/l has settling velocity (30/25 × 60) =) 0.02 cm/s at 30 °C. The corresponding diameter of particle is 0.00133 cm with $n = 0.013$, $d = 0.00133$ cm, $k = 0.8$ and $s = 2.65$.

$$v_{\text{mean}} = \frac{1.49}{0.013}\left(\frac{0.00133}{4}\right)^{\frac{1}{6}}(0.8(2.65 - 1)0.00133)^{\frac{1}{2}} = 1.26402\ cm/sec$$

(iii) *Limitation of spacing distance h:*

Spacing distance h should be well within the depth over which initial discrete settling is exhibited by the settling of settleables in settling column test data. This distance from Fig. 19.3 is 30 cm. To keep down the impairment in settling, let Reynolds' number = 2000 be maintained, i.e.

$$\frac{v_{\text{mean}}(2h)}{\nu_{30^0\text{c}}} = 2000, \text{ i.e. } h = \frac{2000 \times 0.8 \times 10^{-2}}{2 \times 1.26402}$$
$$= 6.32901\ cm$$

Adopt $h = 6.0$ cm.

Although the spacing h=6.0 cm may prevent the impairment in settling (De 2009b) (Chap. 15), it may not prevent the conversion of flow into two-dimensional at $R = 2000$.

In case of two-dimensional flow, the present way of assessment of removal will not provide a reasonable estimate.

To limit to one-dimensional flow, use $R = 1000$ when $h = 3.0$ cm.

Allowable flow rate per unit width

$$= 3 \times 1 \times 1.26402 \frac{\text{cm}^3}{\text{s}}.$$

To remove solids from 5 MLD waste water, the width of settler required

$$= \frac{5 \times 10^6 \times 1000}{(24 \times 60 \times 60) \times 3 \times 1 \times 1.26402}$$
$$= 15260.43215\, cm,\ use\ width = 15200\, cm$$
$$i.e.\ (76 \times 10) \times 20 cm$$

(iv) *Critical length* (from Eq. 19.3)

$$l_C = \left(\frac{v_{\text{mean}}}{v_c} - \sin\theta\right)\frac{h}{\cos\theta}$$

$$= \left(\frac{1.26402}{0.02} - \sin 10^0\right)\frac{3}{\cos 10^0}$$

$$= 191.99895\, cm; \quad Adopt\ l_c = 192\, cm$$

Thickness of the Plates

Use PVC plates of sufficient flexural rigidity to limit the central deflection reasonably.

Designed Module and the Settler

Use 20 nos of modules, to be used in parallel, each of which is made up of 11 nos of 1920 mm × 760 mm × 6 mm thick (arbitrarily chosen) PVC plates arranged one above the other at a distance of separation of 30 mm, the plates being sloping upwards along the length at an angle of 10° with the horizontal.

Effluent concentration of solids through the settler is 92 mg/l.

19.6 Conclusion

The observations may be summarised as follows:

(i) Couette flow settler can be a successful application of shallow-depth sedimentation.

(ii) The theory of couette flow settling has been worked out and presented.
(iii) The system adjustment for the minor quantitative changes in the system parameters can be controlled as indicated by the expressions derived herein.
(iv) The basis and the procedure of design of couette flow settling module have been presented and illustrated by working out a problem.

Notations

x, y, α	Instantaneous coordinates of the settling particle
Q	Flow rate per unit width
h	Distance of separation between consecutive parallel plates
θ	Angle of inclination
v_s	Settling velocity of the particle
v_{cr}, v_c	Critical fall velocity
L, l_c	Length or critical length of the plate
y_s	The vertical distance from the bottom at which a particle enters and just reaches the bottom travelling through the distance L
C_s	Concentration of solids with particles having settling velocity v_s
C_0	Concentration of solids with particles having settling velocities $v_s \geq v_{cr}$
n	Friction factor
k	Characteristic constant of the sediment
S	Specific gravity of the material of solids
d	Diameter of the particle
ν	Kinematic viscosity

References

De A (1998) Revised mode of analysis of column settling data. J Indian Chem Engr Section B 40

De A (2009a) Velocity profile theorem – concept for solving settling problem analysis. J IPHE 2009–10(1):25

De A (2009b) Experimental verification of the theory of 'tube settling'. 2009–10(3):41

De A (2009c) Theoretic study on the control of design parameters for tube settling. Jour IPHE 2009–10(2): 32

De A (2010a) Analysis of the effect of short-circuiting on settling – An application of velocity profile 3. De Theorem. J IPHE 2009–10(4):20

De A (2010b) Design of simple couette flow module for the removal of solids. J IPHE, India 2010–2011(2)

Chapter 20
Design of Thickeners

Abstract This chapter presents different theories of thickener design. These theories have been applied to the solution of the same problem, wherever applicable, to reveal the relative outcome of the solutions for comparison.

Keywords Coe and Clevenger's method • Robert's derivation • Kynch's theory • Talmadge and Fitch • Flux flow method

20.1 Introduction

Gravity settling of solids produces clarification of the influent at the top and forms deposited sludge at the bottom of the settling tank. Clarification aims at the quality of the clarified water and is controlled by overflow rate and weir loading. The deposited sludge looks for the sludge thickening to its desired consistency, and this depends on the sludge loading per unit area of the thickener.

A settling tank may, therefore, function both as a clarifier and thickener. The desired thickening may not be achieved economically when both the functions are aimed at. For the efficient and economic sludge handling and disposal, the thickening of the sludge may be desired to such an extent that a separate thickener may have to be employed for the purpose. Design of gravity thickener is the concern of this chapter.

20.2 Methods of Thickener Design

20.2.1 Coe and Clevenger's Method (1916)

Coe and Clevenger (Coe and Clevenger 1916) demonstrated the characteristic settling of sludge in a glass cylinder. The earliest design method for the design of a continuous thickener was devised by them from the interface height versus time graph of settling sludge, and this remained as the only method for such design for quite some time.

A. De, *Sedimentation Process and Design of Settling Systems*,
Springer Transactions in Civil and Environmental Engineering,
DOI 10.1007/978-81-322-3634-4_20

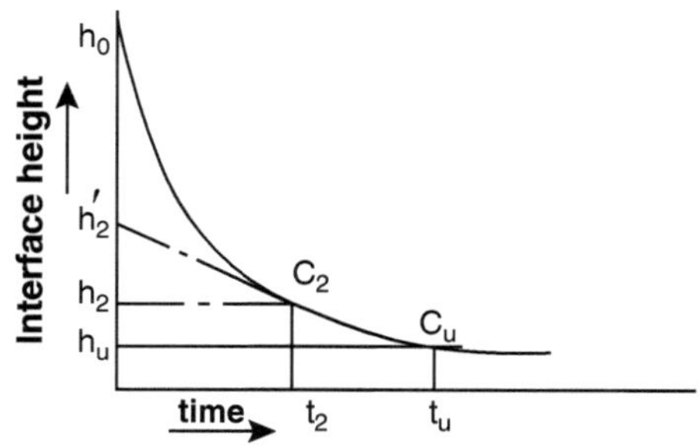

Fig. 20.1 Interface height versus time plot

Figure 20.1 shows a plot of interface height versus time curve of a settling sludge in a settling cylinder initially filled up with sludge concentration C_0 to a height h_0.

At interface height h_2, the average concentration of solids over the height h_2 is C_2 where $C_0h_0 = C_2h_2$.

At interface height h_2, the settling rate of the interface is

$$\overline{u_2} = \frac{h_2' - h_2}{t_2}.$$

This implies that if the water passes out of the interface at the rate $\overline{u_2}$ or less, no solids will be lost into the flow across it. To thicken the sludge from concentration C_2 to concentration C_u, the underflow concentration, the volume of water to be removed per unit mass of solids

= Water per unit mass at concentration C_2–water volume per unit mass at concentration C_u.
$= (1/C_2 - 1/C_\mathrm{u})$.

If such a sludge has to be thickened in a continuous thickener at volumetric flow rate of Q with interface height at average concentration C_2 over the height, sufficient settling time t_u is to be allowed; for the sludge to be thickened to the underflow concentration C_u, the volume of water to overflow per unit time from the entering sludge QC_0 per unit time

$$= QC_0(1/C_2 - 1/C_u).$$

The limiting overflow rate for no solids passing into the overflowing water being

$\overline{u_2} = \frac{h_2' - h_2}{t_2}$ per unit area, the area of the thickener
$A = \frac{QC_0}{\overline{u_2}}\left(\frac{1}{C_2} - \frac{1}{C_u}\right)$, i.e unit area (UA) $\frac{A}{QC_0} = \frac{1}{\overline{u_2}}\left(\frac{1}{C_2} - \frac{1}{C_u}\right)$.

In reality, all possible combinations of different C_2 and C_u are to be tried over all possible range of concentrations that are expected to be thickened in the continuous thickener. The largest unit area is to be selected for design.

To provide settling time t_u to thicken the sludge to underflow concentration C_u, the volume of thickener should be $= Qt_u$.

The depth of thickener $= \frac{Qt_u}{A}$

$$= \frac{Qt_u}{\frac{QC_0}{\overline{u_2}}\left(\frac{1}{C_2} - \frac{1}{C_u}\right)}, \text{ i.e. } \frac{\overline{u_2}t_u}{C_0\left(\frac{1}{C_2} - \frac{1}{C_u}\right)},$$

and the diameter $= \sqrt{\frac{4}{\pi} \cdot \frac{QC_0}{\overline{u_2}}\left(\frac{1}{C_2} - \frac{1}{C_u}\right)}$

The foregoing deductions are based on the following assumptions:

1. Settling velocity of any concentration layer is a function of the concentration.
2. Characteristic settling exhibited by a settling sludge in batch settling is independent of the initial concentration of sludge in the column. Roberts (1946) expressed that this assumption may not be true.

Problem 20.1 A sludge at concentration of 2475 mg/l is to be thickened to 11000 mg/l from the flow of sludge at the rate of 8000 m^3/d.

Batch settling study of the sludge was performed in a glass cylinder. The interface heights in cm at different times in min are noted and plotted to draw the interface height in cm versus time in min (for convenience of scaling) graph as shown in Fig. 20.2 (the typical graph is drawn arbitrarily for the purpose of illustration and is not from any practical data).

Design the thickener.

Solution Scaled from the graph are the following values presented in Table 20.1, where $C_0h_0 = C_uh_u = C_2h$ and $C_0 = 2475$ mg/l, $h_0 = 40$ cm, $C_u = 11000$ mg/l, $h_u = 9$ cm

Unit areas are calculated with data presented in Table 20.1 after converting into units presented in Table 20.2.

From Table 20.2

Unit area (UA) for the thickener $= 0.12874$ m^2/Kg/hr.

Flow rate $= 8000$ m^3/d with solids concentration 2.475 Kg/m^3.

The sludge entering per hour $= \frac{8000}{24} \times 2.475\,\text{Kg} = 825\,\text{Kg}$.

The area of the thickener $= 825$ Kg/hr $\times\ 0.12874$ m^2/Kg/hr $= 106.21$ m^2.

If Q_0 be the flow rate of sludge at concentration C_0 and Q_u be the rate of underflowing of sludge thickened to concentration C_u, the volume rate of water overflowing is

$$(Q_0 - Q_u) = Q_0\left(1 - \frac{Q_u}{Q_0}\right)$$

$$= Q_0\left(1 - \frac{C_0}{C_u}\right)$$

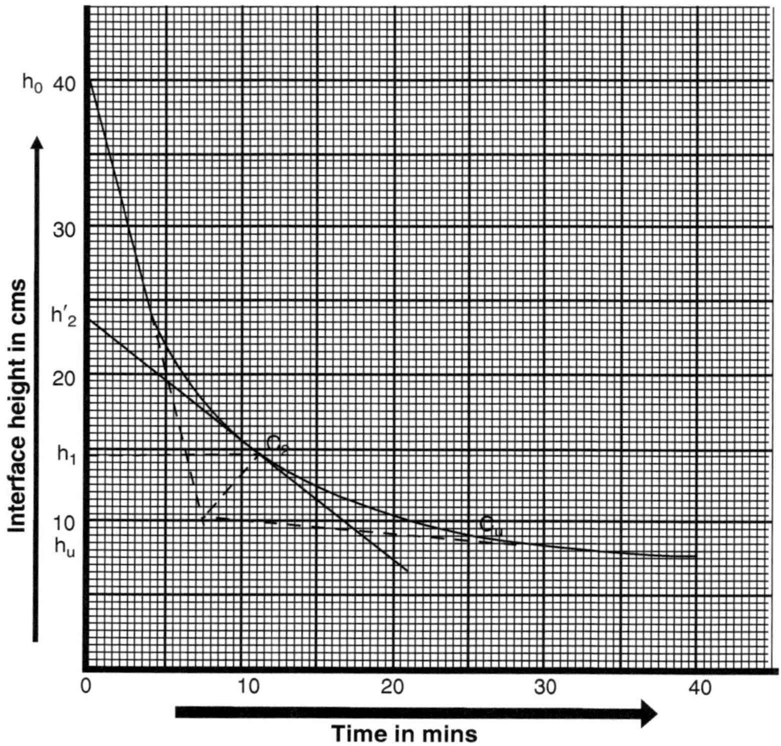

Fig. 20.2 Interface height versus time graph of settling sludge

$$= \frac{8000}{24}\left(1 - \frac{2475}{11000}\right)$$

$$= 258.33\frac{\text{m}^3}{\text{hr}}.$$

Hindered settling rate being 2.40 m/hr.
The clarifier area needed to be $= \frac{258.33}{2.4}$, i.e. $107.6\,\text{m}^2$.
Area 107.6 m^2 governs.
The diameter of the thickener $= \sqrt{\frac{107.40\times 4}{\pi}}$, i.e. 11.7 m;.

From Table 20.1, the average concentration of the sludge reaches 11000 mg/l when interface reaches 9 cm height at $t_u = 25$ min.
To provide detention time of 25 min,
the volume of the thickener required $= \frac{8000}{24} \times \frac{25}{60}$, i.e. $138.89\,\text{m}^3$.
The depth of the tank $= \frac{138.89}{\pi(11.7/2)^2}$, i.e. 1.29 m.

That is, the thickener is 11.7 m dia × 1.29 m depth (SWD).

Table 20.1 Scaled and derived data from Fig. 20.2

Time t min	0	4	7.5	10	11.5	12.5	15	20	22.3	25
Interface ht h cm		24	18	15.5	14.5	13.5	12.5	10	9.5	9
Intercept ht of the tangent at the interface point h' cm		40	27	24	23.5	21	19.5	16.5	15	14
Settling vel. of the interface, $\overline{u_2} = \text{slope} = \frac{h'-h}{t}$ cm/min.	3.75	4	1.2	0.8500	0.7826	0.6000	0.4667	0.3250	0.2444	0.2000
Average concentration $C_2 = \frac{C_0 h_0}{h} = \frac{99000}{h}$ mg/l	2475	4125	5500	6387	6828	7333	7920	9900	10421	11000

Table 20.2 Calculation of unit area by Coe and Clevenger's method with data extracted from Table 20.1

t min	C_2 Kg/m^3	C_u Kg/m^3	$\overline{u_2}\frac{\mathrm{m}}{\mathrm{h}}$	$UA = \frac{1}{u_2}\left(\frac{1}{C2} - \frac{1}{C_u}\right)\mathrm{m}^2/\mathrm{Kg/hr}$
4.0	4.125	11.00	2.40000	0.06313
7.5	5.50	11.00	0.72000	0.12626
10	6.387	11.00	0.51000	0.12874.
11.5	6.828	11.00	0.46956	0.11830
12.5	7.333	11.00	0.36000	0.12628
15.0	7.920	11.00	0.28000	0.12626
20.0	9.900	11.00	0.19500	0.05180
22.5	10.421	11.00	0.14664	0.03444
25.0	11.000			

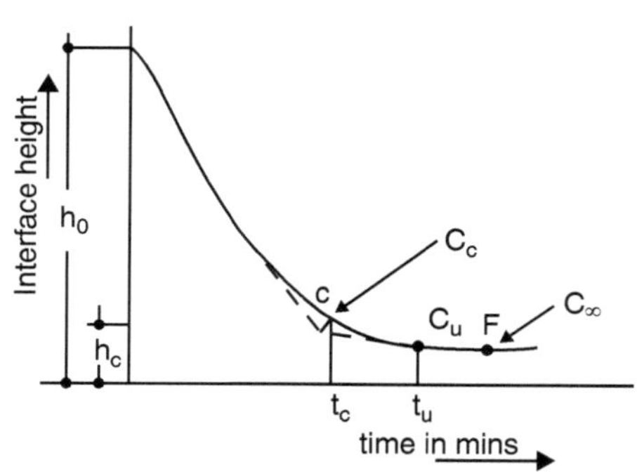

Fig. 20.3 Interface height versus time curve

20.2.2 *Method of Design Based on Robert's Derivation (1946)*

Robert (1946) concentrated on the compression thickening of sludge. Batch settling of sludge was carried out in a glass cylinder. The initial height of sludge in cylinder was h_0 at concentration C_0 throughout. The interface heights at progressive intervals of time were noted and plotted to draw the curve shown in Fig. 20.3.

At the point c, the settling sludge enters into compression phase when the underlying settling particles start sharing weights of the overlying settling solids due to oozing out of water from the interstices in the sludge solids.

Interface height at c is h_c at time t_c and concentration C_c. From the point F, the curve runs almost parallel to the time axis. The height of the sludge at F is h_∞, i.e the ultimate compressed height of the sludge at time t_∞ and at concentration of solids C_∞. F identifies the completion of sludge settling.

The subsidence rate $\left(-\frac{dh}{dt}\right)$ at time t in compression phase may be written proportional to the compressible height $(h - h_\infty)$ at time t as

$$\left(-\frac{dh}{dt}\right) = k(h - h_\infty) \tag{20.1}$$

where k is the rate constant.

Integrating from $h = h_c$ at $t = t_c$ to $h = h$ at $t = t$

$$(-)\int_{h_c}^{h} \frac{dh}{h - h_\infty} = k\int_{t_c}^{t} dt,$$

$$\text{i.e.}\log_e \frac{h_c - h_\infty}{h - h_\infty} = k(t - t_c) \tag{20.2}$$

At $h = h_l$, $t = t_1$ and k can be known as

$$k = \frac{1}{(t_1 - t_c)}\log_e \frac{h_c - h_\infty}{h_1 - h_\infty} \tag{20.3}$$

Sludge per unit area

$$C_0 h_0 = C_c h_c = C_\infty h_\infty = Ch \tag{20.4}$$

From Eqs. 20.2 and 20.4

$$\log_e \frac{(C_\infty - C_c)C}{(C_\infty - C)C_c} = k(t - t_c). \tag{20.5}$$

At $t = t_u$, $C = C_u$ and
Equation (20.5) may be written as

$$\log_e \frac{(C_\infty - C_c)C_u}{(C_\infty - C_u)C_c} = k(t_u - t_c) \tag{20.6}$$

Design steps:

(i) From interface height versus time curve from batch settling, find h_c, h_∞, h_1, t_c, t_∞ and t_1, and k known from Eq. 20.2

$$k = \frac{1}{(t_1 - t_c)}\log_e \frac{h_c - h_\infty}{h_1 - h_\infty}.$$

(ii) From the chosen value of C_u, find

$$(t_u - t_c) = \frac{1}{k}\log_e \frac{(C_\infty - C_c)C_u}{(C_\infty - C_u)C_c},$$

and t_u is known from the known value of t_c from the batch settling curve.

(iii) To provide the necessary retention time for thickening, the volume of the thickener $= Qt_u$.

(iv) To limit the overflow rate for no solid passing into the overflowing water, it should be the slope of the batch settling curve at the compression point $\overline{u_2} = \frac{h_2' - h_2}{t_2}$ (Ref. Fig. 20.2).

(v) To thicken the sludge from C_c to C_u, volume of water to be eliminated from the unit sludge mass

$$= \left(\frac{1}{C_c} - \frac{1}{C_u}\right).$$

(vi) For the total sludge mass QC_0, the area A of the thickener

$$= \frac{QC_0}{\overline{u_2}}\left(\frac{1}{C_c} - \frac{1}{C_u}\right).$$

(vii) Check for clarifier area

$$= \frac{QC_0}{\overline{u_H}}\left(1 - \frac{C_0}{C_u}\right),\ \overline{u_H} \text{ is hindered settling velocity.}$$

Problem 20.2 Same as Problem 20.1.

Solution From the graph in Fig. 20.2

$$h_c = 14.5\,\text{cm},\ t_c = 11\,\text{min.}$$

$$h_\infty = 7.5\,\text{cm},\ t_\infty = 37\,\text{min.}$$

$$h_1 = 8.0\,\text{cm},\ t_1 = 31\,\text{min.}$$

$$k = \frac{1}{(t_1 - t_c)}\log_e \frac{h_c - h_\infty}{h_1 - h_\infty}$$

$$= \frac{1}{(31 - 11)}\log_e \frac{14.5 - 7.5}{8 - 7.5}$$

$$= 0.13195\,\text{per min.}$$

$$C_c = \frac{2475 \times 40}{14.5},\ \text{i.e.} 6827\,\text{mg/l.}$$

$$C_\infty = \frac{2475 \times 40}{7.5},\ \text{i.e.} 13200\,\text{mg/l.}$$

$$C_u = 11000\,\text{mg/l}.$$

$$t_u = \frac{1}{k}\log_e\frac{(C_\infty - C_c)C_u}{(C_\infty - C_u)C_c} + \text{t}_\text{c}$$

$$= \frac{1}{0.13195}\log_e\frac{(13200 - 6827)11000}{(13200 - 11000)6827} + 11,\ \text{i.e}\,22.68\,\text{min}.$$

Volume of the thickener $Qt_u = \frac{8000}{24} \times \frac{22.68}{60}$, i.e. $126\,\text{m}^3$;.

From Fig. 20.2

$$h' = 24\,\text{cm},\ h = 14.5\,\text{cm},\ t = 11\,\text{min}.$$

Slope at the compression point $\overline{u_2} = \frac{24-14.5}{11}$, i.e. 0.86364 cm/min, i.e. 0.51818 m/hr.

Area of the thickener $= \frac{QC_0}{\overline{u_2}}\left(\frac{1}{C_c} - \frac{1}{C_u}\right)$

$= \frac{825}{0.51818}\left(\frac{1}{6.827} - \frac{1}{11}\right)\text{m}^2$ i.e. $0.10724\,\frac{\text{m}^2}{\text{Kg}} \times 825\,\text{Kg}$

$= 88.47\ \text{m}^2$.

Hindered settling velocity from Fig. 20.2 $\overline{u_H} = 2.4\frac{\text{m}}{\text{hr}}$.

Clarifier area $= \frac{QC_0}{\overline{u_H}}\left(1 - \frac{C_0}{C_u}\right)$

$= \frac{8000}{24\times2.4}\left(1 - \frac{2.475}{11}\right)$, i.e. $107.64\,\text{m}^2$.

Clarifier area governs.

Diameter of the thickener $= \sqrt{\frac{4\times107.6}{\pi}}$, i.e. 11.7 m dia.

Surface water depth $= \frac{126}{107.64}$, i.e. 1.17 m.

The thickener is 11.7 m dia × 1.17 m.

20.2.3 *Method of Design Based on Kynch's Theory of Sedimentation (1952)*

Kynch (Kynch 1952) analysed zone settling layers that are formed. No particle can settle across the layers. Each layer releases particles at its bottom to add to the adjoining bottom layer and receives solids at greater rate from the layer lying just above. The concentration value, in consequence, rises upwards. All concentration values from the initial concentration of solids to the concentration of finally settled sludge originate almost simultaneously at the bottom and rise upwards to end journey at the interface. Different interfaces have different concentrations which increase in the interfaces at greater depths.

Kynch utilised one assumption made by Coe and Clevenger (Talmadge and Fitch 1955) that the settling velocity of solids leaving a layer is a function of the concentration of the layer.

Equating the rate of solids received by a concentration layer from a layer just above with the rate of solids released by it to the layer just lying below, it was shown (Chap.7) that each concentration value rises upwards with uniform velocity, characteristic of the concentration, throughout its journey right from the bottom to its journey's end at the interface.

It was also derived that $C_i h' = C_0 h_0$, where C_i is interface concentration and h' is the intercept on the interface height axis made by the tangent at the interface point on the interface height versus time curve. C_0 is the initial concentration of sludge filling the batch settling column to height h_0.

Design steps:

(i) Batch settling of sludge is to be carried out to generate the interface height versus time curve.
(ii) From the curve are scaled the necessary data for the design of the continuous thickener as presented in Table 20.1.
(iii) Table 20.1 includes the data computed such as interface concentration C_i, average concentration of solids over the interface height, settling rate $\overline{u_2}$, etc.
(iv) Retention time for thickening t_u is scaled from the graph from the point of intersection of the line parallel to the time axis drawn at the interface height $h' = C_0 h_o / C_u$ with the graph.
(v) Unit areas are calculated for various interface concentrations at different points of the curve. The maximum value is to be selected to protect against the loss of solids through any interface using the relationship

$$UA = \frac{1}{\overline{u_2}}\left(\frac{1}{C_i} - \frac{1}{C_u}\right).$$

Problem 20.3 Same problem as in Problem 20.1.

Solution Unit areas are calculated with data presented in Table 20.1 after converting into units presented in Table 20.3.

From Table 20.3

Unit area (UA) of the thickener $= 0.42095\ \mathrm{m^2/Kg/hr}$.

Flow rate $= 8000\ \mathrm{m^3/d}$ with solids concentration $2.475\ \mathrm{Kg/m^3}$.

The sludge entering per hour $= \frac{8000 \times 2.475}{24}$, i.e.825 Kg.

The area of the thickener $= 825 \times 0.42095\ \mathrm{m^2} = 347.28\ \mathrm{m^2}$.

Clarifier area (worked out in previous problems) $= 107.6\ \mathrm{m^2}$.

Area $347.28\ \mathrm{m^2}$ governs.

The diameter of the thickener $= \sqrt{\frac{4 \times 347.28}{\pi}}$, i.e.21 m.

To provide retention time $t_u = 25$ min (scaled from the graph as given in design step), the volume of the thickener $= \frac{8000}{24} \times \frac{25}{60}$, i.e.$138.9\ \mathrm{m^3}$.

Table 20.3 Calculation of unit area by Kynch's method with data extracted from Table 20.1

t min	C_i Kg/m^3	C_u Kg/m^3	$\overline{u_2}\frac{\text{m}}{\text{h}}$	$UA = \frac{1}{\overline{u_2}}\left(\frac{1}{C_i} - \frac{1}{C_u}\right)\text{m}^2/\text{Kg/hr}$
4.0	2.475	11.00	2.40000	0.13047
7.5	3.667	11.00	0.72000	0.25322
10.0	4.125	11.00	0.51000	0.29709
11.5	4.213	11.00	0.46956	0.31189
12.5	4.714	11.00	0.36000	0.33674
15.0	5.077	11.00	0.28000	0.37878
20.0	6.000	11.00	0.19500	0.38850
22.5	6.600	11.00	0.14664	0.41330
25.0	7.071	11.00	0.12000	0.42095

The surface water depth of the thickener $= \frac{138.9}{347.28}$, i.e.0.4 m.
The thickener is 21 m dia × 0.4 m SWD.

20.2.4 Talmadge and Fitch's Method of Thickener Design (1955)

Talmadge and Fitch (1955) by geometrical construction determines the retention time in continuous flow thickener required to thicken the sludge to the desired consistency from the interface versus time curve obtained from batch settling of sludge.

Following Kynch's theory of sedimentation, they could find out the unit area for thickening the sludge mass.

The sludge at concentration C_0 filling a settling cylinder of cross-sectional area A to a height h_0 is subjected to batch settling. The interface height versus time curve was plotted as shown in Fig. 20.4.

To thicken the sludge to concentration C_u, the sludge height is to be reduced to $h_u = C_0 h_0 / C_u$. The volume of water to be eliminated $= A\ (h_0 - h_u)$.

This water should move through the interface. If compression point at concentration C_2 at height h_2 at time t_2 be the interface, the settling velocity through the interface is the slope of the tangent $= \left(h_2' - h_2\right)/t_2$, and the volume rate of flow through it is $\frac{A\left(h_2' - h_2\right)}{t_2}$.

From interface height h_0 to h_2, volume of water $A(h_0 - h_2)$ flows out through the interface in time t_2.

The further volume of $A(h_2 - h_u)$ will flow across the interface at the rate $\frac{A\left(h_2' - h_2\right)}{t_2}$ in further interval of time (t_u-t_2).

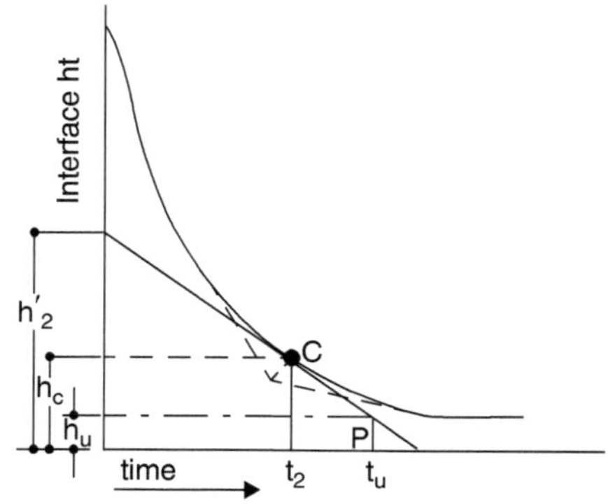

Fig. 20.4 Interface height versus time curve of settling sludge

That is, $(t_u - t_2) = \frac{A(h_2-h_u)}{\frac{A(h'_2-h_2)}{t_2}}$, i.e. $\frac{(h_2-h_u)}{(t_u-t_2)} = \frac{(h'_2-h_2)}{t_2}$.

By geometry 't_u' is corresponding to the point P on the time axis (Fig. 20.4).

Now total solids in the column $= C_0 h_0 A$.

To move through the compression point interface in time t_u, the rate of settling of solids through it $= \frac{C_0 h_0 A}{t_u}$.

So in a continuous thickener of area A, the input of sludge per unit time QC_0 at flow rate Q should be equal to $C_0 h_0 A / t_u$.

$$\text{That is, } QC_0 = \frac{C_0 h_0 A}{t_u}.$$

$$\text{That is, } A = \frac{Q t_u}{h_0}.$$

Design steps:

(i) From the batch settling of sludge, the interface height versus time curve is prepared.
(ii) At interface height $h_u = \frac{C_0 h_0}{C_u}$, a line parallel to the time axis is drawn.
(iii) Compression point is determined as the point of intersection of the curve with the bisector of the angle between two near tangents.
(iv) Tangent is drawn at the compression point to intersect the line through h_u interface height. The point of intersection determines t_u.

Depending upon the value of C_u, i.e.h_u, the line through the interface height may run:

1. Below the compression point (No. 1 in Fig. 20.5)
2. Through the compression point (No. 2 in Fig. 20.5)
3. Above the compression point (No. 3 in Fig. 20.5);

t_{u1}, t_{u2} and t_{u3} can be scaled from the figure accordingly.

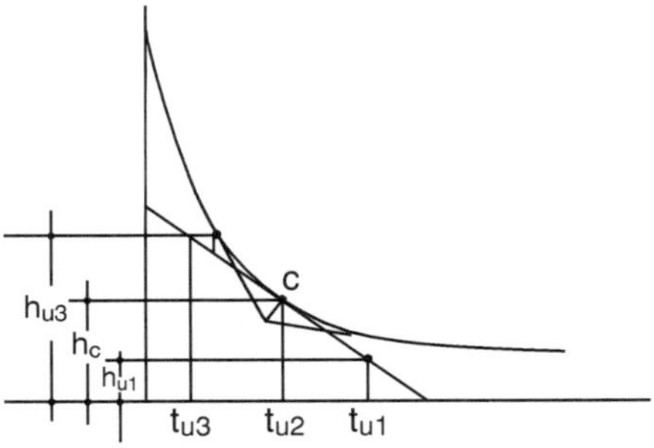

Fig. 20.5 Interface height versus time curve of settling sludge

Problem 20.4 Same as in Problem No. 20.1.

Solution From Fig. 20.2, the line parallel to the time axis intersects the tangent at the compression point at a point corresponding to which t_u= 18 min can be scaled.
Flow rate of sludge = 8000 m^3/d.
Area of the continuous thickener = Qt_u/h_0.

$$= \frac{8000 \times 18 \times 100}{24 \times 60 \times 40} \text{m}^2 = 250\,\text{m}^2.$$

Clarifier area = 107.6 m^2.
250 m^2 area governs.
Diameter = $\sqrt{\frac{4 \times 250}{\pi}}$, i.e. 17.8 m.
To provide the retention time t_u = 18 min.
The volume of the continuous thickener = $\frac{8000}{24} \times \frac{18}{60}$, i.e. 100.00 m^3.
Surface water depth $\frac{100}{250.0}$, i.e. 0.40 m.
The thickener is 17.3 m dia × 0.40 m SWD.

20.2.5 Flux Flow Method of Thickener Design

20.2.5.1 Batch Settling of Sludge

Interface versus time curve obtained from the observation with batch settling of sludge forms the very basis for the thickener design in continuous operation.

Kynch's analysis can find the settling velocities of sludge at higher concentration from a single batch settling of sludge.

Dick and Ewing (1967) reported that settling velocities at higher concentrations according to Kynch's theory may not give the true picture in case of real sludges and that different batch settling tests are necessary for such determinations with different concentrations of solids.

Vesilind (1968) stressed on the effect of diameter of the batch settling column on the settling velocities of the sludge possibly due to the friction exerted by the wall of the settling column, the extent being dependent on the nature of the wall and that of the sludge.

The effect of the wall may be reduced with stirring at very low rate of 250 cm/min (Vasilind 1968).

The bigger length of sludge in column increases the settling velocity of sludge (Dick and Ewing 1967).

According to Dick (Dick and Ewing 1967), batch settling should be conducted with sludge length in the column equal to the effective sludge depth in the thickener in continuous operation to be designed.

20.2.5.2 Batch Flux Curve

Batch settling of sludge forms solid-liquid interface by the settling of solids. This interface will be settling, and the settling velocity of the interface can be measured by the slope of the interface height versus time curve at the point of its movement.

If the initial concentration of the sludge is in flocculating phase, the interface settling velocity will be increasing. With progressively increased initial concentration of solids in the sludge, the settling nature of the interface settling rate will be changing. Firstly, with completion of the flocculation interface, the settling velocity will cease to increase, and the solids are latticed to settle as zone releasing solids at the bottom of a layer and receiving solids at its top. Initially both of the release and reception of solids being same solids settle with constant velocity. Characterising the phase as 'hindered settling', this hindered settling velocity u_H is the slope of the linear portion of the interface height versus time curve. Hindered settling occurs at concentration more than 500 mg/l (Schroeder). Concentration and the velocity of its settling solids and how they change with different initial concentrations need observations for the design of thickener in continuous operations.

Solids crossing per unit area per unit time, henceforth, be called solid flux $F_{sb} = Cu_H$; the subscript 'sb' stands for solid flux in batch settling.

As shown in Fig. 20.6, the 'flux curve' initiates with the point A. At increased concentrations, solids settling velocity decreases being nearer to each other and experiencing increased drag. Increased concentration increases the flux but decreased settling velocity pulls down the due increase.

From A to B, thus, the flux increases but at progressively decreasing rate and at B, the increase of flux is completely offset by its decrease in settling velocity.

Subsequently from B the flux decreases due to predominative decrease of flux for decreasing velocity over the increase of flux due to increased concentration. At a point C, the concentration is in compression phase and the flux will be rapidly declining. The flux will show asymptotic diminish. Theoretically the curve should pass through (0.0) shown by dotted extension.

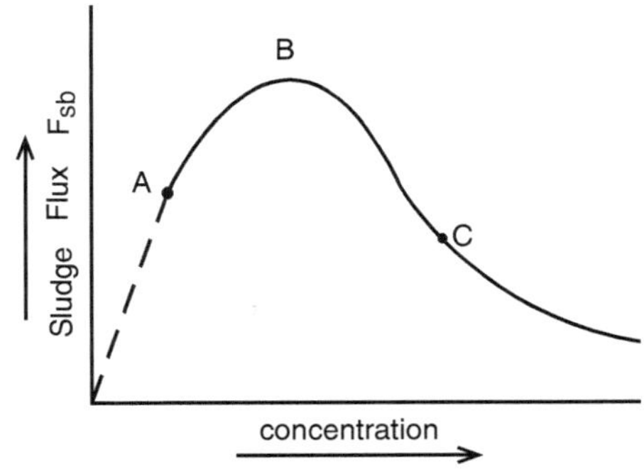

Fig. 20.6 Batch flux curve

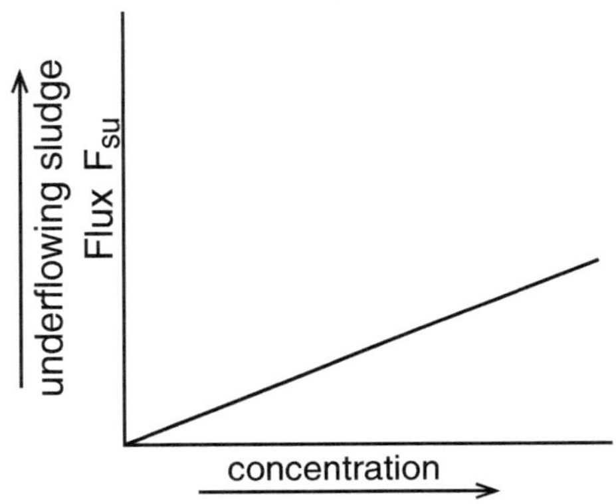

Fig. 20.7 Underflow flux curve

20.2.5.3 Underflow Flux Curve

Underflow rate of Q_u at concentration C_u will pull down underflow flux through area A:

$$F_{su} = C_u(Q_u/A)$$
$$= C_u u_u.$$

The underflow velocity 'u_u' being constant, F_{su} versus C_u curve is a straight line as shown in Fig. 20.7.

20.2.5.4 Total Flux F_T

In Fig. 20.8 is shown the batch flux curve $F_{sb} = Cu_H$ and the underflow flux curve $F_{su} = C_u u_u$. Corresponding to each concentration, the ordinates of the batch flux curve $F_{sb} = Cu_H$ and that of the underflow flux curve $F_{su} = C_u u_u$ are added and plotted as

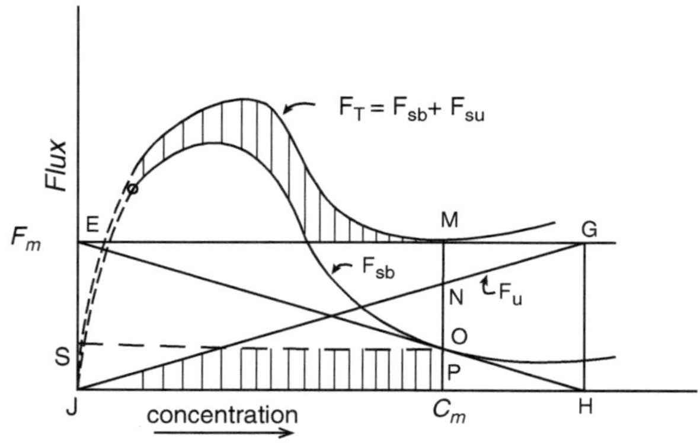

Fig. 20.8 Total flux curve and geometry of Yoshioka's modification

$$F_T = F_{sb} + F_{su}$$
$$= Cu_H + C_u u_u$$

to get the total flux curve.

20.2.5.5 Flux Flow Method

The total flux curve F_T in Fig. 20.8 passes through the minimum at M. This shows that the flux is minimum at concentration C_m between feed concentration C_0 and underflow concentration C_u. For the flow rate of Q_0 at feed concentration C_0, the thickener area

$$A = \frac{Q_0 C_0}{F_m} = \frac{Q_u C_u}{F_m}.$$

20.2.5.6 Yoshioka's Modification

In Fig. 20.8, the triangles GEJ and HJE are geometrically equal and so also are MN and 0P.

When measured from F_T curve

$$\begin{aligned} F_m &= \text{MP} \\ &= (\text{MN} + \text{NO}) + \text{OP} \\ &= (\text{PO} + \text{NO}) + \text{OP} \\ &= \text{PN(underflow flux)} + \text{OP(batch flow flux)} \end{aligned}$$

Alternatively if the point measuring C_u on the abscissa, i.e. the point H, is joined by a straight line with a point measuring F_m on the flux axis, i.e. the point E in Fig. 20.8, the straight line HE will be tangent to the F_{sb} curve at O lying on MP since MN = OP.

Measured from F_{sb} curve - PN + OP

$$= \text{PN} + \text{MN}$$
$$= F_m.$$

This shows that tangent to the batch flux curve drawn through the desired value of C_u on concentration axis will intersect the flux axis to show the minimum allowable flux for design.

This modification provides flexibility to the design unlike the graphical method that requires different graphical constructions for each of the differently chosen underflow concentrations and flow rate of sludge.

20.2.5.7 Design Steps

To find the minimum flux for design, the following steps are to be followed:

1. A suitable depth of thickener to be designed is chosen.
2. In a cylinder batch settling, study is conducted with suitably and carefully collected sludge sample providing sludge column of depth equal to that chosen for the thickener to be designed. The diameter of the column should be more than 7.5 cm.
3. The interface height versus time curves for different feed concentrations are to be prepared.
4. From the slope of the linear portion of each of the curves, the hindered settling velocity u_H is to be found out for each feed concentration.
5. From the data obtained in step 4, flux values for different feed concentrations are calculated and batch flux curve is prepared.
6. The desired thickened concentration of sludge is marked on the abscissa.
7. From the point, a tangent is drawn to the batch flux curve drawn in step 5.
8. The tangent intercept on flux axis will show the minimum flux for design.

Problem 20.5 Waste water containing sludge concentration of 2500 mg/l has to be processed @10,000 m^3/d to thicken the sludge to a concentration of 12,500 mg/l. Sludge settling studies carried with 3.0 m × 10 cm sludge column and the results are reported in the following Table 20.4.

Table 20.4 Batch settling of sludge

Concentration Kg/m^3	1.490	2.600	3.94	5.425	6.930	9.100	12.0
Sett.vel. m/h	5.50	3.23	1.95	1.01	0.55	0.26	0.14

(a) Design the thickener.
(b) The designed thickener is subjected to the processing of 6000 m^3/d at concentration of 3000 mg/l of solids. What are the underflow concentration and the underflow rate through the thickener?

Solution From the given data, flux flows at different concentrations were calculated and presented in the following Table 20.5.

From the values given in Table 20.5, flux versus feed concentration curve is drawn as shown in Fig. 20.9.

From the point indicating concentration 12.5 Kg/m^3 on the abscissa, tangent is drawn to the flux versus concentration curve in Fig. 20.9, and this tangent intersects the flux axis at a point showing minimum flux of 8.2 Kg/m^2.h.

Flow rate = 10000 m^3/d, i.e. 416.67 m^3/h.

Feed concentration = 2.5 Kg/m^3.

Sludge per hour = $\frac{10000 \times 2.5}{24}$ Kg.

To maintain minimum flux of 8.2 Kg/m^2.h,

Table 20.5 Flux flow at different concentrations

Flux Kg/m^2h	8.2	8.4	7.7	5.5	3.8	2.4	1.7
Concentration Kg/m^3	1.5	2.6	3.9	5.4	6.9	9.1	12.0

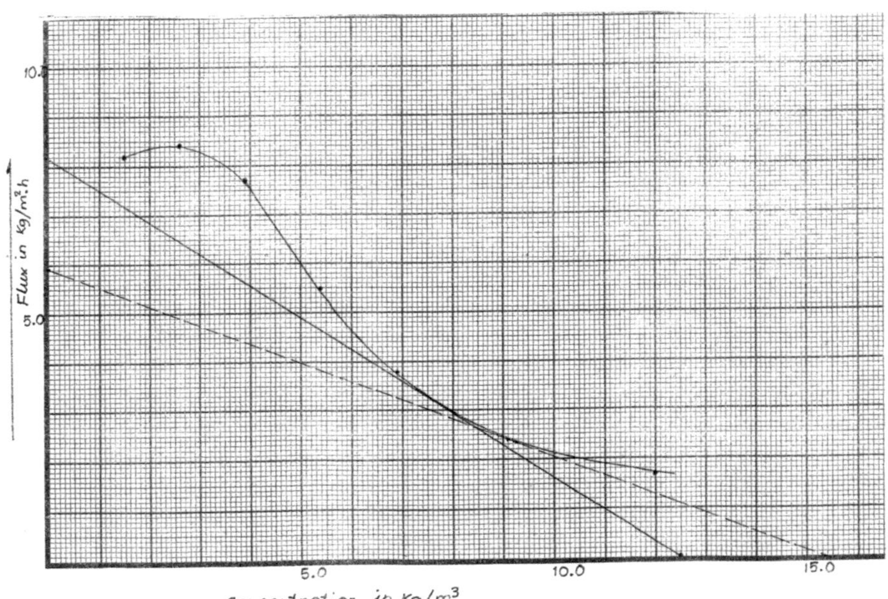

Fig. 20.9 Concentration versus flux curve for Problem 20.5

the area of the thickener $= \frac{10000\times2.5}{24\times8.2}$, i.e. $127.03\,m^2$;

that is, the diameter of the thickener $= \sqrt{\frac{4\times127.03}{\pi}}$, i.e. 12.7 m.

Of the flow rate of 416.67 m^3/h, the flow rate carried with the underflow

$$= \frac{\frac{10000\times2.5}{24\times12.5}\,m^3}{hr}, \text{ i.e. } 83.33\,m^3/h.$$

The water that will overflow is at the rate of $= 416.67 - 83.33$, i.e. $333.34\,\frac{m^3}{h}$.

From the flux versus feed concentration curve corresponding to concentration 2.5 Kg/m^3, flux is 8.4 Kg/m^2.h.
Hence hindered settling velocity at feed concentration 2.5 Kg/m^3

$$= \frac{8.4}{2.5}, \text{ i.e. } 3.36\frac{m}{h}.$$

(Approximate value of hindered velocity may also be found out from the interpolation of the data presented in Table 20.4 for the problem.)
The clarification area required $= \frac{333.34}{3.36}$, i.e. $99.2\,m^2$.
Area 127.03 m^2 governs.

Hence the thickener is 12.7 m (dia) × 3.0 m (SWD).
(b) Flow rate = 6000 m^3/d

$$= \frac{6000}{24}, \text{ i.e. } 250\frac{m^3}{h}.$$

Feed concentration = 3.0 Kg/m^3.
Sludge input per hour = 3.0 × 250 Kg/h, i.e. 750 Kg/h.
Thickener area = 127.03 m^2

$$\text{That is, flux} = \frac{750}{127.03}\frac{Kg}{m^2}.h, \text{ i.e. } 5.9\frac{Kg}{m^2}.h.$$

A point on the flux axis showing flux = 5.9 Kg/m^2.h is located. Through the point, a tangent to the batch flux is drawn. This tangent intersects abscissa at concentration 15.4 Kg/m^3, i.e. 15400 mg/l.
Volume rate of water carried with the underflow $= \frac{6000}{24} \times \frac{\frac{3.0}{15.4}m^3}{hr}$, i.e. $48.7\,\frac{m^3}{h}$.
Volume rate of water that will overflow $= \frac{6000}{24} - 48.7$, i.e. $201.3\,\frac{m^3}{h}$.
Clarification rate $= \frac{201.3}{127.03}$, i.e. $1.58467\frac{m}{h}$.

This is less than hindered settling velocity that should lie between 3.23 m/h and 1.95 m/h as evident in Table 20.4.

Notations

C_0 Feed concentration of solids
h_0 Sludge column height
C_2 Concentration of solids at the interface
h_2 Interface height
h'_2 Intersected height on interface height axis of interface height versus time curve by the tangent drawn at interface height h_2 of the curve
$\overline{u_2}$ Interface settling rate
C_u Concentration of underflowing sludge
t_u Settling time to thicken the sludge to underflow concentration C_u
C_∞ Ultimate concentration of settled sludge
t_∞ Least time to thicken the sludge to its ultimate concentration
C_c Concentration of sludge at compression point
t_c Time to reach the compression point
$\overline{u_H}$ Hindered settling velocity
h_u Interface height at underflow concentration C_u
F_T Total flux flow
F_{sb} Batch sludge Flux
F_{su} Underflow sludge flux
F_m Minimum total flux

References

Coe HS, Clevenger GS (1916) Methods of determining the capacities of slime settling. Trans Am Inst Mining Met Engg 5(5):336

Dick RI, Ewing BB (1967) Evaluation of activated sludge thickening theories. J San Engg Div ASCE 93(SA4):9

Dick RI (1970) Thickening advances in water quality improvement physical and chemical processes. In: Gloyna EF, Eckenfelder WW Jr (eds) University of Texas Press, Austin, p 358

Kynch GJ (1952) A theory of sedimentation. Trans Faraday Soc 48:166

Robert's EJ (1946) Thickening – art or science? Mining Engg 1:61

Schroeder ED. Water and waste water treatment, International Student's Edition. Kogakusha, Ltd, McGraw Hill, p 154

Talmadge WP, Fitch EB (1955) Ind Engg Chem 47:38–41

Vasilind PA (1968) The influence of stirring in the thickening of biological sludges. Ph.D thesis, University of North Carolina at Chapel Hill

Yoshioka N et al (1957) Continuous thickening of homogeneous flocculated slurries (English abstract). Chem Eng 21, Tokyo

Author Index

B

C

D

E

F

G

H

I

K

A. De, *Sedimentation Process and Design of Settling Systems*,
Springer Transactions in Civil and Environmental Engineering,
DOI 10.1007/978-81-322-3634-4

Subject Index

A. De, *Sedimentation Process and Design of Settling Systems*,
Springer Transactions in Civil and Environmental Engineering,
DOI 10.1007/978-81-322-3634-4